Applied Probability and Statistics (*Continued*)

continued on back

Survival Distributions:

Reliability Applications in

the Biomedical Sciences

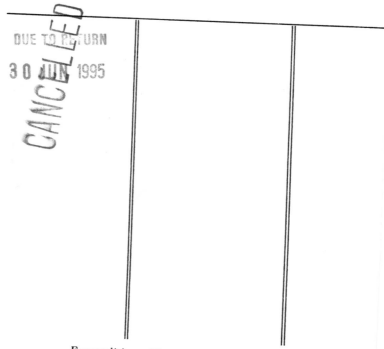

Survival Distributions: Reliability Applications in the Biomedical Sciences

ALAN J. GROSS,

Division of Public Health
School of Health Sciences
University of Massachusetts
Amherst, Massachusetts

and

VIRGINIA A. CLARK,

Division of Biostatistics
School of Public Health
University of California
Los Angeles, California

JOHN WILEY & SONS New York · London · Sydney · Toronto

Library of Congress Cataloging in Publication Data:

Gross, Alan J   1934-
  Survival distributions.

  (Wiley series in probability and mathematical statistics)
  Bibliography: p.
  Includes index.
  1. Medical statistics. 2. Distribution (Probability theory) 3. Life expectancy. 4. Reliability (Engineering) 5. Biometry. I. Clark, Virginia A.
II. Title. [DNLM: 1. Biometry. 2. Statistics. HA29 G878s]
RA409.G74   612.6′8′0182   75-6808
ISBN 0-471-32817-0

Printed in the United States of America
10 9 8 7 6 5 4 3

To Carl E. Hopkins

Preface

Reliability and survivability theory have an important common component. They are both concerned with the measurement of length of life, whether it is the life of a mechanism, a system, an animal, or a human being. There are, of course, differences. An important problem in reliability theory is the optimization of system reliability in multicomponent systems. Such optimization depends on the relation between the number and the location of components in a system. No analogous problem exists in survivability theory, since the system in this case is living and functioning, which makes the rearrangement of components (organs in this case) an impractical alternative. On the other hand, a very important problem in survivability theory is the comparison of survival times of patients or experimental animals receiving two or more different treatments. Similar problems arise in reliability theory when it is necessary to compare life distributions before and after engineering changes; by and large, however, they are not as important as the comparable survivability problems.

This book is concerned with the similarities between the disciplines, emphasizing the aspects of survivability, such as comparing the survival times of two groups of patients entering the study over time with the same acute illness, each group receiving a different treatment. Prerequisites include a year of mathematical statistics at the level of Hoel [1962] or Mood and Graybill [1963]. Also, it is assumed the student has a basic knowledge of matrix algebra and its application to solving systems of linear equations.

The aim is to study many of the parametric models used in both reliability and survivability, illustrating their application in survival studies. Estimation of the requisite parameters for censored and noncensored distributions is a primary consideration. The distributions studied in detail are the exponential, gamma, Weibull, and Rayleigh. The Rayleigh distribution is derived by assuming a linear hazard rate. Parameters of the distributions are estimated in the absence and presence of censoring. Other distributions that are discussed in less detail are the lognormal, normal, and Gompertz.

In addition to the study of parametric survival distributions and models, we consider applied nonparametric procedures. These include graphical plots of observed hazard rates, Kaplan and Meier [1958] estimates for censored data, censored two-sample Wilcoxon tests, and life table methods for analyzing survival data. These applied nonparametric procedures are not covered in the more theoretical nonparametric reliability text by Barlow and Proschan [1965]. In addition to parameter estimation, the nonparametric graphical procedure for estimating the cumulative hazard rate is discussed. This procedure has recently been developed by Nelson [1972].

Chapter 1 serves as an introduction in which we define a death density function, a survivorship function, and a hazard function. We then show how these functions are related. Some of the common death density functions are introduced with discussions of their elementary properties.

Chapter 2 is concerned with life tables. The emphasis is on the clinical life table and the derivation of formulas showing the relationship between life tables and their concomitant survival distributions. Much of the material that appears in Gehan [1969] is discussed in this chapter. The product-limit estimates of Kaplan and Meier are introduced.

Chapters 3 and 4 treat the estimation of parameters in some of the more commonly used survival distributions, including the exponential, gamma, Weibull, and Rayleigh. Estimation is considered for both the censored and noncensored survival data. For the censored case, both singly censored and progressively censored data are included. Parameter estimation in the presence of concomitant information is taken up in Chapter 3 with particular reference to the work of Feigl and Zelen [1965] and Zippin and Armitage [1966]. Competing risk models are also briefly introduced in Chapter 3.

In clinical trials of new therapeutic agents, the likelihood of success in treating patients may be low initially. However, as a trial continues, modifications in the drug and/or mode of treatment may enhance the chances of treatment success. Chapter 5 discusses the applicability of reliability growth models to the measurement of the increase in the probability of treatment success in clinical trials.

In Chapters 3, 4, and 5 maximum likelihood estimation plays an important role. Maximum likelihood estimators are often not obtainable in closed form, particularly when observations are censored. Hence iterative numerical procedures are required to obtain the maximum likelihood estimates. Chapter 6, therefore, covers iterative procedures for solving nonlinear systems of equations.

In Chapter 7 procedures are presented for the comparison of two survival distributions. Parametric and nonparametric tests are considered for the censored and noncensored cases. Chapter 8, the concluding chapter, deals with the sample size problem and its application to clinical trials. The treatment includes a brief introduction to Bayesian procedures and play-the-winner rules.

If the instructor uses this book for a one-semester course, the following sections can be omitted: Section 3.5 on the use of concomitant information, Section 3.7 on competing risks, Section 4.5 on truncated distributions, Section 4.6 on nonmaximum likelihood techniques for Weibull and gamma distributions, the last part of Section 4.8 covering the general regression procedure for estimating survival parameters, Chapter 5 on growth and assessment of survivability, Section 8.4 on sample size for a known population size, and Section 8.5 on sample size allocation to two treatment groups.

There is a rich source of literature in both survival and reliability theory; thus an effort has been made to reference many of the important papers and books on both subjects. The references are annotated and have been placed at the end of the book so they can be readily located.

Finally, problems for solution are included at the end of each section within the chapters, permitting the reader to test his understanding of the material by section. There are theoretical problems, which serve to introduce topics not discussed in detail in the book, and applied problems, designed to give the reader some practical experience with survival data, and often making use of data from actual studies.

ALAN J. GROSS
VIRGINIA A. CLARK

Amherst, Massachusetts
Los Angeles, California
March 1975

Acknowledgments

The work of Alan J. Gross was supported in part by grant 8-P02-HS-00234 from the National Center for Health Services Research (Lester Breslow, M.D. Principal Investigator), while he was with the California Center for Health Services Research, School of Public Health, University of California at Los Angeles.

We are indebted to the following students: Stanley Lemeshow, Carl Pierchala, Rocco Brunelle, Min-An Huang, and David Tyler for their technical assistance and their critical reading of early drafts. The major portion of the typing was performed by Mrs. Anne Eiseman. Editorial assistance was contributed by Mrs. Leona Povondra. Additional typing was carried out by Mrs. Muriel Nemiroff, Ms. Lois Kaplan, and Mrs. Barbara Richer.

A.J.G.
V.A.C.

Contents

Survival Distributions:
Reliability Applications in
the Biomedical Sciences

CHAPTER 1

Introduction

1.1 INTRODUCTORY REMARKS

The theory of reliability has grown primarily out of military applications and experiences with complex military equipment. Since the reliability of a complex piece of equipment can be modeled by a probability statement concerning its lifetime operation, there is a very close connection between reliability theory and survivability theory, where it is desired to make probability statements about survival or remission times of acutely ill patients or animals undergoing biological experimentation.

There is much literature in both areas—reliability and survivability (length of life studies), and given the extent of this literature, a complete listing in the bibliography is not feasible. However, this book contains an annotated bibliography listing many important references. This feature is designed to aid the reader in his search for appropriate references.

This book presents methodology useful in dealing with survival data either from laboratory studies on animals or survival studies of acutely ill humans. We draw quite extensively on reliability theory, showing how some of the estimation and test of hypothesis techniques developed for problems in reliability can be applied to corresponding problems with biological application. In addition, other procedures developed specifically to deal with biological or medical problems are included. For example, survival time for patients with acute leukemia is correlated with the initial white blood count. The procedure to estimate survival time in patients with acute leukemia accounting for initial white blood counts was originally studied by Feigl and Zelen [1965] and is dealt with in Section 3.5.

In animal survival studies, the researcher usually starts with a fixed number of animals. He subjects them to a treatment or treatments and determines the length of life of the animals from time of treatment(s). Often because of time and/or cost constraints, the researcher cannot wait until all the animals have died. Typically he waits until a fixed number of the animals has died (e.g., 50 out of 60), or he determines the length of life for a fixed period of time (e.g., 6 months), after which the surviving animals are sacrificed. The time recorded for the animals that are sacrificed is then the period of the study. Some statisticians—for example, Cohen [1965]—refer to such data as singly censored data. The statistical procedures accounting for censoring of data have been utilized extensively in reliability applications.

In most patient survival studies the patients arrive for treatment during the course of their illness; thus the length of follow-up after entry to the study is usually different for each patient. In industry the situation would be analogous if the date of installation of equipment varied for each piece of equipment: after a period of time we could determine the failure time for the items that have failed and the length of time they have run or survival for those still surviving. Nonparametric procedures for analyzing this type of data are discussed in Chapters 2 and 7, and parametric procedures are given in Chapters 3, 4, and 7. Such data are called progressively censored samples.

Examples of survival data are not hard to find in the medical and biological literature. Examples 1.1.1 to 1.1.5 cited here are from the extensive literature of cancer and heart disease.

Example 1.1.1. MacDonald [1963] reports on lengths of survival of 256 males with malignant melanoma who had metastases (spread of the disease beyond the original site) on admission to the M. D. Anderson Tumor Clinic. The period of admission was 1944 to 1960. Table 1.1 shows the survival pattern of these patients. Using patient records, it is not difficult to obtain their exact survival times. In Chapter 3 we use such a complete sample of survival times in the estimation of survival distribution parameters, such as mean survival time.

Example 1.1.2. Data on the survival times of patients with angina pectoris are given by Gehan [1969]. These patients are part of a larger group of patients examined at the Mayo Clinic during the 10-year period January 1, 1927–December 31, 1936. Some of the patients in the study were lost to follow-up, as Table 1.2 indicates. Methodology for analyzing this type of survival data is developed in Chapter 2, which treats clinical life tables.

Example 1.1.3. Data on survival time of patients who have advanced cancer of the bladder are compared for two treatment groups; one group had a standard treatment, and one had the standard treatment plus immunological therapy. In Chapter 7 we study parametric and nonparametric methods of comparing such treatment groups.

Table 1.1

Survival Time (years)	Number of Patients Surviving at Beginning of Interval	Number of Patients Dying in Interval
0–1	256	167
1–2	89	48
2–3	41	23
3–4	18	6
4–5	12	3
5–6	9	6
6–7	3	1
7–8	2	1
8–9	1	1
9+	0	0

Table 1.2

Survival Time (years)	Number of Patients Known to Survive at Beginning of Interval	Patients Lost to Follow-up
0–1	2418	0
1–2	1962	39
2–3	1697	22
3–4	1523	23
4–5	1329	24
5–6	1170	107
6–7	938	133
7–8	722	102
8–9	546	68
9–10	427	64
10–11	321	45
11–12	233	53
12–13	146	33
13–14	95	27
14–15	59	23
15+	30	0

Example 1.1.4. If the study outlined in Table 1.1 were terminated at the end of 3 years, 18 melanoma patients would still be alive at the end of the study. We study techniques for analyzing such data (singly and progressively censored data) in Sections 3.3 and 4.4.

Example 1.1.5. Clinical trials can be conducted in stages. Refinements in treatment are made between successive stages, and the probability of treatment success is therefore expected to improve from stage to stage. For example, a clinical trial conducted in five stages shows the following success ratios for each stage of the trial: 6/10, 7/10, 7/10, 8/10, and 9/10. Here there are two problems of interest to the clinician. The first is to predict with some level of assurance the probability of treatment success at the next stage of testing. Also required is assessment of the overall success probability after the completion of the five stages. These problems are addressed in Chapter 5.

Examples 1.1.1 through 1.1.5 are given to preview the kinds of problems whose methodology of analysis we consider. We note here that real data examples are used whenever possible to illustrate the theory which is developed.

1.2 EQUIVALENT FUNCTIONS DESCRIBING SURVIVAL

Survival data such as we discussed in Examples 1.1.1 through 1.1.5 are data that measure the times to death of individuals or animals. As such, these times to death are random variables having frequency distributions. Figure 1.1 is a histogram of the number of deaths in each interval for the data in Example 1.1.1.

In addition to using such elementary procedures as histograms, we can characterize any distribution of survival times by three equivalent functions.

1. *Death Density Function* $f(t)$: the probability that an individual dies during the time interval $t < x < t + \Delta t$, no matter how small Δt is. This is the probability density function or frequency function, where the random variable is time. The death density function, which is sometimes also called the unconditional failure rate, is then mathematically $f(t) = \lim_{\Delta t \to 0} \text{pr}\{t < x < t + \Delta t\}/\Delta t$. The death density function possesses the following properties:

$$f(t) \geqslant 0 \quad \text{for all } t$$

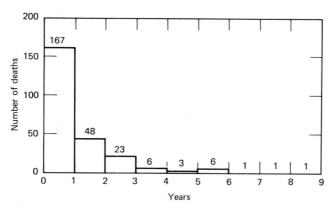

Figure 1.1 Histogram of number of deaths due to malignant melanoma by years after admission to M. D. Anderson Clinic (data from Example 1.1.1).

and

$$\int_{-\infty}^{\infty} f(t)\,dt = 1.$$

Furthermore, because survival time is only measured for positive values of t,

$$f(t) = 0 \qquad \text{for } t < 0;$$

hence

$$\int_{0}^{\infty} f(t)\,dt = 1.$$

Figure 1.2 shows a typical death density function. In Figure 1.3 the data from the malignant melanoma example (Example 1.1.1) are graphed as a frequency polygon by connecting the midpoints of the histogram in Figure 1.1. Thus $f(0.5) = 167/256 = 0.652$, $f(1.5) = 48/256 = 0.188$, etc.

    2. *Survivorship Function S(t)*: the probability that an individual survives at least time $t(t > 0)$. That is, if T is a random variable that represents the survival time of an individual, $S(t)$ is the probability that T is at least a fixed time $t(t > 0)$. In statistics texts the cumulative distribution function $F(t)$ is usually defined as

$$F(t) = \int_{-\infty}^{t} f(T)\,dT$$

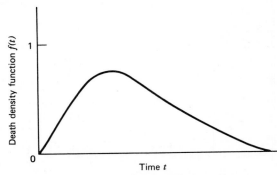

Figure 1.2 A typical death density curve $f(t)$ as a function of t.

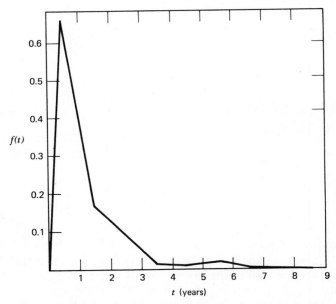

Figure 1.3 Frequency polygon for patients with malignant melanoma (data from Example 1.1.1).

or, since $t > 0$,

$$F(t) = \int_0^t f(T) \, dT.$$

Thus

$$S(t) = 1 - F(t) = \int_t^\infty f(T) \, dT = \operatorname{pr}\{t \leqslant T\}. \tag{1.2.1}$$

Throughout the book we utilize the probability an individual survives at least time t rather than the probability he dies before time t. That is, $S(t)$ is used usually instead of $F(t)$. Assuming that $F(t)$ is differentiable in t, $t \geqslant 0$, the death density is given by

$$f(t) = \lim_{\Delta t \to 0} \left[\frac{\text{prob individual dies in } (t, t + \Delta t)}{\Delta t} \right]$$

$$= F'(t) = -S'(t). \tag{1.2.2}$$

Figure 1.4 shows a typical theoretical survival curve. Figure 1.5 gives the estimated survival distribution $\hat{S}(t)$ for 256 patients with malignant melanoma from the data in Example 1.1.1. Clearly, $\hat{S}(0) = 1$, at $t = 0$, since all 256 patients are alive. After one year of disease, 89 patients are alive. Thus the probability a given patient survives at least one year is estimated from Table 1.1, $\hat{S}(1) = 89/256 = 0.347$. After 2 years of disease 41 patients are alive. The estimated probability a given patient survives at least 2 years is $\hat{S}(2) = 41/256 = 0.160$. Continuing in this way it is easy to obtain the points $\hat{S}(3)$ through $\hat{S}(9)$ from Table 1.1. These points appear in Figure 1.5. Note the decreasing nature of $S(t)$ and $\hat{S}(t)$ in Figures 1.4 and 1.5, respectively.

It follows from the definition of $S(t)$ that $S(0) = 1$, $S(t)$ is a decreasing function of t, and $S(\infty) = 0$.

3. *Hazard Function* $\lambda(t)$: the probability an individual dies in the time interval $t < x < t + \Delta t$, *no matter how small* Δt *is*, given he has survived to time t, is defined as $\lambda(t)\Delta t$. The hazard function, also termed the failure rate, the instantaneous death rate, or the force of mortality, is defined mathematically as

$$\lambda(t) = \lim_{\Delta t \to 0} \left[\frac{\operatorname{pr}\{t < x < t + \Delta t \mid x > t\}}{\Delta t} \right]. \tag{1.2.3}$$

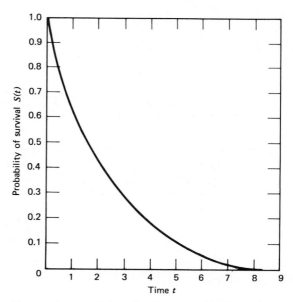

Figure 1.4 A typical survival curve $S(t)$ as a function of t.

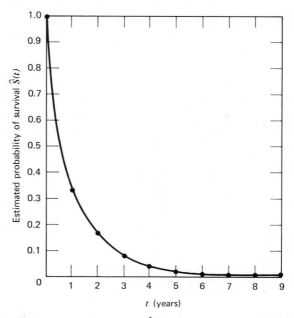

Figure 1.5 Estimated survival distribution $\hat{S}(t)$ for patients with malignant melanoma (estimated using data from Example 1.1.1).

We define $\lambda(x|t)$ to be the conditional death density function for an individual who dies at time $x > t$, given he has lived to t, for a fixed value of t. Since $\lambda(x|t)$ is a death density function,

$$\int_{t}^{\infty} \lambda(x|t)\,dx = 1. \tag{1.2.4}$$

Furthermore, since $\lambda(x|t)$ is a death density function, it is related to the unconditional death density function as follows:

$$\lambda(x|t) = k(t)f(x), \qquad x \geqslant t$$
$$= 0, \qquad x < t, \tag{1.2.5}$$

where $k(t)$ is a constant of proportionality depending only on t. By (1.2.4), it then follows that $k(t) = 1/S(t)$. Note that

$$\frac{1}{S(t)} \int_{t}^{\infty} \lambda(x|t)\,dx = \frac{S(t)}{S(t)} = 1. \tag{1.2.6}$$

If we note that $\lambda(t)$, the conditional failure rate at time t, is the same as $\lambda(t|t)$, it follows that

$$\lambda(t) = \frac{f(t)}{S(t)}. \tag{1.2.7}$$

Graphically, we have $f(t)$ with a total area of 1 and also $\lambda(t)$ with a total area of 1, as in Figure 1.6. Figure 1.7 plots increasing, constant, J-shaped, and decreasing hazard rates as functions of time.

As examples of populations having each of these four hazard functions the following are cited:

1. Individuals who have reached retirement age (i.e., 65 or older) exhibit an increasing hazard rate. For example, the probability an individual survives to age 71, given he has lived to age 70, is greater than the probability an individual survives to age 72, given he has lived to age 71.

2. Individuals from a population whose only risks of death are accidents or rare illness show a constant failure rate.

3. The J-shaped failure rate is typical of the results from a population life table in which an entire population is followed to death. Such is the life table on which life insurance companies base their policy premiums. We have a high failure rate initially followed by a period of few failures, then an increasing failure rate associated with old age or wearing out.

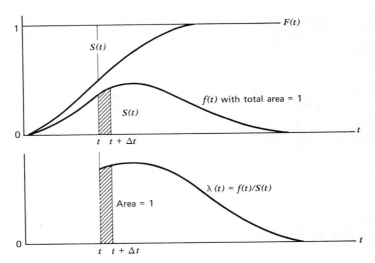

Figure 1.6   Display of $f(t)$, $S(t)$, and $\lambda(t)$.

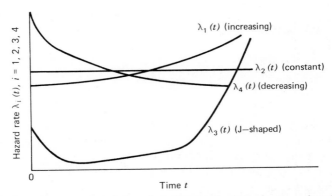

Figure 1.7   Increasing, constant, J-shaped, and decreasing hazard rates as functions of time.

4. Children who have undergone an operative procedure to correct a congenital condition such as a heart defect represent a population exhibiting a decreasing hazard rate. This follows because the principal risk of death is the surgery or complications immediately thereafter.

Figure 1.8 shows the hazard rate estimate for patients with malignant melanoma in Example 1.1.1. To estimate $\lambda(t)$ from the data in Example 1.1.1, we have used the actuarial estimate given in Chapter 2, which is the number of deaths per unit time in the interval divided by the average number of survivors at the midpoint of the interval. Thus the estimated hazard rate average during the first year is $\hat{\lambda}(0.5) = 167/(256 - 167/2) = 0.968$. Similarly the estimated hazard rate average during the second year is $\hat{\lambda}(1.5) = 48/(89 - 48/2) = 0.738$. In a similar way, the other points $\hat{\lambda}(2.5)$ through $\hat{\lambda}(8.5)$ are computed. We note that the hazard rate curve here shows a basic decreasing pattern, although there are reversals. This is not surprising because patients with malignant melanoma tend to die quickly after diagnosis. More details on this computation are given in Chapter 2.

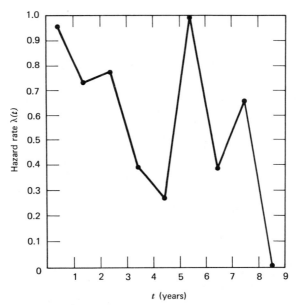

Figure 1.8 Hazard rate $\lambda(t)$ curve for patients with malignant melanoma (data from Example 1.1.1).

This actuarial estimate is somewhat different from the estimate of $\lambda(t)$ defined in (1.2.7) which is used by most reliability engineers. The usual estimate of $\lambda(t)$ is the number of deaths per unit time in the interval divided by the number of survivors at the *beginning* of the interval. Thus $\hat{\lambda}(1) = 167/256 = 0.652$ is the estimated hazard rate at the end of the first year; $\hat{\lambda}(2) = 48/89 = 0.539$ is the estimated hazard rate at the end of the second year. In a similar way, the other points $\hat{\lambda}(3)$ through $\hat{\lambda}(9)$ are computed. Either method of plotting $\hat{\lambda}(t)$ is valid as long as one distinguishes midpoints from end points of intervals. Many medical investigators are more accustomed to the actuarial estimates, but the estimate obtained from (1.2.7), familiar to reliability engineers, is theoretically easier to describe.

Finally, it is useful to display the relationships among $f(t)$, $S(t)$, and $\lambda(t)$. Using (1.2.2), we can rewrite (1.2.7) as

$$\lambda(t) = -\frac{S'(t)}{S(t)} = -\frac{d\log_e S(t)}{dt}. \tag{1.2.8}$$

Hence

$$S(t) = \exp\left[-\int_0^t \lambda(u)\,du\right]. \tag{1.2.9}$$

We have thus the three connecting relationships among $\lambda(t)$, $f(t)$, and $S(t)$, which are

$$\text{(i)} \quad f(t) = -S'(t),$$

$$\text{(ii)} \quad \lambda(t) = \frac{f(t)}{S(t)},$$

and

$$\text{(iii)} \quad S(t) = \exp\left[-\int_0^t \lambda(u)\,du\right]. \tag{1.2.10}$$

Example 1.2.1. Although any of the three functions $\lambda(t)$, $f(t)$, or $S(t)$ defines uniquely a specific survival distribution, each provides the researcher a different view of the data. The shape of $\lambda(t)$ indicates the type of risk to which the population under study is exposed as a function of time. For example, if $\lambda(t)$ is increasing, there is an aging process that increases the rate of death in the population. The density $f(t)$ may be used to assess the peak period of death in a population. For example, if the researcher requires the time when 50% of his sample survives, he would use $\hat{S}(t)$, the estimated survival probability. To demonstrate the relationship among the three functions, suppose that $\lambda(t) = \lambda$ independent of t. Then it follows from (iii) of (1.2.10) that

$$S(t) = \exp(-\lambda t), \qquad t > 0. \tag{1.2.11}$$

Thus if the hazard rate is constant as a function of time, the survival distribution is the negative exponential survival distribution. Furthermore, from (i) of (1.2.10) we have

$$f(t) = \lambda \exp(-\lambda t), \qquad t > 0. \tag{1.2.12}$$

Figure 1.9 presents plots of $f(t)$ and $S(t)$ for the constant hazard rate situation, where $S(t)$ and $f(t)$ are called the exponential survival function and the exponential death density function, respectively.

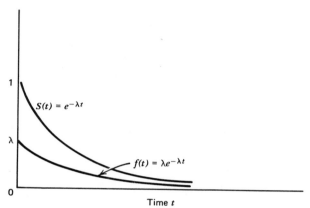

Figure 1.9 Typical exponential survival and density functions for $0 < \lambda < 1$.

It would be reasonable to assume an exponential death density if the cause of death occurs according to a Poisson process with a constant rate. If an individual is subject to random events, such as a blood clot or thrombosis that causes the body to "fail" if and only if that event occurs, we would expect the exponential death density to govern the length of life of the individual.

EXERCISES

The data in Table 1.3 are survival times in years of 100 patients suffering from coronary thrombosis.

1. Plot a histogram of the number of deaths in each interval for the data in Table 1.3.
2. Plot graphs of $\hat{S}(t)$, $\hat{\lambda}(t)$, and $\hat{f}(t)$, the survival distribution, hazard rate, and death density curves, respectively, for the data in Table 1.3. Obtain both the actuarial and reliability estimates of $\lambda(t)$.

Table 1.3

Survival Time (Years)	Number of Patients Surviving to Beginning of Interval
0–1	100
1–2	65
2–3	56
3–4	41
4–5	29
5–6	20
6–7	14
7–8	6
8–9	5
9+	3

3. Suppose the probability that an individual lives longer than $t_1 + t_2$ years, given that he has already lived longer than t_1 years, is the same as the probability he lives longer than t_2 years (unconditionally). Show that the survival distribution for this is of the form $S(t) = \exp(-\lambda t)$.

4. Prove the converse to the theorem in Exercise 3. That is, if an individual's survival distribution is $S(t) = \exp(-\lambda t)$, the probability he lives longer than $t_1 + t_2$ years, given he has already lived longer than t_1 years, is the same as the unconditional probability he lives longer than t_2 years.

5. For an individual to survive, it is necessary that all n of his vital organs (heart, liver, brain, etc.) function or survive. Suppose that $t_1, \ldots, t_n$ are the survival times for the n organs, which are independently and identically distributed with survival distribution $S(t) = \exp(-\lambda t)$. Find the survival distribution for the individual. Can you extend this result if the survival distribution for the ith organ is $S_i(t) = \exp(-\lambda_i t)$, $i = 1, 2, \ldots, n$?

6. Use a limit argument to show that $\lambda(t) = -S'(t)/S(t)$. *Hint*: pr(death in $(t < x < t + \Delta t)$|survival to t) =

$$[F(t + \Delta t) - F(t)]/S(t).$$

7. Using the identity $S(t) \equiv e^{\log_e S(t)}$, derive (1.2.9). *Hint*: $\int du/u = \log_e u$.

1.3 SOME SPECIAL SURVIVAL DISTRIBUTIONS

As noted in Figure 1.4 the hazard function $\lambda(t)$ can take many different forms—each specific form of the hazard function leads to a specific survival distribution $S(t)$. In Example 1.2.1 as well as in the exercises which follow Section 1.2, we described the survival distribution associated with a constant hazard rate $\lambda(t)=\lambda$: namely, the exponential survival distribution. In Chapter 3 the exponential survival distribution is examined in more detail.

In this Section we briefly introduce four other statistical distributions that are used to describe survival times. These are the gamma, Weibull, normal, and lognormal survival distributions. The gamma and Weibull distributions receive detailed treatment in Chapter 4. In this section we want to acquaint the reader with these distributions and to give some reasons how they arise in theory and practice.

Let us suppose that $t_1, t_2, \ldots, t_n$ are independent survival times for n individuals, each of whom has a constant hazard rate λ. Let $y = \sum_{i=1}^{n} t_i$. That is, y is the total survival time for all n individuals. It is instructive to determine the death density function, hence hazard rate and survival distribution of y, since the average time to death $\bar{t}$, is y/n. We proceed by mathematical induction. If $n=2$, the joint death density function for t_1 and t_2 is

$$f(t_1, t_2) = \lambda^2 \exp[-\lambda(t_1 + t_2)], \qquad t_1 \geq 0, t_2 \geq 0. \tag{1.3.1}$$

Letting $y = t_1 + t_2$, we see that

$$f(t_1, y) = \lambda^2 \exp(-\lambda y), \qquad 0 \leq t_1 \leq y, y \geq 0. \tag{1.3.2}$$

Integrating out t_1, we have

$$f(y) = \lambda^2 y \exp(-\lambda y), \qquad y \geq 0. \tag{1.3.3}$$

Now assume for $n = k$, that

$$f(y_k) = \frac{\lambda^k y_k^{k-1} \exp(-\lambda y_k)}{(k-1)!}, \qquad y \geq 0. \tag{1.3.4}$$

That is, $f(y_k)$ is the assumed (under the induction hypothesis) death density function for $y_k = \sum_{i=1}^{k} t_i$. Equations 1.3.3 and 1.3.4 are special cases of gamma density function, which is defined by (1.3.8). Now

$$f(y_k, t_{k+1}) = \frac{\lambda^{k+1} y_k^{k-1} \exp[-\lambda(y_k + t_{k+1})]}{(k-1)!}, \qquad y_k \geq 0, t_{k+1} \geq 0. \tag{1.3.5}$$

Letting $t = y_k + t_{k+1}$, we see that

$$f(t_{k+1}, t) = \frac{\lambda^{k+1} (t - t_{k+1})^{k-1} \exp(-\lambda t)}{(k-1)!}, \qquad 0 \leq t_{k+1} \leq t, t \geq 0. \tag{1.3.6}$$

Integrating out t_{k+1} we finally have

$$f(t) = \frac{\lambda^{k+1} t^k \exp(-\lambda t)}{k!}, \qquad t \geqslant 0. \qquad (1.3.7)$$

Thus we have shown that if $t = t_1 + t_2 + \cdots + t_n$ where each time t_i has an exponential death density function $\lambda \exp(-\lambda t_i)$, and the times $t_1, t_2, \ldots, t_n$ are mutually statistically independent, the death density function of t is given by

$$f(t) = \frac{\lambda^n t^{n-1} \exp(-\lambda t)}{(n-1)!}, \qquad t \geqslant 0. \qquad (1.3.7')$$

This death density function is a special case of the gamma density function whose general form is

$$f(t) = \frac{\lambda^\gamma t^{\gamma-1} \exp(-\lambda t)}{\Gamma(\gamma)}, \qquad t \geqslant 0, \qquad (1.3.8)$$

where $\lambda > 0$, and $\gamma > 0$ are the scale and shape parameters of the gamma density function, respectively, and

$$\Gamma(\gamma) = \int_0^\infty x^{\gamma-1} e^{-x} dx. \qquad (1.3.9)$$

If we integrate (1.3.9) by parts, we obtain the recursion formula $\Gamma(\gamma) = (\gamma - 1)\Gamma(\gamma - 1)$. Hence if $\gamma = n$, an integer, (1.3.8) and (1.3.7') are equivalent.

The gamma death density function is thus a generalization of the exponential death density, the latter having a shape parameter $\gamma = 1$. In fact, if $\gamma > 1(<1)$, the hazard rate for the gamma death density increases (decreases) as a function of time.

For the special case when $\gamma = n$, an integer, it is not difficult to show (again successive integration by parts) that $S(t)$, the gamma survival distribution, is given by

$$S(t) = \sum_{\nu=0}^{n-1} \frac{(\lambda t)^\nu}{\nu!} \exp(-\lambda t), \qquad t \geqslant 0. \qquad (1.3.10)$$

Hence in this special case the hazard rate is given by

$$\lambda(t) = \frac{f(t)}{S(t)} = \frac{\lambda^n t^{n-1}/(n-1)!}{\sum_{\nu=0}^{n-1} (\lambda t)^\nu / \nu!}, \qquad (1.3.11)$$

which is an increasing hazard rate. (See Exercise 4.)

An application of the hazard rate (1.3.11) to a real-life situation can be seen. Suppose we are studying a group of patients suffering from a kidney disease in which the failure rate of each kidney is constant and equals λ. For a patient to die from the disease, both kidneys must fail. Hence the hazard rate of these patients with this disease is $\lambda[\lambda t/(1+\lambda t)]$, which is easily seen by setting $n=2$ in (1.3.11). It is clear that this hazard rate is an increasing function of time for $\lambda>0$.

Another widely used hazard rate is the Weibull hazard rate given by

$$\lambda(t)=\lambda\gamma(t-\delta)^{\gamma-1}, \tag{1.3.12}$$

where λ, γ, and δ are all real parameters greater than zero. Weibull [1939] proposed this hazard rate because the function $\lambda(t)$ under fairly general mathematical conditions can be expressed as a polynomial in t. Furthermore, it has been found that the Weibull hazard rate often yields an excellent fit to survival data. We usually assume δ, the location or threshold parameter for the Weibull hazard function, is known or is zero. When δ is known, we make a transformation $t'=t-\delta$. Note that since δ must be less than $\min_i t_i$, if one has small values of t_i taking $\delta=0$ is a sensible choice. (An estimator of δ that is independent of γ and λ is given by Dubey [1967] for the case in which the smallest t_i are much larger than zero.) If the prime notation is dropped, we can write the Weibull hazard rate $\lambda(t)$ as

$$\lambda(t)=\lambda\gamma t^{\gamma-1}. \tag{1.3.12'}$$

It then follows that $f(t)$ and $S(t)$ corresponding to (1.3.12') are, respectively,

$$f(t)=\lambda\gamma t^{\gamma-1}\exp(-\lambda t^\gamma) \tag{1.3.13}$$

and

$$S(t)=\exp(-\lambda t^\gamma). \tag{1.3.14}$$

For the Weibull distribution, as for the gamma distribution, λ and γ are the scale and shape parameters, respectively. Like the gamma hazard rate, the Weibull hazard rate is increasing, constant, or decreasing according to whether $\gamma>1$, $\gamma=1$ (in which case we obtain the exponential); or $\gamma<1$, respectively.

Reviewing, we see that the gamma and Weibull death densities are generalizations of the exponential death density. Although it is often difficult to distinguish between them, we see that the gamma often arises from physical considerations (the sum of survival times). Since these distributions are difficult to distinguish in practice, it is helpful if the researcher can pinpoint the way in which his data are obtained. For example, can he ascertain that the survival time of a patient is the sum of survival times, each one having a constant hazard rate? Further discussion on this problem is given in Chapter 4.

Although the normal distribution is widely used in many areas of statistical research, its use as a death density function is limited. One difficulty is that if the random variable t is assumed to follow a normal density function with mean μ and variance σ^2, then

$$f(t) = \frac{1}{\sqrt{2\pi}\,\sigma} \exp\left[-\frac{(t-\mu)^2}{2\sigma^2} \right], \tag{1.3.15}$$

with $-\infty < \mu < \infty$, $\sigma > 0$, and $-\infty < t < \infty$. Thus theoretically both the observed time to death and the mean time to death can take on negative values. However, if $\mu > 0$ and $\mu/\sigma \geqslant 3$, it is virtually impossible for t to be negative.

The normal distribution has been used in reliability applications when it is assumed that failure or death is due to accumulated wear. Thus if k failures in total are required for a death, then for k large (as a consequence of the central limit theorem), the death density would approach normality. Or if the amount of some substance needed by the body is normally distributed and failure occurs because it totally consumed, we would expect a normal death density.

The survival function then is

$$S(t) = \frac{1}{\sqrt{2\pi}\,\sigma} \int_t^\infty \exp\left[-\frac{(x-\mu)^2}{2\sigma^2} \right] dx. \tag{1.3.16}$$

The hazard rate $\lambda(t)$ is

$$\lambda(t) = \frac{f(t)}{S(t)} = \frac{\exp\left\{ -\dfrac{(t-\mu)^2}{2\sigma^2} \right\}}{\displaystyle\int_t^\infty \exp\left\{ -\dfrac{(x-\mu)^2}{2\sigma^2} \right\} dx}. \tag{1.3.17}$$

The inverse of the hazard rate for the normal distribution is called Mills' ratio and is tabulated in Sheppard [1939]. Kendall and Stuart [1963] provide an expression for Mills' ratio in terms of gamma functions and in terms of continued fractions. The hazard function is a monotonically increasing function and has been graphed in Gumbel [1958].

Customarily, when t has a normal distribution with mean μ and variance σ^2, we write $t \sim N(\mu, \sigma^2)$. This notation is used in Chapters 7 and 8.

Another death density function in common use closely associated with the normal is the lognormal density function. This density function has the form

$$f(t) = \frac{1}{t\sigma\sqrt{2\pi}} \exp\left[\frac{-1}{2\sigma^2}(\log_e t - \mu)^2\right], \qquad -\infty < \mu < \infty, \sigma > 0, t \geq 0,$$

(1.3.18)

$$S(t) = \frac{1}{\sigma\sqrt{2\pi}} \int_t^\infty \frac{1}{x} \exp\left[\frac{-1}{2\sigma^2}(\log_e x - \mu)^2\right] dx, \qquad (1.3.19)$$

and

$$\lambda(t) = \frac{f(t)}{S(t)}, \qquad (1.3.20)$$

where $f(t)$ and $S(t)$ are given by (1.3.18) and (1.3.19), respectively.

Boag [1949] gives an application of the lognormal survival distribution to patients with cancer. He compares fitting patient survival to both the lognormal and exponential survival distributions. We also note that the exponential is *not* a special case of the lognormal. In fact $\lambda(t)$ defined by (1.3.20) can be shown to increase to a maximum value and then decrease as t increases to infinity. Furthermore, $f(t)$ defined by (1.3.18) is skewed to the right and has a long tail.

As a practical example of the use of the lognormal survival distribution, length of times to recovery from injuries or surgery often follows a lognormal survival distribution. Most recoveries are quick (within a few months) but some take considerably longer.

Other more complex types of distributions can be considered. For example, we could have a J-shaped or bathtub-shaped hazard rate characterized by a period of decreasing risk followed by a relatively constant failure rate and concluding with a rapidly increasing failure rate. This would correspond to a high risk of death in early infancy, a fairly constant and low rate from childhood through middle age, and an increasing rate in old age (or, for equipment, a breaking-in period, useful life, and a final wear-out period). Murthy, Swartz, and Yuen [1973] discuss this type of distribution where the hazard rate can be expressed as

$$\lambda(t) = \frac{a}{1+bt} + c\,dt^{d-1}. \tag{1.3.21}$$

Finally we consider the death density function whose hazard rate is constant except for jump changes. That is,

$$\lambda(t) = \begin{cases} \lambda_1 & 0 \leqslant t < t_1, \\ \lambda_2 & t_1 \leqslant t < t_2, \\ \cdot & \cdot \\ \cdot & \cdot \\ \cdot & \cdot \\ \lambda_{k-1} & t_{k-2} \leqslant t < t_{k-1}, \\ \lambda_k & t \geqslant t_{k-1}, \end{cases} \tag{1.3.22}$$

where $t_1, t_2, \ldots, t_{k-1}$ are known points of change of the parameter. Figure 1.10 shows a typical hazard rate that is constant except for jump changes.

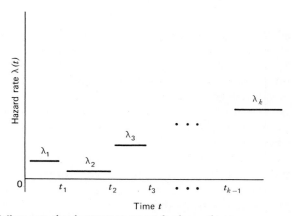

Time t

Figure 1.10 Failure rate that is constant except for jump changes.

There is an application of (1.3.22) to life tables. Namely, the hazard rate or instantaneous death rate of people is assumed to be constant in each time interval defined in the table. However, the overall death rate is not constant in life tables in general. See Table 2.3 (Chapter 2), which is a clinical life table for patients with malignant melanoma.

Expressions for $S(t)$ and $f(t)$, the survival probability function and the death density function, respectively, are obtainable from $\lambda(t)$. Thus,

$$
S(t) = \begin{cases}
\exp(-\lambda_1 t), & 0 \leqslant t < t_1, \\
\exp[-\lambda_1 t_1 - \lambda_2(t - t_1)], & t_1 \leqslant t < t_2, \\
\quad \cdot & \\
\quad \cdot & \\
\quad \cdot & \\
\exp[-\lambda_1 t_1 - \lambda_2(t_2 - t_1) - \cdots - \lambda_k(t - t_{k-1})], & t \geqslant t_{k-1},
\end{cases}
$$

$$(1.3.23)$$

and

$$
f(t) = \begin{cases}
\lambda_1 \exp(-\lambda_1 t), & 0 \leqslant t < t_1, \\
\lambda_2 \exp[-\lambda_1 t_1 - \lambda_2(t - t_1)], & t_1 \leqslant t < t_2, \\
\quad \cdot & \\
\quad \cdot & \\
\quad \cdot & \\
\lambda_k \exp[-\lambda_1 t_1 - \lambda_2(t_2 - t_1) - \cdots - \lambda_k(t - t_{k-1})], & t \geqslant t_{k-1}.
\end{cases}
$$

$$(1.3.24)$$

The density function $f(t)$ in (1.3.24) is termed the piecewise exponential death density function.

Chapter 1 affords the reader a menu of survival distributions. Admittedly the list is not complete; however, in many practical applications the experimenter will be able to plot a graph of $\lambda(t)$ (see Figure 1.7) or $S(t)$ (see Figure 1.4). Such a graph may allow the experimenter to choose one of the survival distributions described in this chapter. For example, if $\lambda(t)$ is plotted and is found to be fairly constant with respect to t, this is a good indication that the exponential survival distribution reasonably describes the survival data. At this point we note that the hazard rate $\lambda(t)$, is a better determinant of the appropriate survival distribution because its form is more distinct than the corresponding values, the death density function $f(t)$, or the survival function, $S(t)$. For example, a constant $\lambda(t)$ is easier to recognize than the corresponding exponential death density or survival distribution.

Finally we note that techniques for estimating a survival distribution from data (particularly life table data) are well documented. Berkson and

Gage [1950] and Cutler and Ederer [1958] use the actuarial method; Chiang [1960], and Kaplan and Meier [1958] develop maximum likelihood estimates. Kaplan and Meier's estimate is asymptotically equivalent to the actuarial or life table method and has the added feature of being distribution free. Life tables including the population and clinical life table and a brief discussion of Kaplan and Meier's results constitute Chapter 2.

EXERCISES

1. Suppose failure rate of each kidney for patients suffering from a kidney disease is $\lambda = 0.14$ per year. Plot the hazard rate $\lambda(t) = \lambda[\lambda t/(1+\lambda t)]$ for this disease as a function of t.

2. For the piecewise exponential death density function, suppose $\lambda_1 = 0.2$, $0 \leqslant t < 1$; $\lambda_2 = 0.4, 1 \leqslant t < 2$; $\lambda_3 = 0.6, 2 \leqslant t < 3$; $\lambda_4 = 0.8, 3 \leqslant t < 4$; and $\lambda_5 = 1.0, t \geqslant 4$. Plot a graph showing $S(t)$, $f(t)$, and $\lambda(t)$. Compute the mean and variance of t.

3. Suppose t_1 and t_2 are random variables, each having the same Weibull death density function. Find the death density function of $t = (t_1^\gamma + t_2^\gamma)^{1/\gamma}$.

4. Show that the hazard rate (1.3.11) is an increasing function of t, hence that the distribution of the sum of death times, each having a constant hazard rate, has a hazard rate that is increasing over time.

5. Obtain the means and variances for all the death density functions discussed in this chapter: namely, exponential, gamma, Weibull, normal, lognormal, and piecewise exponential death density functions.

6. Suppose t_1 and t_2 are independent random variables with death density functions $\gamma_1 \exp(-\gamma_1 t_1)$ and $\gamma_2 \exp(-\gamma_2 t_2)$, respectively, $\gamma_1 \neq \gamma_2$. Find the death density function of the random variable $t = t_1 + t_2$.

7. From (1.3.22) and the fundamental equations relating $\lambda(t)$, $f(t)$, and $S(t)$, derive (1.3.23) and (1.3.24), the survival function and the death density function, respectively, for the piecewise exponential.

8. Suppose patients with a kidney disease show $\lambda = 0.5$ failures/year. Plot the failure rate $\lambda(t) = 0.5[0.5t/(1+0.5t)]$ as a function of time. Also plot $\lambda(t) = 0.5$ as the constant failure rate of each kidney, and compare the graphs of $\lambda(t)$ and λ.

9. Suppose $\lambda(t)$ is the hazard rate for gamma death density function (1.3.8). Show that $\lambda(t)$ increases (decreases) with t for $\gamma > 1 (<1)$, and $\lim_{t \to \infty} \lambda(t) = \lambda$.

CHAPTER 2

Life Table Analysis of Survival Data

2.1 INTRODUCTORY REMARKS: POPULATION LIFE TABLE

Chapter 2 introduces the study of nonparametric analysis of survival data. Ordinarily these more descriptive analyses are utilized prior to the parametric techniques given in subsequent chapters. In many actual studies, they are sufficient to analyze the data. In fact, only a very small percentage of clinical studies is presently analyzed by techniques other than those presented here. Section 2.1 discusses population life tables, Section 2.2 includes clinical life tables, Section 2.3 provides formulas for the variances of the estimates used in clinical life tables and formulas for the median, and Section 2.4 introduces the work of Kaplan and Meier [1958].

One of the basic tools in the description of the mortality experience of a population, the life table, was first developed by E. Halley [1693] in England; it has been used extensively by biostatisticians and actuaries to portray the pattern of survival in populations. Although by survival we usually signify time to death, the technique is useful for describing other data. For example, we could be describing the length of stay in a mental hospital, wherein "birth" is entry into the hospital and "death" is discharge. As a second example, we may be considering the length of remission in acute leukemia; here "birth" is the time at which the patient goes into remission, and "death" is the time at which he relapses. Admittedly, birth and death are more difficult to discern in the second example.

There are three types of life table in common use—the population life table, the clinical life table, and the cohort life table. We discuss population life tables because they were developed first historically; the reasoning and notation has derived from them. Second, we describe the clinical life table because this alternative form of analysis to that given in the rest of the book is useful in deciding what hazard rate to assume. Cohort life

23

tables describe the actual survival experience of a group or "cohort" of individuals who were born at about the same time. (See Chiang [1968] for a discussion of the differences between population and cohort tables.) We do not discuss cohort tables further because they are not revelant to the rest of the book, and they are the least used of the life tables.

To construct a population life table, two sources of data are required. These are: (1) census data on the number of living persons at each age (i.e., 0–1, 1–2, etc.) for a given year measured at midyear (i.e., June 30), and (2) vital statistics on the number of deaths in the given year for each age. We now define the following terms:

1. $_hd_i$ is number of persons ages t_i to t_{i+h} registered dead in a given year, year z.

2. $_hP_i^{z+1/2}$ is the total population whose age is t_i to t_{i+h} at the midyear of year z.

3. $_hm_i$ is the age-specific death rate for persons whose age is t_i to t_{i+h} in the given year z (see Shryock and Siegel [1973]).

It follows from these definitions that

$$_hm_i = \frac{_hd_i}{_hP_i^{z+1/2}} . \tag{2.1.1}$$

To construct a current population life table, we need to obtain the probability that an individual dies between ages t_i and t_{i+h} if he is alive in year z. Let $_h\hat{q}_i$ be this probability. The quantity $_h\hat{q}_i$ is not the same as $_hm_i$ but is related to $_hm_i$. Note that $_h\hat{q}_i$ relates to an individual's birthdate, whereas $_hm_i$ relates to the regular calendar year. Both Chiang [1968] and Spiegelman [1968] show that

$$_h\hat{q}_i = \frac{_hm_i}{1 + (1 - {_ha_i})_hm_i} , \tag{2.1.2}$$

where $_ha_i$ is the fraction of a year that a person lives who is alive at t_i and dead by t_{i+h}. By assuming that both births and deaths are uniformly distributed over the year, Spiegelman shows that $_ha_i = \frac{1}{2}$. This is a reasonable assumption except for the first few years of life and for the very elderly, where other values of $_ha_i$ are used. (See Jaffe [1951] for values to use at the extremes of life.) When $h = 1$, it is commonly omitted from the notation. Thus for $h = 1$ and ignoring the very young or old,

$$\hat{q}_i = \frac{m_i}{1 + \frac{1}{2}m_i} . \tag{2.1.3}$$

Thus $\hat{q}_i$ gives us the conditional probability of dying at an age between t_i and t_{i+1}, given that an individual has lived to age t_i.

Usually, we start with $s_0 = 100,000$ persons alive at age 0. This value of s_0 along with the $\hat{q}_i$ are the only statistics used in computing a population life table. If we define d_i as the estimated number of individuals who die between ages t_i and t_{i+1} and s_i as the number of individuals who are alive at age t_i, we obtain

$$d_0 = s_0\hat{q}_0 \tag{2.1.4}$$

and

$$s_1 = s_0 - d_0. \tag{2.1.5}$$

In general, at age t_i

$$d_i = s_i\hat{q}_i \tag{2.1.6}$$

and

$$s_{i+1} = s_i - d_i = s_i(1 - \hat{q}_i). \tag{2.1.7}$$

Let L_i be the average number of years of life in the ith interval. For persons over age 1 we write

$$_hL_i = \frac{h}{2}(s_i + s_{i+h}), \tag{2.1.8}$$

or for $h = 1$

$$L_i = \tfrac{1}{2}(s_i + s_{i+1}), \tag{2.1.9}$$

or equivalently

$$L_i = \left(s_i - \frac{d_i}{2}\right). \tag{2.1.9'}$$

The expected length of life for a person already at age t_i is

$$\overset{o}{e}_i = \frac{\sum_{j=i}^{\infty} L_j}{s_i} = \frac{T_i}{s_i}. \tag{2.1.10}$$

The statistic, $\overset{o}{e}_i$ is used by actuaries in determining rates and benefits for life insurance and by biostatisticians in comparing survivorship from several populations. A summary of the usual data contained in a life table is given in Table 2.1.

If we graphed $_h\hat{q}_i$, we would have a J-shaped graph similar to that given in Figure 1.7. A graph of s_i would show a curve that decreases rapidly soon after birth, is almost flat during early adulthood when few deaths occur, and again decreases rapidly in old age. The values of $_h\hat{q}_i$ and s_i from Table 2.1 are graphed in Figure 2.1. For a less well-developed country the graph of $_h\hat{q}_i$ would show a higher tail at the left end because of high infant mortality. The sharp and prolonged rise in $_h\hat{q}_i$ after age 60 or 70 indicates that one should consider hazard rates, which can be increasing functions if the patients under study are elderly and the time interval is more than 2 or 3 years. The values of $_h\hat{q}_i$ should be graphed at the midpoint of the age interval and s_i at the beginning point of the interval.

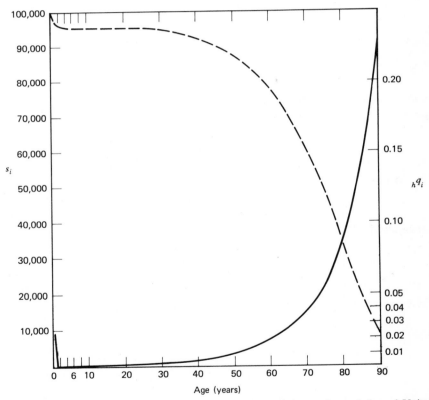

Figure 2.1 Graph of s_i (dashed curve) and $_h q_i$ (solid curve) for total population of United States in 1959–1961.

Table 2.1 Complete life table for the total population of the United States: 1959–1961

Age Interval: Period of Life Between Two Exact Ages (years) i to $i+h$	Proportion Dying: Proportion of Persons Alive at Beginning of Age Interval Dying During Interval $_hq_i$	Of 100,000 Born Alive		Stationary Population		Average Remaining Lifetime: Average Number of Years of Life Remaining at Beginning of Age Interval $\overset{o}{e}_i$
		Number Living at Beginning of Age Interval s_i	Number Dying During Age Interval $_hd_i$	In the Age Interval $_hL_i$	In this and All Subsequent Age Intervals T_i	
	(1)	(2)	(3)	(4)	(5)	(6)
0–1........	.02593	100,000	2,593	97,815	6,989,030	69.89
1–2........	.00170	97,407	165	97,324	6,891,215	70.75
2–3........	.00104	97,242	101	97,192	6,793,891	69.87
3–4........	.00080	97,141	78	97,102	6,696,699	68.94
4–5........	.00067	97,063	65	97,031	6,599,597	67.99
5–6........	.00059	96,998	57	96,969	6,502,566	67.04
6–7........	.00052	96,941	50	96,916	6,405,597	66.08
7–8........	.00047	96,891	46	96,868	6,308,681	65.11
8–9........	.00043	96,845	42	96,824	6,211,813	64.14
9–10.......	.00039	96,803	38	96,784	6,114,989	63.17
10–11......	.00037	96,765	36	96,747	6,018,205	62.19
11–12......	.00037	96,729	36	96,711	5,921,458	61.22
12–13......	.00040	96,693	39	96,674	5,824,747	60.24
13–14......	.00048	96,654	46	96,630	5,728,073	59.26
14–15......	.00059	96,608	57	96,580	5,631,443	58.29
15–16......	.00071	96,551	68	96,517	5,534,863	57.33
16–17......	.00082	96,483	80	96,443	5,438,346	56.37
17–18......	.00093	96,403	89	96,358	5,341,903	55.41
18–19......	.00102	96,314	98	96,265	5,245,545	54.46
19–20......	.00108	96,216	105	96,163	5,149,280	53.52
20–21......	.00115	96,111	110	96,056	5,053,117	52.58
21–22......	.00122	96,001	118	95,942	4,957,061	51.64
22–23......	.00127	95,883	122	95,822	4,861,119	50.70
23–24......	.00128	95,761	123	95,700	4,765,297	49.76
24–25......	.00127	95,638	121	95,578	4,669,597	48.83
25–26......	.00126	95,517	120	95,456	4,574,019	47.89
26–27......	.00125	95,397	120	95,337	4,478,563	46.95
27–28......	.00126	95,277	120	95,217	4,383,226	46.00
28–29......	.00130	95,157	123	95,095	4,288,009	45.06
29–30......	.00136	95,034	129	94,970	4,192,914	44.12

Table 2.1 (*Continued*)

Age Interval: Period of Life Between Two Exact Ages (years) i to $i+h$	Proportion Dying: Proportion of Persons Alive at Beginning of Age Interval Dying During Interval $_hq_i$	Of 100,000 Born Alive		Stationary Population		Average Remaining Lifetime: Average Number of Years of Life Remaining at Beginning of Age Interval $\overset{o}{e}_i$
		Number Living at Beginning of Age Interval s_i	Number Dying During Age Interval $_hd_i$	In the Age Interval $_hL_i$	In this and All Subsequent Age Intervals T_i	
	(1)	(2)	(3)	(4)	(5)	(6)
30–31......	.00143	94,905	136	94,836	4,097,944	43.18
31–32......	.00151	94,769	143	94,698	4,003,108	42.24
32–33......	.00160	94,626	151	94,551	3,908,410	41.30
33–34......	.00170	94,475	160	94,395	3,813,859	40.37
34–35......	.00181	94,315	171	94,229	3,719,464	39.44
35–36......	.00194	94,144	183	94,053	3,625,235	38.51
36–37......	.00209	93,961	196	93,863	3,531,182	37.58
37–38......	.00228	93,765	214	93,658	3,437,319	36.66
38–39......	.00249	93,551	232	93,435	3,343,661	35.74
39–40......	.00273	93,319	255	93,191	3,250,226	34.83
40–41......	.00300	93,064	279	92,925	3,157,035	33.92
41–42......	.00330	92,785	306	92,632	3,064,110	33.02
42–43......	.00362	92,479	335	92,311	2,971,478	32.13
43–44......	.00397	92,144	366	91,961	2,879,167	31.25
44–45......	.00435	91,778	400	91,578	2,787,206	30.37
45–46......	.00476	91,378	435	91,161	2,695,628	29.50
46–47......	.00521	90,943	473	90,707	2,604,467	28.64
47–48......	.00573	90,470	519	90,210	2,513,760	27.79
48–49......	.00633	89,951	569	89,667	2,423,550	26.94
49–50......	.00700	89,382	626	89,069	2,333,883	26.11
50–51......	.00774	88,756	687	88,412	2,244,814	25.29
51–52......	.00852	88,069	751	87,693	2,156,402	24.49
52–53......	.00929	87,318	811	86,913	2,068,709	23.69
53–54......	.01005	86,507	870	86,072	1,981,796	22.91
54–55......	.01082	85,637	926	85,174	1,895,724	22.14

Table **2.1** (*Continued*)

Age Interval: Period of Life Between Two Exact Ages (years) i to $i+h$	Proportion Dying: Proportion of Persons Alive at Beginning of Age Interval Dying During Interval $_hq_i$	Of 100,000 Born Alive Number Living at Beginning of Age Interval s_i	Number Dying During Age Interval $_hd_i$	Stationary Population In the Age Interval $_hL_i$	In this and All Subsequent Age Intervals T_i	Average Remaining Lifetime: Average Number of Years of Life Remaining at Beginning of Age Interval $\overset{o}{e}_i$
(1)	(2)	(3)	(4)	(5)	(6)	
55–56......	.01161	84,711	983	84,220	1,810,550	21.37
56–57......	.01249	83,728	1,047	83,204	1,726,330	20.62
57–58......	.01352	82,681	1,117	82,123	1,643,126	19.87
58–59......	.01473	81,564	1,202	80,962	1,561,003	19.14
59–60......	.01611	80,362	1,295	79,715	1,480,041	18.42
60–61......	.01761	79,067	1,392	78,371	1,400,326	17.71
61–62......	.01917	77,675	1,489	76,930	1,321,955	17.02
62–63......	.02082	76,186	1,586	75,393	1,245,025	16.34
63–64......	.02252	74,600	1,680	73,760	1,169,632	15.68
64–65......	.02431	72,920	1,773	72,033	1,095,872	15.03
65–66......	.02622	71,147	1,866	70,214	1,023,839	14.39
66–67......	.02828	69,281	1,959	68,302	953,625	13.76
67–68......	.03053	67,322	2,055	66,295	885,323	13.15
68–69......	.03301	65,267	2,155	64,189	819,028	12.55
69–70......	.03573	63,112	2,255	61,985	754,839	11.96
70–71......	.03866	60,857	2,352	59,681	692,854	11.38
71–72......	.04182	58,505	2,447	57,282	633,173	10.82
72–73......	.04530	56,058	2,539	54,788	575,891	10.27
73–74......	.04915	53,519	2,631	52,204	521,103	9.74
74–75......	.05342	50,888	2,718	49,529	468,899	9.21
75–76......	.05799	48,170	2,794	46,773	419,370	8.71
76–77......	.06296	45,376	2,857	43,948	372,597	8.21
77–78......	.06867	42,519	2,920	41,059	328,649	7.73
78–79......	.07535	39,599	2,983	38,108	287,590	7.26
79–80......	.08302	36,616	3,040	35,096	249,482	6.81
80–81......	.09208	33,576	3,092	32,030	214,386	6.39
81–82......	.10219	30,484	3,115	28,926	182,356	5.98
82–83......	.11244	27,369	3,078	25,830	153,430	5.61
83–84......	.12195	24,291	2,962	22,811	127,600	5.25
84–85......	.13067	21,329	2,787	19,935	104,789	4.91

Table 2.1 (*Continued*)

Age Interval: Period of Life Between Two Exact Ages (years) i to $i+h$	Proportion Dying: Proportion of Persons Alive at Beginning of Age Interval Dying During Interval $_hq_i$	Of 100,000 Born Alive		Stationary Population		Average Remaining Lifetime: Average Number of Years of Life Remaining at Beginning of Age Interval $\overset{o}{e}_i$
		Number Living at Beginning of Age Interval s_i	Number Dying During Age Interval $_hd_i$	In the Age Interval $_hL_i$	In this and All Subsequent Age Intervals T_i	
	(1)	(2)	(3)	(4)	(5)	(6)
85–86......	.14380	18,542	2,666	17,209	84,854	4.58
86–87......	.15816	15,876	2,511	14,620	67,645	4.26
87–88......	.17355	13,365	2,320	12,205	53,025	3.97
88–89......	.19032	11,045	2,102	9,995	40,820	3.70
89–90......	.20835	8,943	1,863	8,011	30,825	3.45
90–91......	.22709	7,080	1,608	6,276	22,814	3.22
91–92......	.24598	5,472	1,346	4,799	16,538	3.02
92–93......	.26477	4,126	1,092	3,580	11,739	2.85
93–94......	.28284	3,034	858	2,605	8,159	2.69
94–95......	.29952	2,176	652	1,849	5,554	2.55
95–96......	.31416	1,524	479	1,285	3,705	2.43
96–97......	.32915	1,045	344	873	2,420	2.32
97–98......	.34450	701	241	580	1,547	2.21
98–99......	.36018	460	166	377	967	2.10
99–100.....	.37616	294	111	239	590	2.01
100–101....	.39242	183	72	147	351	1.91
101–102....	.40891	111	45	89	204	1.83
102–103....	.42562	66	28	52	115	1.75
103–104....	.44250	38	17	29	63	1.67
104–105....	.45951	21	10	17	34	1.60
105–106....	.47662	11	5	8	17	1.53
106–107....	.49378	6	3	5	9	1.46
107–108....	.51095	3	2	2	4	1.40
108–109....	.52810	1	0	1	2	1.35
109–110....	.54519	1	1	1	1	1.29

Source: U. S. National Center for Health Statistics, Life Tables, 1959–61, Vol. 1, No. 1, "United States Life Tables: 1959–61," December 1964, pp. 8–9.

Current population life tables present a picture of mortality as of a moment of time (usually 1 or 5 years). Note that one uses census and vital statistics data taken at a single time period for all age groups. The life table shows a fictitious pattern of mortality conditions because no combination of cohorts (persons born at the same time) has actually experienced or will experience this particular pattern of mortality. For example, if we look at the expected length of life of a person age 25 in 1961 ($\overset{o}{e}_{25} = 48$ years), we will be using data from persons who are currently older than 25. In 20 years the life span of a person now 25 years old may be different from those of persons now 45 years old. A population life table shows a combined current experience of many different cohorts.

2.2 CLINICAL LIFE TABLES

Clinical life tables reflect the thinking and notation of population life tables but use data from clinical studies of patients instead of census vital statistics data. To understand the procedure, let us examine a small set of data from a given clinic. In a study of a disease condition such as lung cancer, it is seldom possible to admit all the patients to the study sample at the same time because of the paucity of patients. We have to accept the patients as they enter for treatment. We then follow each patient to death. Unfortunately, a portion of the patients may be lost to follow-up because the investigator can neither find the patient nor determine that he has died.

Figure 2.2 illustrates some typical cases. For example, patient 1 entered shortly after the start of the study and soon died. Patient 2 entered later and died after a long follow-up. Patient 3 was lost to follow-up during the study period. Patient 4 entered later but died during the study period, whereas patient 5 entered later and was withdrawn alive because the study

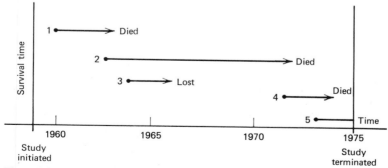

Figure 2.2 Patients entering and leaving study.

ended. In working with these data, we assume that neither the treatment nor the characteristics of the patients change during the period of study; hence the starting point for each patient (this is illustrated in Figure 2.3 for the same patients shown in Figure 2.2) can be moved to the beginning of the study.

Thus in clinical trials when the length of time until death or some other event, such as the absence of clinical symptoms (remission), can be long, and when the disease is so rare that we enter patients one at a time over a long period, we obtain progressively censored samples.

The simplest technique for analyzing these data is the T-year survival rate (see Berkson and Gage [1950]). Only patients who entered the study early (i.e., so that they have a known exposure to the risk of dying in T years) are used. Of this reduced number of patients, the T year survival rate is the proportion surviving T years. For example, suppose we wish to compute the 5-year survival rate for the patients in Figure 2.2. We would have

$$5\text{-year survival rate} = \frac{\text{patient 2}}{\text{patient 1} + \text{patient2}}$$

$$= \tfrac{1}{2} = 0.5 \text{ or } 50\%.$$

Patients 1 and 2 can be used because they entered the study early enough to be exposed to the risk of dying for more than 5 years and they were not lost before 5 years. Patients 3, 4, and 5 could not be used in computing the rate, since they were not exposed to risk of dying for 5 years. (Patient 3 was lost before 5 years and patients 4 and 5 entered too late.) Thus all the information available cannot be used, and larger sample sizes are needed to compare a new treatment to previous results or for the simultaneous

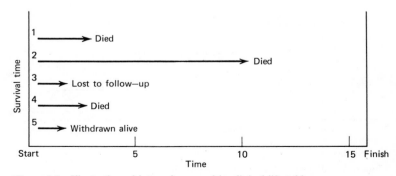

Figure 2.3 Illustration of intervals as used in clinical life tables.

comparison of two treatments. Gehan [1967] showed that in conducting a clinical trial to compare the effects of two treatments, appreciably fewer patients would be needed if an exponential distribution were used instead of the T-year survival rate. Since obtaining a large sample is difficult in many medical studies, techniques that allow the investigator to reach conclusions with a smaller sample size (i.e., parametric techniques) are emphasized in this book. In Chapter 8 techniques for determining necessary sample size are given.

For the patient who is not lost nor withdrawn, two points determine the length of time a patient spends in the study—the starting point and the time of death. The time of death is obvious, but the choice of a proper starting point can be quite difficult. Often comparisons between studies are complicated by different choices of starting points. In a population life table the starting point is birth at year 0. Analogously, to compute a clinical life table, we need to define the starting point from which we calculate survival. For example, consider using onset of the disease as a starting point. For some diseases where the onset is clear-cut and recent, such as a skiing accident resulting in a broken leg, this may be adequate. But for conditions in which the onset is insidious, it is difficult to establish an equivalent point of onset for all patients. Other possibilities are first visit to a physician, diagnosis, admission to hospital, start of treatment (e.g., drugs or surgery), completion of treatment (e.g., surgery), and discharge from hospital. (If the beginning of surgery is the starting point, patients may die from the remedy and not the disease.) We want a starting point that will be similar in the course of the disease for all patients. In a study of patients with peripheral arterial occlusions, it was suggested that a good starting time would be admission to the hospital, which is easily defined for each patient. It was decided not to use this criterion, however, because many of the patients had been admitted for some other reason and the occlusion had occurred later in the hospital.

In addition to deciding on a starting point, the investigator also needs to determine the frequency and manner of follow-up. Each patient could be followed at regular intervals—say one year from his date of entry to the study. Another method is to attempt to trace all the subjects at a particular time, such as January of each year, regardless of when during the year the patient entered the study. In the latter case information often is discarded on the subject during the last period. Elveback [1958] discusses the two methods of follow-up and concludes that the loss due to the discarded information is slight. In either case, we *know* at the end of each period whether a subject has lived through the period, has died, or has been lost to follow-up either by the study closing or because of failure to trace the subject. One of the advantages of life table analysis is that data on subjects lost to follow-up can be used up to the period in which they are lost.

We use the rather extensive notation for clinical life tables of Gehan [1969] because it relates to later discussions of survival distributions. Definitions of the notation are as follows:

1. $[t_{i-1}, t_i)$, $i = 1,\dots,s+1, t_{s+1} \equiv +\infty$ is the interval. This column gives the groupings into which the survival times and times to loss or withdrawal are distributed. The notation $[t_{i-1}, t_i)$ represents the half-open time interval $t_{i-1} \leqslant t < t_i, i = 1,\dots,s+1$. The last interval is infinite in length.

2. $t_{m,i}$ is the midpoint of the interval $[t_{i-1}, t_i), i = 1,\dots,s$. The midpoints are used in plotting the hazard and death density functions.

3. $h_i = t_i - t_{i-1}$, $i = 1, 2, \dots, s$ is the width of the ith interval. The widths are required to calculate the hazard and death density functions. Since the width of the last interval is infinite, no estimate of either the hazard or death density function can be obtained for this interval.

4. l_i is the total number of individuals who are lost to follow-up during the ith interval $[t_{i-1}, t_i)$. Individuals may be lost to observation if they move or fail to return for a follow-up visit, for example.

5. w_i is the total number of individuals who are withdrawn from the study alive during the ith interval $[t_{i-1}, t_i)$. These individuals have not been lost, but since they started late in the study, we have incomplete information on them.

6. d_i is the number of individuals who die in the ith interval. The time to death for each individual is measured from his own zero point.

7. n_i' is the total number of individuals who enter the interval $[t_{i-1}, t_i), i = 1, 2, \dots, s+1$. Thus the total sample size for the study is n_1', the number of individuals who enter the study at t_0. Clearly $n_i' = n_{i-1}' - l_{i-1} - w_{i-1} - d_{i-1}$. That is, the number of individuals who remain on study at the beginning of the ith interval equals the number of individuals who were on study at the beginning of the $(i-1)$th interval minus the number who were lost, withdrawn alive, or who died during the $(i-1)$th interval.

8. $n_i = n_i' - \frac{1}{2}(l_i + w_i)$ is the number of individuals exposed to risk during the ith interval. If there are no losses due to follow-up or withdrawal of patients, then clearly $n_i' = n_i$ in the ith interval. It is assumed that times to loss or withdrawal are uniformly distributed. Thus on the average individuals who are lost to follow-up or who withdraw from the study are lost for half the total interval.

9. $\hat{q}_i = d_i / n_i$, $i = 1,\dots,s$, is the conditional proportion dying in the ith interval. $\hat{q}_{s+1} = 1$, clearly.

10. $\hat{p}_i = (1 - \hat{q}_i)$ is the conditional proportion surviving the ith interval, $i = 1,\dots,s$.

11. The cumulative proportion $\hat{P}_i$ or $\hat{S}(t_i)$ surviving until time t_i is often referred to as the cumulative survival rate. The estimate is $\hat{P}_i = \hat{p}_{i-1}\hat{P}_{i-1}$, $i = 1, 2, \ldots, s$ and $\hat{P}_0 = 1.00$. This is the usual life table estimate. The result is based on the fact that surviving to the start of the ith interval means surviving to the start of the $(i-1)$th interval and *given* survival to the start of the $(i-1)$th interval, surviving the $(i-1)$th interval. Frequently $\hat{P}_i$ is compared with the cumulative proportion surviving in a population life table starting at the average age of the patients in the study to determine roughly how seriously the disease affects life expectancy.

The foregoing information is summarized in clinical life tables such as Table 2.2.

A natural estimate for the death density function $f(t)$ is given by

$$\hat{f}(t_{mi}) = \frac{\hat{P}_i - \hat{P}_{i+1}}{h_i} = \frac{\hat{P}_i \hat{q}_i}{h_i}, \qquad i = 1, \ldots, s. \tag{2.2.1}$$

Thus $\hat{f}(t_{mi})$ is the probability of dying in the ith interval per unit width— the definition of the probability density function.

The hazard rate $\hat{\lambda}(t_{mi})$ is *not* directly obtained as $\hat{f}(t_{mi})/\hat{P}_i$, since $\hat{P}_i$ is the probability of survival at t_i *not* t_{mi}. We thus must find $\hat{P}(t_{mi})$, the probability of survival at t_{mi}, the midpoint of the ith interval. Clearly

$$\hat{P}(t_{mi}) = \frac{\hat{P}_{i+1} + \hat{P}_i}{2} = \frac{\hat{P}_i(1 + \hat{p}_i)}{2}. \tag{2.2.2}$$

Thus,

$$\hat{\lambda}(t_{mi}) = \frac{2\hat{q}_i}{h_i(1 + \hat{p}_i)}. \tag{2.2.3}$$

To exhibit computations for the clinical life table, we use the clinical data gathered on 913 male and female patients with malignant melanoma who were treated at the M. D. Anderson Tumor Clinic between 1944 and 1960. These data, given by MacDonald [1963], comprise Table 2.3. We note that the computed failure rate for these patients is decreasing. That is, if an individual survives 3 years with the disease it appears he has an excellent chance to survive 5 years or longer. These 913 patients include 256 males (see Table 1.1) and 213 females who had metastases when they were first seen, plus an additional 444 patients in whom metastases had not occurred or could not be established.

Table 2.2 Clinical life table

Interval	Mid Point	Width	Number Entering Interval	Number Lost To Follow-up	Number Withdrawn Alive	Number Exposed to Risk	Number Dying	Conditional Proportion Dying	Conditional Proportion Surviving	Cumulative Proportion Surviving	$\hat{f}(t_{mi})$	$\hat{\lambda}(t_{mi})$
$[t_0, t_1)$	t_{m1}	h_1	n_1'	l_1	w_1	n_1	d_1	$\hat{q}_1$	$\hat{p}_1$	$\hat{P}_1 = 1.00$	$\hat{f}(t_{m1})$	$\hat{\lambda}(t_{m1})$
$[t_1, t_2)$	t_{m2}	h_2	n_2'	l_2	w_2	n_2	d_2	$\hat{q}_2$	$\hat{p}_2$	$\hat{P}_2$	$\hat{f}(t_{m2})$	$\hat{\lambda}(t_{m2})$
.	.	.	.	.	.	.	.	.	.	.	.	.
.	.	.	.	.	.	.	.	.	.	.	.	.
$[t_{s-1}, t_s)$	$t_{m,s}$	h_s	n_s'	l_s	w_s	n_s	d_s	$\hat{q}_s$	$\hat{p}_s$	$\hat{P}_s$	$\hat{f}(t_{m,s})$	$\hat{\lambda}(t_{m,s})$
$[t_s, \infty)$	—	—	n_{s+1}'	—	—	n_{s+1}'	d_{s+1}	1.00	0.00	$\hat{P}_{s+1}$	—	—

$h_i = t_i - t_{i-1}, \quad i = 1, 2, \ldots, s$

$n_i = n_i' - \frac{1}{2}(l_i + w_i), \quad i = 1, 2, \ldots, s$

$\hat{q}_i = d_i / n_i, \quad i = 1, \ldots, s$

$\hat{q}_{s+1} = 1.00$

$\hat{p}_i = 1 - \hat{q}_i, \quad i = 1, 2, \ldots, s$

$\hat{S}(t_i) = \hat{P}_i$

$\hat{P}_i = \hat{p}_{i-1}\hat{P}_{i-1}, \quad i = 2, \ldots, s$

$\hat{P}_1 = 1.00$

$\hat{f}(t_{mi}) = \frac{\hat{P}_i - \hat{P}_{i+1}}{h_i}, \quad i = 1, \ldots, s$

$\hat{\lambda}(t_{mi}) = \frac{2(1 - \hat{p}_i)}{h_i(1 + \hat{p}_i)} = \frac{2\hat{q}_i}{h_i(1 + \hat{p}_i)}, \quad i = 1, 2, \ldots, s$

Table 2.3 Malignant melanoma at the M. D. Anderson Tumor Clinic, 1944–1960[a]

Interval (years)	Mid point	Width	Number Entering Interval	Number Lost to Follow-up	Number Withdrawn Alive	Number Exposed to Risk	Number Dying	Conditional Proportion Surviving	Cumulative Proportion Surviving	$\hat{f}(t_{mi})$	$\hat{\lambda}(t_{mi})$
[0, 1)	0.5	1	913	19	77	865.0	312	0.639	1.000	0.361	0.441
[1, 2)	1.5	1	505	3	71	468.0	96	0.795	0.639	0.131	0.228
[2, 3)	2.5	1	335	4	58	304.0	45	0.852	0.508	0.075	0.160
[3, 4)	3.5	1	228	3	27	213.0	29	0.864	0.433	0.059	0.146
[4, 5)	4.5	1	169	5	35	149.0	7	0.953	0.374	0.018	0.048
[5, 6)	5.5	1	122	1	36	103.5	9	0.913	0.356	0.031	0.091
[6, 7)	6.5	1	76	0	17	67.5	3	0.956	0.325	0.014	0.045
[7, 8)	7.5	1	56	2	10	50.0	1	0.980	0.311	0.006	0.020
[8, 9)	8.5	1	43	0	8	39.0	3	0.923	0.305	0.024	0.080
[9, ∞)[b]	—	—	32	—	—	32.0	32	0.000	0.281	—	—

[a]Patient years are measured from time of initial diagnosis. It is also noted that these 913 patients did not all enter study at the same time.

[b]All 32 patients must die in the last interval.

EXERCISES

1. Verify the computations for the data in Table 2.3.
2. In the life table (Table 2.4), fill in the blank spaces.
3. Sacher [1956] proposes as an estimate of the hazard function

$$\hat{\lambda}(t_{mi})^+ = \frac{-\log_e \hat{p}_i}{h_i}.$$

   Derive this estimate on the assumption that the hazard rate is constant within an interval but varies among intervals. *Hint*: given survival to the ith interval, the chance of death in the ith interval is $[1 - \exp(-\lambda_i h_i)]$.

4. Compare numerically, $\hat{\lambda}(t_{mi})$ and $\hat{\lambda}(t_{mi})^+$ in Tables 2.3 and 2.4.
5. Beginning with the hazard function estimate $\hat{\lambda}(t_{mi}) = 2\hat{q}_i/(h_i(1 + \hat{p}_i))$, show that

$$\hat{\lambda}(t_{mi}) = \frac{d_i}{h_i(n_i - d_i/2)}.$$

6. Consider a life table in which no patients are withdrawn alive and no patients are lost to follow-up. If n patients are entered in the life table at the beginning of the first interval, all intervals fixed, prove that the numbers of deaths in the s intervals have a multinomial distribution. Specifically, show that $f(d_1,\ldots,d_s)$ has the form

$$f(d_1,\ldots,d_s) = \frac{n!}{d_1! d_2! \cdots d_s!} \pi_1^{d_1} \cdots \pi_s^{d_s}, \tag{2.2.4}$$

   where

$$\sum_{i=1}^{s} d_i = n, \qquad \sum_{i=1}^{s} \pi_i = 1, \tag{2.2.5}$$

   $\pi_i = \mathrm{pr}\{\text{an individual dies during the } i\text{th interval}\}.$

2.3 VARIANCES AND MEDIANS COMPUTED FROM CLINICAL LIFE TABLES

Approximate variances for $\hat{\lambda}(t_{mi})$, $\hat{f}(t_{mi})$, and $\hat{P}_i$ are given next, along with a brief discussion of how they were obtained. In these formulas it is assumed that a fixed sample size n is followed until all subjects die at a known time in one of the s intervals. No patients are withdrawn alive or lost to follow-up. In addition, instead of obtaining variances of the functions directly, a Taylor series expansion was made of the function and then the variance of the linear terms was found. The properties of variances of linear functions are well-known; thus by approximating an arbitrary function by a linear one, its variance can be approximated. If x is a random

Table 2.4 Clinical life table

Interval	Mid point	Width	Number Entering Interval	Number Lost to Follow-up	Number Withdrawn Alive	Number Exposed to Risk	Number Dying	Conditional Proportion Surviving	Cumulative Proportion Surviving	$\hat{f}(t_{mi})$	$\hat{\lambda}(t_{mi})$
[0,1)	0.5	1	1000	7	3	—	22	—	—	—	—
[1,2)	1.5	1	—	5	3	—	19	—	—	—	—
[2,3)	2.5	1	—	12	9	—	26	—	—	—	—
[3,4)	3.5	1	—	15	11	—	31	—	—	—	—
[4,5)	4.5	1	—	17	20	—	27	—	—	—	—
[5,∞)	—[a]	—[a]	—	—[a]	—[a]	—	773	—	—	—[a]	—[a]

[a] leave blank

variable with finite variance σ^2 and $f(x)$ is an arbitrary function of x which is differentiable, then expanding $f(x)$ in a Taylor series about μ, the mean of x, we can approximate $\mathrm{Var} f(x)$ as $\mathrm{Var} f(x) \doteq [f'(\mu)]^2 \sigma^2$, ignoring all terms in the Taylor series beyond the linear term. This is discussed in Bennett and Franklin [1954] and Kendall and Stuart [1963].

Using $[f'(\mu)]^2 \sigma^2$ as the approximation for $\mathrm{Var} f(x)$, the approximate variance formulas for $\hat{\lambda}(t_{mi})$, $\hat{f}(t_{mi})$, and $\hat{P}_i$ are obtained. For example, if we write

$$\hat{\lambda}(t_{mi}) = \frac{d_i}{h_i(n_i - d_i/2)} \tag{2.3.1}$$

(see Exercise 5, Section 2.2), and carry through some rather intricate algebraic manipulations, we can show that an approximate formula for the variance of $\hat{\lambda}(t_{mi})$ is given by

$$\mathrm{Var}\,\hat{\lambda}(t_{mi}) \doteq \frac{\pi_i}{n h_i^2 (1 - \phi_i - \pi_i/2)^2} \left\{ 1 - \left[\frac{\pi_i}{2(1 - \phi_i - \pi_i/2)} \right]^2 \right\} \tag{2.3.2}$$

where $\phi_i = \sum_{j=1}^{i-1} \pi_j$ and π_j is the probability that an individual dies during the jth interval, $j = 1, 2, \ldots, i$.

Equation 2.3.2 is valid when the survival times of all n'_1 patients on study are recorded. For incomplete samples (i.e., when not all the patients on study have expired or are not followed to death), we estimate π_i by $\hat{P}_i \hat{q}_i$, ϕ_i by $1 - \hat{P}_i$, and n_i by $n'_1 \hat{P}_i$, where the actual number starting on study is n'_1. The value n_i takes into account patients lost to follow-up and withdrawn alive from study. With these assumptions it is not difficult to show that (2.3.2) becomes

$$\mathrm{Var}\{\hat{\lambda}(t_{mi})\} \doteq \frac{\left[\hat{\lambda}(t_{mi})\right]^2}{n_i \hat{q}_i} \left\{ 1 - \left[\frac{h_i \hat{\lambda}(t_{mi})}{2} \right]^2 \right\}. \tag{2.3.3}$$

If n_i is small, (2.3.2) and (2.3.3) lose accuracy. This implies that if there are very few patients in the later intervals of a clinical life table, the computation of $\mathrm{Var}\{\hat{\lambda}(t_{mi})\}$ is not worthwhile for these later stages.

The derivation of an approximation to $\mathrm{Var} \hat{f}(t_{mi})$ is accomplished by observing that

$$\hat{f}(t_{mi}) = \frac{(1 - \hat{p}_i)\hat{p}_{i-1} \cdots \hat{p}_1}{h_i}, \qquad i = 1, 2, \ldots, s. \tag{2.3.4}$$

Equation 2.3.4 follows from (2.2.1) and the definition of $\hat{p}_i$.

Applying the formulas for variances of arbitrary functions, we have

$$\text{Var}\{\hat{f}\,(t_{mi})\} \doteq \left(\frac{\hat{P}_i \hat{q}_i}{h_i}\right)^2 \left(\sum_{j=1}^{i-1} \frac{\hat{q}_j}{n_j \hat{p}_j} + \frac{\hat{p}_i}{n_i \hat{q}_i}\right), \qquad (2.3.5)$$

observing that $\text{cov}(\hat{p}_j, \hat{p}_l) = 0$, for $j \neq l$.* Note that (2.3.4) is defined for $n_i > 0$, $i = 1, 2, \ldots, s-1$, and this approximation is better as n_i increases.

Finally, Greenwood [1926] computed the approximation to $\text{Var}\,\hat{P}_i$. This is given by

$$\text{Var}\,\hat{P}_i \doteq \hat{P}_i^2 \sum_{j=1}^{i-1} \frac{d_j}{n_j(n_j - d_j)}. \qquad (2.3.6)$$

Again it should be pointed out that for small values for n_i (2.3.6) is not a good approximation to the true variance. Unfortunately (2.3.6) is likely to underestimate the variance of $\hat{P}_i$, because we treat n_i as a constant for $i > 1$ when it is a random variable. Kuzma [1967] made some numerical comparisons between Greenwood's formula and Chiang's [1968] results. He found that the variance formula of Greenwood considerably underestimated the variance when the withdrawal rate was high.

The parametric techniques discussed in the subsequent chapters provide reliable variance formulas for their estimates, although they also should be viewed cautiously if a considerable proportion of the patients has not been followed until death. In particular, the reasons for the lack of complete follow-up should be explored to determine whether a biasing effect exists. For example, if the patients who were lost to follow-up are in a lower socioeconomic class than those followed, the results of the study may be quite biased, especially if the study is at a teaching hospital located in a slum area where many of the patients are poor.

To estimate mean lifetime most accurately, one must wait until all individuals have expired in the last interval. Otherwise, the estimate of mean lifetime is based on incomplete information. We consider the problem of estimating mean lifetimes based on incomplete information in the next section and also in the next two chapters.

Often the median lifetime is a better estimate of survival time than the mean lifetime. This is because one or two exceptionally long-term survivors will cause the mean lifetime to be larger than the median lifetime. Ordinarily, a mean lifetime substantially larger than the median lifetime indicates that there is at least one long-term survivor in the life table.

---

*Chiang shows that $\text{cov}(\hat{p}_j, \hat{p}_l) = 0$. This, however, does not imply that $\hat{p}_j$ and $\hat{p}_l$ are independent. In fact they are not independent in a clinical life table.

We present two median measures in this section—the median lifetime and the median remaining lifetime. Let $[t_j, t_{j+1})$ be the interval such that $\hat{P}_j \geq \frac{1}{2}$ and $\hat{P}_{j+1} < \frac{1}{2}$. Then the median lifetime $\tilde{t}$ is given by

$$\tilde{t} = (t_j - t_0) + \frac{(\hat{P}_j - \frac{1}{2})}{\hat{f}(t_{mj})}, \tag{2.3.7}$$

where t_0 is the time at which measurements for the life table begin and $\hat{f}(t_{mj})$ is defined by (2.2.1). A more general measure is the median remaining lifetime. That is, if $\hat{P}_i$ is the proportion of individuals surviving to time $\tilde{t}_i$, we wish to find that point $\tilde{t}_i$ such that the proportion of individuals surviving to $\tilde{t}_i$ is $\hat{P}_i/2$. Let $[t_j, t_{j+1})$ be the interval, such that $\hat{P}_j \geq \hat{P}_i/2$ and $\hat{P}_{j+1} < \hat{P}_i/2$. Then

$$\tilde{t}_i = (t_j - t_i) + \frac{(\hat{P}_j - \hat{P}_i/2)}{\hat{f}(t_{mj})}. \tag{2.3.8}$$

The approximate variances of (2.3.7) and (2.3.8) are, respectively,

$$\mathrm{Var}(\tilde{t}) \doteq \left\{ 4n_1 \left[\hat{f}(t_{mj}) \right]^2 \right\}^{-1} \tag{2.3.9}$$

and

$$\mathrm{Var}(\tilde{t}_i) \doteq \frac{\hat{P}_i^2}{4n_i \left[\hat{f}(t_{mj}) \right]^2}. \tag{2.3.10}$$

We compute the median remaining lifetimes $\tilde{t}_i$ and their variances and standard deviations in Table 2.5 from the data in Table 2.3. Note that the median is computed by (2.3.8) for only the first two time periods in Table 2.3, since the median remaining life for all time periods beyond the second is in the last time period, which is infinite in length. For this reason, only the variances and standard deviations are given for the first two time periods.

The life table approach for the analysis of clinical data is quite useful because it is nonparametric, it uses the information that is available, and it provides the type of information that medical investigators often desire.

Table 2.5 Median remaining lifetime estimates and their variances and standard deviations of melanoma patients

Year After Diagnosis	Midpoint	Width	$\tilde{t}_i$	$\mathrm{Var}\,\tilde{t}_i$	$\sqrt{\mathrm{Var}\,\tilde{t}_i}$
0	0.5	1.0	2.107	2.22×10^{-3}	0.0471
1	1.5	1.0	5.392	0.0127	0.1127
2	2.5	1.0	7.000+	—	—
3	3.5	1.0	6.000+	—	—
4	4.5	1.0	5.000+	—	—
5	5.5	1.0	4.000+	—	—
6	6.5	1.0	3.000+	—	—
7	7.5	1.0	2.000+	—	—
8	8.5	1.0	1.000+	—	—
9	9.5	1.0	0+	—	—
10	—	—	—	—	—

The cumulative proportion surviving is frequently graphed. This picture can be compared graphically with data from an appropriate population life table to assess how the length of life is affected by the disease condition being studied. Suppose the average age of the patients in Table 2.3 were 60 years. We could plot s_{60+i}/s_{60} for $i = 0, \ldots, 9$ from the population life table (Table 2.1) to assess the difference in length of survival and shape of the curve between patients with malignant melanoma and the population in general. Questions such as what are the chances of living 5 years can be answered directly from the data in Table 2.3. The variance formulas allow the investigator to assess the dispersion of his results. The length of a typical patient's survival can be addressed by computing the median survival time.

Furthermore, the results from the clinical life table are useful in determining which parametric distribution to assume if a parametric approach is to be used. By plotting $\hat{\lambda}(t_{mi})$ we can often obtain considerable information about possible choices of a death density function. For example, from Table 2.3, where $\hat{\lambda}(t_{mi})$ appears to be decreasing over time, we would not want to choose a death density function that has a constant or increasing hazard rate such as exponential or normal, respectively. By grouping the data into intervals, we can assess the shape of our distributions in a fashion analogous to the use of a histogram. Thus regardless of later analysis this is often a first step in data analysis. Computer programs are widely available (e.g., BMD01S and BMD011S, Dixon [1973]), or if the sample size is moderate, a clinical life table can be computed easily with a desk calculator. Theoretical properties of the statistics available from clinical life tables have been extensively studied by Breslow and Crowley [1974].

It is often useful to refer to the population life table (Table 2.1) when attempting to determine an appropriate parametric distribution. The table provides an estimate of the hazard rate that exists minus the effects of the disease. For example, if the investigator is examining a disease of young infants and the death rate from the disease is expected to stay constant or decrease with time, he can postulate a decreasing hazard rate because the death rate for infants, in general, is decreasing.

EXERCISES

1. Verify the variances and standard deviations for the melanoma patients for the second year after diagnosis in Table 2.6 from the data in Table 2.3.

2. Show that (2.3.4) follows from (2.2.1). Hence obtain the approximate value $\text{Var}\{\hat{f}(t_{mi})\}$ given by (2.3.5).

3. Verify the variance computations for the third and fourth year after diagnosis for the data in Table 2.6.

4. Find the approximate variance of Sacher's estimate of the hazard function $\hat{\lambda}(t_{mi})^{+} = (-\log_{e}\hat{p}_{i})/h_{i}$, $i = 1,\ldots,s-1$. Consider both the cases of complete and incomplete sampling. *Hint:* Expand $-\log_{e}\hat{p}_{i}$ in a power series about $1 - \pi_{i}$, where π_{i} is the true proportion dying in the ith interval.

5. How is (2.3.5) defined if $i = 1$? Justify your answer by deriving the formula for $\text{Var}\{\hat{f}(t_{m1})\}$.

Table 2.6 Variance and standard deviations of life table estimates of melanoma patients

Interval (years)	Mid-point	Width	$\text{Var}\,\hat{P}_{i}$	$\text{Var}\,\hat{f}(t_{mi})$	$\text{Var}\,\hat{\lambda}(t_{mi})$	$\sqrt{\text{Var}\,\hat{P}_{i}}$	$\sqrt{\text{Var}\,\hat{f}(t_{mi})}$	$\sqrt{\text{Var}\,\hat{\lambda}(t_{mi})}$
[0,1)	0.5	1.0	—	2.67×10^{-4}	5.92×10^{-4}	—	0.0163	0.0243
[1,2)	1.5	1.0	2.66×10^{-4}	1.53×10^{-4}	5.33×10^{-4}	0.0163	0.0124	0.0231
[2,3)	2.5	1.0	3.11×10^{-4}	1.39×10^{-4}	5.66×10^{-4}	0.0176	0.0118	0.0238
[3,4)]	3.5	1.0	3.33×10^{-4}	1.25×10^{-4}	7.32×10^{-4}	0.0182	0.0112	0.0271
[4,5)	4.5	1.0	3.52×10^{-4}	4.89×10^{-5}	3.29×10^{-4}	0.0188	0.0070	0.0181
[5,6)]	5.5	1.0	3.61×10^{-4}	1.15×10^{-4}	9.18×10^{-4}	0.0190	0.0107	0.0303
[6,7)	6.5	1.0	3.98×10^{-4}	7.62×10^{-5}	6.81×10^{-4}	0.0200	0.0087	0.0261
[7,8)	7.5	1.0	4.31×10^{-4}	4.36×10^{-5}	4.00×10^{-4}	0.0208	0.0066	0.0200
[8,9)	8.5	1.0	4.52×10^{-4}	1.97×10^{-4}	5.33×10^{-4}	0.0213	0.0140	0.0231
[9,∞)	9.5	1.0	5.53×10^{-4}	8.55×10^{-5}	1.11×10^{-3}	0.0235	0.0092	0.0333

For Exercises 6 through 8 we make use of the life table given as Table 2.7.

Table 2.7

Interval	Number Entering	Number Dying	Number Lost to Followup	Number Withdrawn Alive	Cumulative Proportion Surviving
[0, 1)	500	2	0	0	1.000
[1, 2)	498	8	0	0	0.996
[2, 3)	490	24	0	0	0.980
[3, 4)	466	80	0	0	0.932
[4, 5)	386	103	0	0	0.772
[5, 6)	283	128	0	0	0.566
[6, 7)	155	105	0	0	0.310
[7, 8)	50	41	0	0	0.100
[8, 9)	9	9	0	0	0.018

6. Calculate the mean and median lifetimes.
7. Calculate the median remaining lifetimes at $t_i = 2$, 3, 4, 5, 6, 7, and 8.
8. Calculate the approximate variances of the estimate in Exercise 7 and the approximate variance of the median lifetime $\bar{t}$.

2.4 PRODUCT–LIMIT ESTIMATES OF $\hat{P}(t)$

In the estimates given in the previous sections of this chapter, we have utilized grouped data. For example, in Table 2.3 the survival time was grouped by yearly intervals. In the Kaplan and Meier [1958] approach, ordered observations are used instead of grouped data. This method has the advantage of yielding results that are not dependent on the choice of the time intervals. It has been used with small samples, where it is difficult to decide on an appropriate parametric distribution. It is not, however, as helpful in the possible later choice of a parametric distribution as the results of a clinical life table. The present technique can be used when the data are progressively censored.

Kaplan and Meier [1958] introduce the product-limit estimate $\hat{P}(t)$ or $\hat{S}(t)$ of $P(t)$ (the probability an individual survives beyond time t) in a life table. To define the product limit estimate $\hat{P}(t)$, order the times of death such that $t_{(1)} \leq t_{(2)} \leq, \ldots, t_{(i)}$. Then n_i is the number of patients still under observation at time $t_{(i)}$, including the one that died at time $t_{(i)}$. If a death and a loss occur at the same time, Kaplan and Meier suggest treating the death as if it had occurred slightly before $t_{(i)}$ and the loss as if it had occurred slightly after $t_{(i)}$. Then

$$\hat{P}(t) = \prod_{j=1}^{i} \left(\frac{n_j - 1}{n_j} \right). \tag{2.4.1}$$

If there are no patient losses, $\hat{P}(t)$ reduces to the ordinary binomial estimate of the probability of survival at time t (i.e., s/n, where s is the number of patients alive after time t).

Example 2.4.1 Consider a clinical trial in which eight patients are observed to have the following survival pattern (in months): $2, 2^+, 3, 4, 4^+, 5, 6, 7$ (2^+ and 4^+ are individuals lost to follow-up at 2 and 4 months, respectively).

$$\hat{P}(2) = \left(\frac{8-1}{8} \right) = 0.875,$$

$$\hat{P}(3) = \left(\frac{8-1}{8} \right)\left(\frac{6-1}{6} \right) = 0.729,$$

$$\hat{P}(4) = \left(\frac{8-1}{8} \right)\left(\frac{6-1}{6} \right)\left(\frac{5-1}{5} \right) = 0.583,$$

$$\hat{P}(5) = \left(\frac{8-1}{8} \right)\left(\frac{6-1}{6} \right)\left(\frac{5-1}{5} \right)\left(\frac{3-1}{3} \right) = 0.389,$$

$$\hat{P}(6) = \left(\frac{8-1}{8} \right)\left(\frac{6-1}{6} \right)\left(\frac{5-1}{5} \right)\left(\frac{3-1}{3} \right)\left(\frac{2-1}{2} \right) = 0.194,$$

and

$$\hat{P}(7) = 0.$$

Example 2.4.1 demonstrates that if $t_{(i-1)}$ and $t_{(i)}$ are consecutive times to death, the relationship

$$\hat{P}(t_{(i)}) = \hat{P}(t_{(i-1)})\left(\frac{n_i - 1}{n_i} \right) \tag{2.4.2}$$

holds, provided there is no loss between $t_{(i-1)}$ and $t_{(i)}$.

An approximate estimate of the variance of $\hat{P}(t)$ is given as

$$\text{Vâr}\left[\hat{P}(t)\right] \doteq \left[\hat{P}(t)\right]^2 \sum_{j=1}^{i} \frac{1}{n_j(n_j-1)} \cdot$$

In Example 2.4.1, we see that

$$\text{Vâr}\left[\hat{P}(4)\right] = (0.583)^2 \left\{ \frac{1}{7\times8} + \frac{1}{5\times6} + \frac{1}{4\times5} \right\} = 0.0344. \quad (2.4.3)$$

Thus the estimated standard error of $\hat{P}(4)$ is 0.185.

We compute an estimate of the mean survival time and its estimated variance based on the Kaplan-Meier product-limit procedure. Let the observed times to death be ordered as $t'_{(1)} \leqslant t'_{(2)} \leqslant \cdots \leqslant t'_{(d)}$. It is shown by Kaplan and Meier that the estimated mean survival time is

$$\hat{\mu} = 1.000 t'_{(1)} + P(t'_{(1)})(t'_{(2)} - t'_{(1)}) + \cdots + P(t'_{(d-1)})(t'_{(d)} - t'_{(d-1)})). \quad (2.4.4)$$

In the case for which the longest time a patient is on study is his time to death, the estimate $\hat{\mu}$ is generally satisfactory. Otherwise it may be biased on the low side.

The estimate of the variance of $\hat{\mu}$, $\hat{V}(\hat{\mu})$ is given as

$$\hat{V}(\hat{\mu}) = \sum_r \frac{A_r^2}{(N-r)(N-r+1)}, \quad (2.4.5)$$

where r is summed over those integers for which t_r corresponds to a death, and

$$A_r = P(t'_{(r-1)})(t'_{(r)} - t'_{(r-1)}) + \cdots + P(t'_{(d-1)})(t'_{(d)} - t'_{(d-1)}). \quad (2.4.6)$$

The estimated mean survival time for the data in Example 2.4.1, using (2.4.2), is

$$\hat{\mu} = (1.000)(2) + (0.875)(3-2) + (0.729)(4-3) + (0.583)(5-4)$$

$$+ (0.389)(6-5) + (0.194)(7-6)$$

$$= 4.770 \text{ months.} \quad (2.4.7)$$

To obtain the estimated variance $\hat{V}(\hat{\mu})$, we first compute the coefficients A_r from (2.4.6).

$$A_7 = 0.389 + 0.194 = 0.583$$

$$A_6 = 0.583 + 0.389 + 0.194 = 1.166$$

$$A_4 = 0.729 + 0.583 + 0.389 + 0.194 = 1.895$$

$$A_3 = 0.875 + 0.729 + 0.583 + 0.389 + 0.194 = 2.770$$

$$A_1 = 2.000 + 0.875 + 0.729 + 0.583 + 0.389 + 0.194 = 4.770. \qquad (2.4.8)$$

Thus from (2.4.5) we have

$$\hat{V}(\hat{\mu}) = \frac{(4.770)^2}{7 \times 8} + \frac{(2.770)^2}{5 \times 6} + \frac{(1.895)^2}{4 \times 5} + \frac{(1.166)^2}{2 \times 3} + \frac{(.583)^2}{1 \times 2}$$

$$= 1.2382.$$

Thus the estimated standard deviation is 1.1127 months.

Computer programs for this technique (e.g., SURVIVAL KIT at Stanford) can also be used, but these are not available in widely distributed and well documented form.

EXERCISES

1. Consider a clinical trial in which 10 patients are observed to have the following survival pattern (in months) $1, 2, 3, 3^+, 4, 4^+, 5, 5^+, 8, 9^+$. (The "plus" values are patients who are lost to follow-up.) Compute $\hat{P}(t)$, the Kaplan-Meier product-limit estimate of survival, at the values $t = 2, 4, 6, 8$, and 9, respectively. Also, compute the estimated standard error of $\hat{P}(5)$.

2. For the data given in Exercise 1, compute the product-limit estimate of the mean lifetime and the variance of the mean lifetime.

CHAPTER 3

Estimation and Inference in the Exponential Distribution

3.1 INTRODUCTORY REMARKS

In this chapter we discuss in some detail procedures for estimating the parameter of the exponential survival distribution. Section 3.1 reviews the procedures for obtaining maximum likelihood estimators, applying these, along with other estimation procedures, to the exponential distribution. Section 3.2 gives formulas for estimation of λ when the data are censored. Variance formulas and confidence intervals are discussed in Section 3.3, and estimation of the probability of survival for the exponential distribution is presented in Section 3.4. Section 3.5 deals with the estimation of survival time when a concomitant (covariate) variable is measured, for the noncensored and censored cases. In Section 3.6 we discuss graphical as well as formal tests to determine whether an exponential distribution can be fitted to actual survival data for the noncensored and censored cases. Section 3.7 introduces competing risk models for the exponential survival distribution. It may be sensible to cover Section 3.6 before tackling Sections 3.4 and 3.5.

Our principal tool for estimating the exponential survival parameter as well as estimating parameters of other survival distributions is the method of maximum likelihood. The reasons we stress the maximum likelihood method of estimation are threefold: (1) conceptually it is a simple procedure, although the reader will discover the computational problems are not always simple; (2) the asymptotic properties of maximum likelihood estimators (under fairly general conditions) make their use desirable; (3) maximum likelihood estimation affords a rather general method of estimation of parameters of survival distributions. Even when observations are censored, for example, one can in most instances obtain the maximum likelihood estimators of the survival parameters of the distribution from which the observations are assumed to come.

49

We review, briefly, the method of maximum likelihood. Suppose $t_1, t_2, \ldots, t_n$ are n independent survival times whose death density is a function of the parameters $\theta_1, \ldots, \theta_k$, $n > k$. We define $L(\theta_1, \ldots, \theta_k)$, the likelihood function given the sample $t_1, t_2, \ldots, t_n$, as

$$L(\theta_1, \ldots, \theta_k) = \prod_{i=1}^{n} f(t_i; \theta_1, \ldots, \theta_k), \qquad (3.1.1)$$

where $f(t_i; \theta_1, \ldots, \theta_k)$ is the death density. The maximum likelihood estimators $\hat{\theta}_1, \hat{\theta}_2, \ldots, \hat{\theta}_k$ of $\theta_1, \theta_2, \ldots, \theta_k$, respectively, based on the sample $t_1, t_2, \ldots, t_n$ are such that if $\tilde{\theta}_1, \tilde{\theta}_2, \ldots, \tilde{\theta}_k$ is any other set of estimators, then

$$L(\hat{\theta}_1, \ldots, \hat{\theta}_k) \geqslant L(\tilde{\theta}_1, \ldots, \tilde{\theta}_k). \qquad (3.1.2)$$

Under regularity conditions outlined in Cramér [1946] or Wasan [1970], the maximum likelihood estimators $\hat{\theta}_1, \hat{\theta}_2, \ldots, \hat{\theta}_k$ of $\theta_1, \theta_2, \ldots, \theta_k$ are obtained as the solution to the $k \times k$ system of equations

$$\left. \frac{\partial \log_e L(\theta_1, \ldots, \theta_k)}{\partial \theta_i} \right|_{\theta_i = \hat{\theta}_i} = 0, \qquad i = 1, 2, \ldots, k, \qquad (3.1.3)$$

since maximizing a function or its logarithm leads to equivalent results. Furthermore, the estimators $(\hat{\theta}_1, \ldots, \hat{\theta}_k)$ are asymptotically normally, distributed with mean $(\theta_1, \ldots, \theta_k)$ and variance–covariance matrix $V_{\hat{\theta}}$ (see Rao [1952] or Mood and Graybill [1963]), where

$$V_{\hat{\theta}} = \begin{bmatrix} -E\left(\dfrac{\partial^2 \log L}{\partial \theta_1^2} \right) & \cdots & -E\left(\dfrac{\partial^2 \log L}{\partial \theta_1 \, \partial \theta_k} \right) \\ \vdots & \ddots & \vdots \\ -E\left(\dfrac{\partial^2 \log L}{\partial \theta_1 \, \partial \theta_k} \right) & \cdots & -E\left(\dfrac{\partial^2 \log L}{\partial \theta_k^2} \right) \end{bmatrix}^{-1} ; \qquad (3.1.4)$$

that is,

$$
\begin{bmatrix} \hat{\theta}_1 \\ \cdot \\ \cdot \\ \cdot \\ \hat{\theta}_k \end{bmatrix} \sim N \begin{bmatrix} \begin{pmatrix} \theta_1 \\ \cdot \\ \cdot \\ \cdot \\ \theta_k \end{pmatrix} ; & V_{\hat{\theta}} \end{bmatrix}. \tag{3.1.5}
$$

There are distributions for which a survival parameter's maximum likelihood estimator cannot be obtained by (3.1.3). Suppose the death density of t involves only a location parameter θ; that is, if

$$
f(t) = \exp[-(t - \theta)], \qquad t \geqslant \theta, \tag{3.1.6}
$$

(3.1.3) cannot be used to obtain $\hat{\theta}$. In this case if $t_{(1)} < t_{(2)} < \cdots < t_{(n)}$ are n ordered survival times whose death density is given by (3.1.6), the maximum likelihood estimator $\hat{\theta} = t_{(1)}$. Physically, this says that if the survival times of n individuals are guaranteed to survive at least a period of length θ, then the maximum likelihood estimator of θ is the minimum of the n recorded survival times, that is,

$$
\hat{\theta} = min_i t_i.
$$

When the parameters of a death density function are shape parameters or scale parameters, however, the maximum likelihood estimators of these parameters are usually obtained by (3.1.3), and their large sample variance– covariance matrix is given by (3.1.4).

Although the maximum likelihood method of estimation is the principal method of estimation in Sections 3.1 through 3.5, we consider alternative methods for estimating the exponential survival parameter in Section 3.6. This presentation includes a graphical method as well as a procedure for testing grouped data for exponentiality. Gehan [1973] provides least squares estimators as alternatives to maximum likelihood estimators for several survival distributions, including the exponential.

As we showed in Chapter 1, determination of either the hazard function, the death density function, or the survival function, determines all these distributions uniquely. Thus we consider the estimation problem in this chapter in light of a constant hazard function, that is, $\lambda(t)=\lambda$ for all $t>0$. The resulting exponential death density and survival functions are, respectively,

$$f(t)=\lambda\exp(-\lambda t), \qquad \lambda>0, t\geqslant 0, \tag{3.1.7}$$

and

$$S(t)=\exp(-\lambda t), \qquad \lambda>0, t\geqslant 0. \tag{3.1.8}$$

Note that (3.1.7) and (3.1.8) were developed in Example 1.2.1. We estimate the exponential survival parameter λ, by the method of maximum likelihood. We note that μ, the mean of an exponential survival distribution, is λ^{-1}. We show that the maximum likelihood estimator $\hat{\mu}$ is also the best linear unbiased estimator, the methods of moments estimator, and the sufficient statistic for μ.

If $t_1, t_2, \ldots, t_n$ are n times to death of individuals who have the constant hazard function λ, we can view these times as n observations from the population whose density function is the exponential. Thus if we require the maximum likelihood estimator $\hat{\lambda}$ of λ, we observe that $L(\lambda)$, the likelihood function, is

$$L(\lambda)= \prod_{i=1}^{n} \lambda\exp(-\lambda t_i). \tag{3.1.9}$$

Recalling that maximizing $L(\lambda)$ is equivalent to maximizing the logarithm of $L(\lambda)$, we have

$$\log_e L(\lambda)= n\log_e\lambda-\lambda \sum_{i=1}^{n} t_i. \tag{3.1.10}$$

Differentiating with respect to λ, equating to zero, and finding that value of λ, $\hat{\lambda}$ (say), which satisfies the equation, we have

$$\hat{\lambda}=\left(\frac{\sum_{i=1}^{n} t_i}{n} \right)^{-1} \equiv \bar{t}^{-1}, \tag{3.1.11}$$

which is the maximum likelihood estimator of λ—the constant hazard rate. If $\mu=\lambda^{-1}$ is the mean time to death, then since maximum likelihood estimates are invariant under one-to-one transformations,

$$\hat{\mu}=\bar{t} \tag{3.1.12}$$

is the maximum likelihood estimator of μ.

The minimum variance unbiased estimator for $\mu, \tilde{\mu}$ (say) is obtained quite easily. First of all $\tilde{\mu}$ is constructed so that it is an unbiased linear estimator of μ. That is,

$$\tilde{\mu} = \sum_{i=1}^{n} a_i t_i, \tag{3.1.13}$$

where $\sum_{i=1}^{n} a_i = 1$. Then $a_1, \ldots, a_n$ must be chosen so that Var $\tilde{\mu}$ achieves a minimum subject to the constraint $\sum_{i=1}^{n} a_i = 1$. It is not difficult to show that $a_i = 1/n$, for all i; thus

$$\tilde{\mu} = \bar{t}. \tag{3.1.14}$$

That is, the maximum likelihood and minimum variance unbiased estimators of μ coincide when the parent density function is the exponential. In fact, in estimation theory terminology, $\bar{t}$ is the best linear unbiased estimator (BLUE) of μ. Unfortunately, $\bar{t}^{-1}$ is not the BLUE for λ, since BLUEs are not generally preserved under inversing.

The method of moments estimator of μ, μ^* (say), is also $\bar{t}$. This follows immediately because the method of moments estimator μ^* is obtained by setting $E(t) = \bar{t}$.

Often it may be desirable to use a method of moments estimator as a check on the adequacy of the maximum likelihood estimator. For example, when the experimenter has only a small sample size, the maximum likelihood estimator may be quite biased. Furthermore, even when sample sizes are large enough to make effective use of maximum likelihood estimators, moment estimates are often used as initial values in the iterative procedure to determine the maximum likelihood estimate. (See Chapter 6.) Further examples of the use of method of moments can be found in Kendall and Stuart [1963].

To show that $\hat{\mu}$ is the sufficient statistic for μ, we must show that the likelihood function: $L(\mu)$ can be expressed as a function of the parameter μ and the estimator $\hat{\mu}$ alone, multiplied by a function of the sample $t_1, t_2, \ldots, t_n$, alone. But this follows, since

$$L(\mu) = \left(\frac{1}{\mu^n}\right) \exp\left(\frac{-n\bar{t}}{\mu}\right)$$

$$= g(\bar{t}, \mu) h(t_1, \ldots, t_n), \tag{3.1.15}$$

where $h(t_1, \ldots, t_n) \equiv 1$. Thus, in summary, $\bar{t}$ is the maximum likelihood, minimum variance unbiased, method of moments estimator for μ, as well as the sufficient statistic for μ, when $t_1, \ldots, t_n$ is a random sample of size n from a population whose hazard rate is constant.

3.2 ESTIMATION OF λ FOR THE EXPONENTIAL DENSITY WITH CENSORING

Suppose that patients with some serious illness (e.g., carcinoma of the lung) are on a study to determine their average survival time. However, the study lasts only for a limited period. Hence it is likely that not all patients are dead at the end of the study. Here there are two cases to consider. First, the patients enter the study independently at different points in time. Suppose there are n patients on study. Let T_i be the maximum time for which the ith patient can be observed, $i = 1, 2, \ldots, n$. Thus if the study ends at time T and if the ith patient enters the study at time z_i, then $T_i = T - z_i$.* The survival time of the ith patient t_i is known only if $t_i \leqslant T_i$. Thus given a sample of n patients, the information available is the set of maximum times $T_1, T_2, \ldots, T_n$ and those times to death t_j, such that $t_j \leqslant T_j$. The sample size n is fixed in advance of the study; but d, the total number of deaths observed, is a random variable because the study is assumed to end at the fixed time T.

Second, n patients who enter the study are assumed to enter at the same time. For example, suppose we wish to study the length of survival of individuals who are exposed to excessive amounts of radiation at the same time (e.g., persons who survived the atomic bomb attack at Hiroshima). In such a study the survival times arrive in a natural ordering, meaning that the individual with the shortest survival time is recorded first, the individual with the next shortest survival time is recorded second, and so on. If the survival times are recorded for the first $d \leqslant n$ individuals who die, then $t_{(1)} \leqslant t_{(2)} \leqslant \cdots \leqslant t_{(d)}$.

We discuss first progressively censored data and second singly censored data. We assume that each patient has the same death density function

$$f(t) = \lambda \exp(-\lambda t), \qquad \lambda > 0, t \geqslant 0. \qquad (3.2.1)$$

Following Bartholomew's [1957] development for the progressively censored case, the probability the ith individual dies while he is on study is

$$1 - S(T_i) = \int_0^{T_i} \lambda \exp(-\lambda t)\, dt = 1 - \exp(-\lambda T_i) = Q_i. \qquad (3.2.2)$$

Let $\mu = \lambda^{-1}$. To estimate μ (hence λ) we use the method of maximum likelihood. We note that the contribution to the likelihood of the sample,

---

*Since we assume a constant hazard function λ, we can proceed to estimate λ under the assumption that $T_i = T - z_i$.

due to the ith individual, is

$$f(t_i) = \begin{cases} \mu^{-1}\exp(-\mu^{-1}t_i), & 0 \leq t_i \leq T_i, \\ \exp(-\mu^{-1}T_i), & t_i > T_i \end{cases}. \tag{3.2.3}$$

Equation 3.2.3 follows because an individual either dies at $t_i \leq T_i$ with density $\mu^{-1}\exp(-\mu^{-1}t_i)$ or he survives beyond time T_i. Should the latter situation occur, we can only measure the probability of his survival, which is $\exp(-\mu^{-1}T_i)$ or $S(T_i)$. The likelihood function is given by

$$L(\mu) = \prod_{i=1}^{n} \left[\mu^{-1}\exp(-\mu^{-1}t_i)\right]^{\delta_i}\left[\exp(-\mu^{-1}T_i)\right]^{1-\delta_i}. \tag{3.2.4}$$

Similarly, for any survival distribution $S(T)$ and corresponding death density function $f(t)$, we can write

$$L(\mu) = \prod_{i=1}^{n} \left[f(t_i)\right]^{\delta_i}\left[S(T_i)\right]^{1-\delta_i}, \tag{3.2.4'}$$

where

$$\delta_i = \begin{bmatrix} 1, & \text{if the } i\text{th patient dies in the interval } 0 \leq t_i \leq T_i \\ 0, & \text{if he does not die in the interval } 0 \leq t_i \leq T_i \end{bmatrix}. \tag{3.2.5}$$

Taking logarithms, we find

$$\log_e L = -\sum_{i=1}^{n} \left[\delta_i(\log_e\mu + \mu^{-1}t_i) + (1-\delta_i)\mu^{-1}T_i\right]. \tag{3.2.6}$$

Setting $\partial \log_e L/\partial\mu|_{\mu=\hat{\mu}} = 0$ we have

$$\sum_{i=1}^{n} \left[\delta_i(-\hat{\mu}^{-1} + \hat{\mu}^{-2}t_i) + (1-\delta_i)\hat{\mu}^{-2}T_i\right] = 0. \tag{3.2.7}$$

The value $\hat{\mu}$ that satisfies (3.2.7), thus maximizes (3.2.4), is

$$\hat{\mu} = d^{-1}\sum_{i=1}^{n} (\delta_i t_i + (1-\delta_i)T_i), \tag{3.2.8}$$

where $d = \sum_{i=1}^{n}\delta_i$ is assumed to be greater than zero. The $\hat{\mu}$ given by (3.2.8) is thus the sum of the times to death of patients who die while on study plus the sum of the patient study times for those patients surviving the duration of the study, divided by the number of patients who die on study. Since maximum likelihood estimators are invariant under one-to-one transformations,

$$\hat{\lambda} = d \left[\sum_{i=1}^{n} (\delta_i t_i + (1 - \delta_i) T_i) \right]^{-1} \tag{3.2.9}$$

is the maximum likelihood estimator for λ. If $d = 0$, we define

$$\hat{\mu} = \sum_{i=1}^{n} T_i. \tag{3.2.10}$$

However, there will be few if any practical situations for which $d = 0$. An approximate variance of $\hat{\mu}$ is given by Bartholomew [1957] as

$$\text{Var}\,\hat{\mu} \doteq \frac{\hat{\mu}^2}{\sum_{i=1}^{n} \hat{Q}_i}, \qquad \text{where } \hat{Q} = 1 - \exp(-\lambda T_i), \qquad i = 1, 2, \ldots, n. \tag{3.2.11}$$

Variance equations are derived in Section 3.3; however, it is needed for the following example.

Example 3.2.1. We consider the example in Bartholomew [1957], assuming that the data represent 10 survival times (in days) of patients with advanced lung cancer and that the study is terminated at a particular point in time.

Table 3.1

Patient number	1	2	3	4	5	6	7	8	9	10
Survival time t_i (Days)	2		51		33	27	14	24	4	
Number of days until end of period T_i	81	72	70	60	41	31	31	30	29	21

We see from Table 3.1 that $\delta_2 = \delta_4 = \delta_{10} = 0$, and all other values of δ_i equal unity. Furthermore, $d = 7$, $\sum_{i=1}^{10} \delta_i t_i = 155$, and $\sum_{i=1}^{10}(1 - \delta_i)T_i = 153$. Thus $\hat{\mu} = (155 + 153)/7 = 44.0$ days, and $\hat{\lambda} = 1/44.0 = 0.023$ death per day. The approximate variance of $\hat{\mu}$ is $\mathrm{Var}\ \hat{\mu} \doteq \hat{\mu}^2/\sum_{i=1}^{n} Q_i = 44^2/6.15 = 314.8$; thus the standard deviation is 17.7 days.

Again we assume that each patient has the same exponential death density function given by (3.2.1). We assume, however, that all patients have the same point of entry into the study and that the study is terminated after the survival time of the dth patient (out of $n \geqslant d$ patients in all) has been recorded. Again n is fixed, and d is assumed to be fixed. Thus $t_{(d)}$, the survival time of the dth patient, is assumed to be a random variable. This case is considered by Halperin [1952] and Epstein and Sobel [1953].

The likelihood function $L(\theta')$ for the general k parameter case, where $\theta' = (\theta_1, \ldots, \theta_k)$ is

$$L(\theta') = \frac{n!}{(n-d)!} \prod_{i=1}^{d} f(t_{(i)}; \theta')[S(t_{(d)}; \theta')]^{n-d}, \qquad (3.2.12)$$

where

$$S(t_{(d)}; \theta') = \int_{t_{(d)}}^{\infty} f(t; \theta')dt. \qquad (3.2.13)$$

For the exponential case, when $\lambda = \mu^{-1}$ is the parameter of interest, Halperin shows

$$L(\lambda) = \frac{n!}{(n-d)!} \lambda^d \exp\left\{ -\lambda \left[\sum_{i=1}^{d-1} t_{(i)} + (n-d+1)t_{(d)} \right] \right\}. \qquad (3.2.14)$$

The maximum likelihood estimator obtained by the standard procedure is

$$\hat{\lambda} = \frac{d}{\displaystyle\sum_{i=1}^{d-1} t_{(i)} + (n-d+1)t_{(d)}} = \frac{d}{y} \qquad \text{(say)}. \qquad (3.2.15)$$

Halperin also obtains the mean and variance of $\hat{\lambda}$, which are, respectively,

$$E(\hat{\lambda}) = \frac{d\lambda}{d-1} \qquad (3.2.16)$$

and

$$\mathrm{Var}\hat{\lambda} = \frac{\lambda^2}{d-1}. \tag{3.2.17}$$

Equations 3.2.16 and 3.2.17 follow because for fixed d, the variable $2\lambda y$ has a chi-square distribution with $2d$ degrees of freedom.

Epstein and Sobel consider μ, the mean time to death, as the parameter of interest vis-à-vis the exponential survival distribution. They obtain $\hat{\mu}$, the maximum likelihood estimator of μ, which is $\hat{\lambda}^{-1}$ where $\hat{\lambda}$ is given by (3.2.15). They then show that $\hat{\mu}$ is the minimum variance unbiased estimator of μ, using the fact that $2y/\mu$ has a chi-square distribution with $2d$ degrees of freedom. It also follows that

$$\mathrm{Var}\hat{\mu} = \frac{\mu^2}{d}. \tag{3.2.18}$$

Finally, Epstein and Sobel obtain the expected value and variance of the time to the dth death:

$$E(t_{(d)}) = \mu \sum_{j=1}^{d} [n-j+1]^{-1} \tag{3.2.19}$$

and

$$\mathrm{Var}(t_{(d)}) = \mu^2 \sum_{j=1}^{d} [n-j+1]^{-2}. \tag{3.2.20}$$

If the time to censoring is fixed so that the number of deaths d is a random variable, (3.2.12) through (3.2.17) still hold; however, the distributions of $2\lambda y$ and $2y/\mu$ are only approximately distributed as chi-squares.

EXERCISES

1. In a clinical trial 10 patients with advanced cancer of the prostate survived (in months) as follows: 7, 19, 13, 12, 21, 16, 4, 9, 21, and 18. Assuming the exponential survival distribution for these patients, compute the mean survival time and the survival rate estimates $\hat{\mu}$ and $\hat{\lambda}$ for these patients.

2. If all the patients in Exercise 1 began their clinical trial at the same time and if the trial is terminated after 15 months, recompute $\hat{\mu}$ and $\hat{\lambda}$. What observations are censored?

3. Suppose that $f(t)$ has the death density function given by (1.3.24). Suppose further that $t_1, \ldots, t_n$ is a sample of size $n > k$ from the population having (1.3.24) as its death density function. Find the maximum likelihood estimators $\hat{\lambda}_1, \ldots, \hat{\lambda}_k$ of $\lambda_1, \ldots, \lambda_k$, respectively.

4. Suppose $k = 2$ in (1.3.20). Calculate the mean and variance of the resulting density function.

5. Suppose $t_1, \ldots, t_n$ is a random sample of size n from a population whose death density function is that in Exercise 3. Use the results in Exercise 3 to obtain $\tilde{\lambda}_1$ and $\tilde{\lambda}_2$ (not necessarily in closed form), which are the method of moments estimators of λ_1 and λ_2, respectively.

6. Suppose $t_1, t_2, \ldots, t_n$ is a random sample of size n from a population whose death density $f(t; \theta)$ is given by (3.1.6). Prove that $L(\theta)$ is maximized if $\hat{\theta} = \min_i t_i$.

7. For singly censored exponentially distributed survival data, obtain $\hat{\lambda}$, show that for fixed d, $2\lambda y$ has a chi-square distribution with $2d$ degrees of freedom; hence obtain $E(\hat{\lambda})$ and $\text{Var}(\hat{\lambda})$.

3.3 LARGE SAMPLE PROPERTIES OF THE MAXIMUM LIKELIHOOD ESTIMATES OF λ: NONCENSORED AND CENSORED

Let us follow n patients, each of whom has a constant hazard function λ, until they have all expired. In Section 3.1 we showed that if $t_1, t_2, \ldots, t_n$ are the survival times of the n patients, $\hat{\lambda}$, the maximum likelihood estimator of λ, is given by (3.1.11). Applying (3.1.4) for a single-parameter situation, the variance of $\hat{\lambda}$ for large n is approximately

$$\text{Var}\hat{\lambda} \doteq \frac{1}{-E(\partial^2 \log_e L / \partial \lambda^2)} = \frac{\lambda^2}{n}. \qquad (3.3.1)$$

The estimator $\hat{\lambda}$ is approximately normally distributed with mean λ and variance λ^2/n. For large n (roughly $n \geq 25$), that is, $\hat{\lambda} \sim N(\lambda, \lambda^2/n)$. Thus a $(1 - \alpha)\,100$ percent approximate confidence interval for λ is given by

$$\hat{\lambda} - \frac{Z(1 - \alpha/2)\hat{\lambda}}{\sqrt{n}} < \lambda < \hat{\lambda} + \frac{Z(1 - \alpha/2)\hat{\lambda}}{\sqrt{n}}, \qquad (3.3.2)$$

where $Z(1 - \alpha/2)$ is the $1 - \alpha/2$ percentile of the standard normal density.

It is not difficult to show that $2n\hat{\mu}/\mu$ has an exact chi-square distribution with $2n$ degrees of freedom where $\hat{\mu} = \bar{t}$; that is, $2n\hat{\mu}/\mu \sim \chi^2(2n)$. (See Epstein and Sobel [1953].) Recalling that $\hat{\lambda} = \hat{\mu}^{-1}$ and $\lambda = \mu^{-1}$, an exact $(1 - \alpha)\,100$ percent confidence interval for λ is

$$\frac{\hat{\lambda}\chi^2(\alpha/2; 2n)}{2n} < \lambda < \frac{\hat{\lambda}\chi^2(1 - \alpha/2; 2n)}{2n}, \qquad (3.3.2')$$

where $\chi^2(\alpha/2; 2n)$ and $\chi^2(1 - \alpha/2; 2n)$ are the $\alpha/2$ and $1 - \alpha/2$ percentiles, respectively, for the chi-square distribution with $2n$ degrees of freedom.

Example 3.3.1. Suppose 500 patients with advanced lung cancer are followed until death. Also suppose that the mean survival time for these patients is 5.4 months and the hazard rate λ is constant. It then follows that $\hat{\lambda} = 0.19/\text{month}$. Using (3.3.2), an approximate 99 percent confidence interval for λ, the true hazard rate is

$$0.19 - \frac{(2.576)0.19}{\sqrt{500}} < \lambda < 0.19 + \frac{(2.576)0.19}{\sqrt{500}}$$

or

$$0.168 < \lambda < 0.212. \tag{3.3.3}$$

Equation 3.3.2′ gives an exact 99 percent confidence interval

$$\frac{0.19(888.45)}{1000} < \lambda < \frac{0.19(1119.10)}{1000}$$

or

$$0.169 < \lambda < 0.213, \tag{3.3.3′}$$

noting that $\chi^2(0.005; 1000) = 888.45$ and $\chi^2(0.995; 1000) = 1119.10$. Hence depending on whether one uses (3.3.2) or (3.3.2′), the true hazard rate (per month) of patients with advanced lung cancer is covered by the interval $(0.168, 0.212)$ or $(0.169, 0.213)$ with probability 0.99. These values are obtained by

$$\chi^2(\alpha; \nu) = \nu \left[1 - \frac{2}{9\nu} + Z(1 - \alpha)\sqrt{\frac{2}{9\nu}} \right]^3, \tag{3.3.4}$$

where $Z(1 - \alpha)$ is the upper $(1 - \alpha)$-th percentage point of the standard normal density. Equation 3.3.4 is used generally for large values of ν that are not tabled (i.e., $\nu > 30$).

For the progressively censored case, suppose that T_i is the maximum time that the ith patient is on study, $i = 1, 2, \ldots, n$. Thus all the patients have not necessarily expired by the end of the study. In Section 3.2 we showed that if d of the n patients, $d > 0$, expire prior to the end of the study, d fixed, the maximum likelihood estimator $\hat{\lambda}$ of the hazard function λ, is given by (3.2.9). The unconditional variance of $\hat{\lambda}$ for large n is approximately

$$\text{Var}\hat{\lambda} = \frac{-1}{E_d E_{\hat{\lambda}}(\partial^2 \log_e L/\partial \lambda^2 | d)} = \frac{\lambda^2}{\sum\limits_{i=1}^{n} Q_i}, \qquad (3.3.5)$$

where

$$Q_i = 1 - \exp(-\lambda T_i), \qquad i = 1, 2, \ldots, n, \qquad (3.3.6)$$

which is (3.2.2).

The maximum likelihood estimator $\hat{\lambda} \sim N(\lambda, \sigma_{\hat{\lambda}}^2)$, approximately for large n, where $\sigma_{\hat{\lambda}}^2 = \lambda^2 / \sum_{i=1}^{n} Q_i$. To obtain a simple approximate $(1 - \alpha)$ 100 percent confidence interval for λ, however, we cannot use $\lambda / \sqrt{\sum_{i=1}^{n} Q_i}$ as the standard deviation of $\hat{\lambda}$ as the equation for determining the appropriate interval, since Q_i is a nonlinear function involving λ [see (3.3.6)]. Instead we may use $\hat{\lambda} / \sqrt{\sum_{i=1}^{n} \hat{Q}_i}$ or $\hat{\lambda} / \sqrt{d}$ as the standard deviation, where

$$\hat{Q}_i = 1 - \exp\left(-\hat{\lambda} T_i\right) \qquad (3.3.7)$$

and d is the number of deaths. Bartholomew [1957, p. 352] refers to $\hat{\lambda} / \sqrt{d}$ as "the quick estimate." Thus

$$\hat{\lambda}\left[1 - \frac{Z(1 - \alpha/2)}{\sqrt{M}}\right] < \lambda < \hat{\lambda}\left[1 + \frac{Z(1 - \alpha/2)}{\sqrt{M}}\right] \qquad (3.3.8)$$

is an approximate $(1 - \alpha)$ 100 percent confidence interval for λ, where

$$M = \left[\begin{array}{ll} \sum\limits_{i=1}^{n} \hat{Q}_i, & \text{if accuracy is important,} \\ d, & \text{if speed is important,} \end{array}\right. \qquad (3.3.9)$$

and $Z(1 - \alpha/2)$ is the $(1 - \alpha/2)100$ percentage point of the standard normal density.

Example 3.3.2. We consider the example in Bartholomew [1957] with the assumption that the data represent 10 survival times (in days) of patients with advanced lung cancer, given that the study is terminated at a particular point in time.

Table 3.2

Patient number	1	2	3	4	5	6	7	8	9	10
Survival time t_i (Months)	2	(119)	51	(77)	33	27	14	24	4	(37)
Number of days until end of period T_i	81		72	70	60	41	31	31	30	29	21

The numbers in parentheses in Table 3.2 are the survival times of those patients who were still alive at the conclusion of the study. Their survival times would not be known, but they are included here to permit comparison of the censored and noncensored cases. First of all, $d = 7$, which means that

$$\hat{\lambda} = \left[\frac{(2 + 51 + 33 + 27 + 14 + 24 + 4) + (72 + 60 + 21)}{7} \right]^{-1} = 0.023/\text{day}.$$

The estimated standard deviation of $\hat{\lambda}$ is

$$\frac{\hat{\lambda}}{\sqrt{\sum_{i=1}^{n} \hat{Q}_i}} = \frac{0.023}{\sqrt{6.15}} = 0.0093,$$

where

$$\sum_{i=1}^{10} \hat{Q}_i = \sum_{i=1}^{10} \left(1 - \exp\left(-\hat{\lambda}T_i\right)\right) = 10 - 3.85 = 6.15.$$

This more precise estimate may be compared with $\hat{\lambda}/\sqrt{d} = 0.023/\sqrt{7} = 0.0087$ for the quick estimate. The associated 95 percent confidence intervals for λ, based on the precise and quick estimates, respectively, are

$$0.005 < \lambda < 0.041 \tag{3.3.10}$$

and

$$0.006 < \lambda < 0.040. \tag{3.3.11}$$

If we had waited until all the patients on study had expired, we would have found $\hat{\lambda} = 0.026/\text{day}$, with an approximate 95 percent confidence interval for λ, given by

$$0.010 < \lambda < 0.042. \tag{3.3.12}$$

Without censoring, the exact 95 percent confidence interval for λ is

$$0.012 < \lambda < 0.049, \tag{3.3.12'}$$

since $\chi^2(0.025; 20) = 9.59$, and $\chi^2(0.975; 20) = 37.57$.

Depending on which method is used and whether we waited until all patients in the study had expired, each confidence interval so generated for λ (the true hazard rate of lung cancer patients) clearly covers λ with probability approximately 0.95.

For singly censored data in which the number of deaths d is fixed in advance of the study, it was demonstrated in Section 3.2 that

$$\mathrm{Var}\,\hat{\lambda} = \frac{\lambda^2}{d-1}, \tag{3.3.13}$$

where $\hat{\lambda}$ is given by (3.2.15). By methods analogous to the complete sample and progressively censored sample cases, it is easy to show that an approximate $(1-\alpha)100$ percent confidence interval for λ is

$$\hat{\lambda} - \frac{Z(1-\alpha/2)\hat{\lambda}}{\sqrt{d-1}} < \lambda < \hat{\lambda} + \frac{Z(1-\alpha/2)\hat{\lambda}}{\sqrt{d-1}}, \tag{3.3.14}$$

where $Z(1-\alpha/2)$ is the $1-\alpha/2$ percentile of the standard normal distribution. Noting that $2\lambda y$ has a chi-square distribution with $2d$ degrees of freedom, an exact $(1-\alpha)$ 100 percent confidence interval for λ is

$$\frac{\hat{\lambda}\chi^2(\alpha/2; 2d)}{2d} < \lambda < \frac{\hat{\lambda}\chi^2(1-\alpha/2; 2d)}{2d}, \tag{3.3.15}$$

where $\chi^2(\alpha/2; 2d)$ and $\chi^2(1-\alpha/2; 2d)$ are the $\alpha/2$ and $1-\alpha/2$ percentiles, respectively, of the chi-square distribution with $2d$ degrees of freedom.

EXERCISES

1. For the data in Exercise 1, Section 3.2, obtain both exact and approximate confidence intervals for λ.

2. For the data in Exercise 2, Section 3.2, obtain both the exact and approximate confidence intervals for λ. For the approximate confidence interval use both the precise and quick estimates.

3. Derive the confidence interval for λ given by (3.3.2).

4. Derive the confidence interval for λ given by (3.3.8).

5. Suppose $t_1, t_2, \ldots, t_n$ is a random sample of survival times from the population having as its death density function (1.3.20) with $k = 2$. Obtain the large sample variance–covariance matrix for the maximum likelihood estimators $\hat{\lambda}_1$ and $\hat{\lambda}_2$; hence obtain an approximate $(1 - \alpha)$ 100 percent simultaneous confidence interval for λ_1 and λ_2. *Hint*: See Section 4.4.

6. (a) Suppose $t_1, t_2, \ldots, t_n$ is a random sample of survival times from a population whose death density is

$$f(t; \mu) = \mu^{-1} \exp(-\mu^{-1} t), \qquad \mu > 0, t \geq 0.$$

Prove that $2n\bar{t}/\mu \sim \chi^2(2n)$, where $\bar{t} = \sum_{i=1}^{n} t_i / n$; that is; show that $2n\bar{t}/\mu$ is distributed as a chi-square variable with $2n$ degrees of freedom. (b) If the n observations in (a) are censored so that only d deaths are observed, where d is selected in advance $(d > 0)$, prove that $2d\hat{\mu}/\mu \sim \chi^2(2d)$, where

$$\hat{\mu} = \frac{\sum_{i=1}^{d-1} t_{(i)} + (n - d + 1) t_{(d)}}{d}.$$

Hence obtain an exact $(1 - \alpha)$ 100 percent confidence interval for $\hat{\mu}$ in the singly censored case.

7. Suppose $t_1, t_2, \ldots, t_n$ is a random sample of size n from a population whose death density is given by the censored exponential death density. Thus t_i is known if and only if $t_i \leq T_i$, $i = 1, 2, \ldots, n$. Derive the large sample variance of $\hat{\mu}$, the maximum likelihood estimator. *Hint*: $E(d)$ and $E(d\hat{\mu})$ are both needed.

3.4 ESTIMATION OF P, THE SURVIVAL PROBABILITY FOR A CONSTANT HAZARD RATE

Consider a clinical trial in which the patients exhibit a constant hazard rate λ and enter the trial at random and uniformly through the period of the trial that is T years (say) long. If there are n patients altogether, for the ith patient we observe his survival time t_i (if he dies during the trial) or his total time on the trial T_i (if he is alive at time T). For a pictorial view the reader should reexamine Figures 2.1 and 2.2.

We consider estimating $P(x)$ the x-year survival probability based on this sample of n patients for two cases: (1) when the values T_i are fixed, and (2) when T_i is a random variable whose density $g(T_i)$ is

$$g(T_i) = \begin{bmatrix} \dfrac{1}{T}, & 0 \leq T_i \leq T, \\ 0, & \text{elsewhere} \end{bmatrix} \qquad (3.4.1)$$

In both cases the data are progressively censored; hence the maximum likelihood estimator for λ is given by (3.2.9). That is,

$$\hat{\lambda} = d\left[\sum_{i=1}^{n} \delta_i t_i + \sum_{i=1}^{n} (1 - \delta_i) T_i \right]^{-1}. \qquad (3.4.2)$$

Where $\delta_i = 1$ if the ith patient dies during the clinical trial, $\delta_i = 0$ if he lives the entire time he is on study, and d is the observed number of deaths, we assume $d > 0$. Thus $\hat{P}(x)$, the estimated probability an individual beginning at time $t = 0$ survives the x years of the T-year clinical trial, is

$$\hat{P}(x) = \exp(-\hat{\lambda}x) = \hat{S}(x). \qquad (3.4.3)$$

Case 1: T_i Fixed

Under the assumption that each T_i is fixed, we compute $\sigma_{\hat{P}}^2$ approximately. To do this we expand $\exp(-\hat{\lambda}x)$ in a Taylor series about λ, ignoring all terms higher than the first order. Thus

$$\exp(-\hat{\lambda}x) \doteq \exp(-\lambda x) - x\exp(-\lambda x)(\hat{\lambda} - \lambda). \qquad (3.4.4)$$

Thus as discussed in Section 2.3, it follows that $\sigma_{\hat{P}}^2$ is approximately

$$\sigma_{\hat{P}}^2 \doteq x^2 \exp(-2\lambda x)\sigma_{\hat{\lambda}}^2. \qquad (3.4.5)$$

Now $\sigma_{\hat{\lambda}}^2$ is given by (3.3.5). Thus

$$\sigma_{\hat{P}}^2 \doteq \frac{(x\lambda)^2 \exp(-2\lambda x)}{\sum_{i=1}^{n} Q_i}, \qquad (3.4.6)$$

where we recall that $Q_i = 1 - \exp(-\lambda T_i)$, which is the probability the ith individual dies during the trial.

Using large sample theory for maximum likelihood estimators, $\hat{P}(x) \sim N(P(x), \sigma_{\hat{P}}^2)$. Thus an approximate $(1 - \alpha)$ 100 percent confidence interval for $P(x)$ is given by

$$\hat{P}(x) - Z\left(1 - \frac{\alpha}{2}\right)\sigma_{\hat{P}} < P(x) < \hat{P}(x) + Z\left(1 - \frac{\alpha}{2}\right)\sigma_{\hat{P}}, \qquad (3.4.7)$$

where $Z(1 - \alpha/2)$ is the $(1 - \alpha/2)$ 100 percentile of the standard normal density. Since $\sigma_{\hat{P}}$ involves the use of the unknown parameter λ, in practical applications of (3.4.7) we replace $\sigma_{\hat{P}}$ by $\hat{\sigma}_{\hat{P}}$, where

$$\hat{\sigma}_{\hat{P}}^2 = \frac{(x\hat{\lambda})^2 \exp(-2\hat{\lambda}x)}{\sum_{i=1}^{n} \hat{Q}_i}, \qquad (3.4.8)$$

and $\hat{Q}_i = 1 - \exp(-\hat{\lambda}T_i)$; $i = 1, 2, \ldots, n$, which is (3.3.7). Thus an approximate $(1 - \alpha)$ 100 percent confidence interval for $P(x)$ is

$$\hat{P}(x) - Z\left(1 - \frac{\alpha}{2}\right)\hat{\sigma}_{\hat{P}} < P(x) < \hat{P}(x) + Z\left(1 - \frac{\alpha}{2}\right)\hat{\sigma}_{\hat{P}}. \qquad (3.4.7')$$

We stress that this confidence interval is approximate. It may be compared, however, to the exact confidence interval for $P(x)$. (See Exercises 5 and 6 at the end of the section.)

Case 2. T_i A Random Variable Whose Density Is (3.4.1)

In this case we compute first of all $\sigma_{\hat{\lambda}}^2$. Recall that

$$\sigma_{\hat{\lambda}}^2 = -1 \Big/ E\left(\frac{\partial^2 \log L}{\partial \lambda^2}\right) \qquad (3.4.9)$$

is the large sample variance of $\hat{\lambda}$. Evaluation of (3.4.9) differs from (3.3.5) in that T_i, the length of time the ith patient is followed during the clinical trial, is uniformly distributed on the interval $(0, T)$. It thus follows that if the ith patient is followed for a uniformly distributed length of time, then

$$\sigma_{\hat{\lambda}}^2 = \frac{\lambda^2}{n\left(1 - \dfrac{1 - \exp(\lambda T)}{\lambda T}\right)}. \qquad (3.4.10)$$

To prove this, consider

$$-\frac{\partial^2 \log L}{\partial \lambda^2} = \frac{d}{\lambda^2}, \qquad (3.4.11)$$

where

$$d = \sum_{i=1}^{n} \delta_i \qquad (3.4.12)$$

and $\delta_i = 1$ if the ith patient dies while on study and is 0 otherwise. Clearly,

$$E_d(\delta_i | T_i) = Q_i \equiv (1 - \exp(-\lambda T_i)), \qquad i = 1, 2, \ldots, n. \qquad (3.4.13)$$

Thus

$$E(d) = \sum_{i=1}^{n} E_{T_i}(Q_i|T_i) = \frac{n}{T} \int_0^T (1 - \exp(-\lambda T_i)) \, dT_i$$

$$= n\left(1 - \frac{1 - \exp(-\lambda T)}{\lambda T}\right). \tag{3.4.14}$$

Hence noting that

$$-E\left(\frac{\partial^2 \log L}{\partial \lambda^2}\right) = \frac{E(d)}{\lambda^2},$$

(3.4.10) follows.

Using (3.4.5) and (3.4.10), an approximate expression for $\sigma_{\hat{p}}^2$ is

$$\sigma_{\hat{p}}^2 \doteq \frac{(\lambda x)^2 \exp(-2\lambda x)}{n\left(1 - \dfrac{1 - \exp(-\lambda T)}{\lambda T}\right)}. \tag{3.4.15}$$

Thus (3.4.7) again provides us with an approximate $(1 - \alpha)$ 100 percent confidence interval for $P(x)$, where $\sigma_{\hat{p}}^2$ is given by (3.4.15) instead of (3.4.6). In practical applications we again would use (3.4.7′) as an approximate $(1 - \alpha)$ 100 percent confidence interval for $P(x)$, where

$$\hat{\sigma}_{\hat{p}}^2 = \frac{(\hat{\lambda} x)^2 \exp(-2\hat{\lambda} x)}{n\left(1 - \dfrac{1 - \exp(-\hat{\lambda} T)}{\hat{\lambda} T}\right)} \tag{3.4.16}$$

replaces (3.4.8).

Example 3.4.1. In a study by Frei et al. [1961] complete remission times for 11 adult patients with acute leukemia were recorded in weeks: 20, 21, 17, 18, 26, 49, 71+, 36, 53, 53+, and 57. The values 53+ and 71+ are censored after 80 weeks of observation. That is, these patients were still in complete remission after 80 weeks. However, their censoring times are measured as 53 and 71 weeks, respectively. We compute a 90 percent confidence interval $P(x)$, the probability of remaining in complete remission at 80 weeks: $\hat{\lambda} = 0.0214$, $x = T = 80$. Thus $\hat{P}(x) = 0.181$. Now $\hat{\sigma}_{\hat{p}} = 0.129$ using either (3.4.8) or (3.4.16). Hence an approximate 90 percent confidence interval for $P(x)$ is*

$$0 < P(x) < 0.393. \tag{3.4.17}$$

We are thus 90 percent confident that the true complete remission rate is covered by an interval whose upper limit is no greater than 0.393 after 80 weeks of the trial.

EXERCISES

1. The following data are simulated survival times in months of patients with cancer:

23	12	2	8	11
10	2	1	21	16
25	5	0	5	
0	3	4	5	
21	0	2	5	
41	21	23	8	

Plot $\lambda(t)$, $S(t)$, and $f(t)$. Would the assumption of a constant hazard rate appear reasonable from these data? (More discussion of graphical techniques of assessing the data are given in Section 3.6.) Compute $\hat{\lambda}$ and $\sigma_{\hat{\lambda}}$, and $\hat{\mu}$ and $\sigma_{\hat{\mu}}$, and compare with your plots.

   If the values 25 and 41 were progressively censored times rather than times to death, compute $\hat{P}(x)$ for $x \leqslant 10$ and 95 percent confidence interval for $\hat{P}(x)$ for $T = 100$.

2. Suppose $t_1, t_2, \ldots, t_n$ are the survival times of n patients in a clinical trial, and all patients are followed to death. If the common death density of these patients is $\lambda \exp(-\lambda t)$, $t \geqslant 0$, find the approximate variance of $\hat{P} = \exp(-\hat{\lambda}T)$, where $\hat{\lambda} = \bar{t}^{-1}$ is the maximum likelihood estimator of λ and T is an arbitrary point in time.

3. Find an approximate $(1 - \alpha)\,100$ percent confidence interval for $P(x) = \exp(-\lambda x)$ in Exercise 2. Apply this result to the data in Example 3.4.1 to obtain an approximate 90 percent confidence interval for $P(x)$ if after following patients on the clinical trial for 90 weeks, the censored times $53+$ and $71+$ are 58 and 80 weeks in complete remission, respectively.

---

*The confidence interval includes values less than 0, but since $P(x) \geqslant 0$ the result is still that given in (3.4.17).

4. Suppose each of n patients entering a clinical trial is followed for at least time $t_0 > 0$ so that

$$g(T_i) = \left[\begin{array}{cc} \dfrac{1}{(T - t_0)}, & t_0 < T_i < T, \\[2ex] 0, & \text{otherwise.} \end{array} \right.$$

Obtain $\text{Var}\,\hat{\lambda}$ and the approximate value of $\text{Var}\,\hat{P}(x)$ when the common death density function of patients is $\lambda \exp(-\lambda t)$, $t > 0$, $\lambda > 0$.

5. Suppose that $t_1, t_2, \ldots, t_n$ is a complete random sample of survival times from a clinical trial for which the death density function is $\lambda \exp(-\lambda t)$, $t > 0$, $\lambda > 0$. Is it possible to obtain an exact confidence interval for $P = \exp(-\lambda T)$, where T is a given point in time? Explain.

6. For the data in Example 3.4.1 is it possible to find an exact 90 percent confidence interval for P? If so, find it. (Answer: $0.065 < P < 0.409$.)

3.5 ESTIMATION OF EXPONENTIAL SURVIVAL PROBABILITIES WITH CONCOMITANT INFORMATION

In the case of a simple linear or multiple linear regression model, estimation of the regression parameters is accomplished by ordinary least squares procedures. In analyzing patient survival data in the presence of concomitant information, however, some modifications of the usual procedures are necessary. In this section, we develop the statistical model put forth by Feigl and Zelen [1965] for one concomitant variate, noting that the extension to the multivariate case is relatively straightforward. We also consider the extension to the censored case treated by Zippin and Armitage [1966].

Full understanding of this section requires a knowledge of least squares theory and the two-dimensional Newton-Raphson procedure. The reader may skip this section on initial reading, but he should be able to return to it after reading Chapter 6. Additional discussion and procedures can be found in Miller [1974].

Suppose n patients are on study and the death density function for the ith patient is

$$f_i(t_i) = \lambda_i \exp(-\lambda_i t_i), \qquad \lambda_i > 0, \; t_i \geqslant 0, \qquad i = 1, 2, \ldots, n. \qquad (3.5.1)$$

Initially we assume that all patients are followed until death. Furthermore, let x_i be observed values of a concomitant variable or covariate, such that

$$E(t_i) = \lambda_i^{-1} = a + bx_i, \qquad i = 1, 2, \ldots, n. \qquad (3.5.2)$$

Thus the mean survival time of patients is assumed to be linearly related to the concomitant variable. For example, if the variates x represent cholesterol levels of patients with acute coronary disease, we might expect small values of the concomitant variate to correspond to relatively large values of the mean survival time. Conversely, we should expect large values of the concomitant variate to correspond to small values of the mean survival time. Note that x_i could refer to a transformed value (e.g., $x_i = z_i^{-1}$). The problem at hand is to estimate the parameters a and b. To this end we use the method of maximum likelihood. Let $L(a,b)$ be the likelihood function for a and b, given the observations $(x_1, t_1), \ldots, (x_n, t_n)$:

$$L(a,b) = \prod_{i=1}^{n} (a + bx_i)^{-1} \exp\left(-(a + bx_i)^{-1} t_i\right). \qquad (3.5.3)$$

Then, differentiating $\log_e L$ with respect to a and b, respectively, the maximum likelihood estimators $\hat{a}$ and $\hat{b}$ are obtained as solutions of

$$\frac{\partial \log_e L}{\partial a} = - \sum_{i=1}^{n} \left(\hat{a} + \hat{b}x_i\right)^{-1} + \sum_{i=1}^{n} t_i \left(\hat{a} + \hat{b}x_i\right)^{-2} = 0 \qquad (3.5.4)$$

and

$$\frac{\partial \log_e L}{\partial b} = - \sum_{i=1}^{n} x_i \left(\hat{a} + \hat{b}x_i\right)^{-1} + \sum_{i=1}^{n} x_i t_i \left(\hat{a} + \hat{b}x_i\right)^{-2} = 0. \qquad (3.5.5)$$

We note that $\hat{a}$ and $\hat{b}$ must satisfy the constraint $\hat{a} + \hat{b}x_i > 0$ for all $i = 1, 2, \ldots, n$. In applications, however, this constraint is usually satisfied. The problem of constrained maximum likelihood estimators is discussed in Chapter 6.

Feigl and Zelen use the Newton-Raphson technique discussed in Chapter 6 for obtaining $\hat{a}$ and $\hat{b}$. The discussion of the two-dimensional Newton-Raphson procedure, which is required to obtain $\hat{a}$ and $\hat{b}$, appears in Section 6.2. The formulas to compute $\hat{a}$ and $\hat{b}$ are given in this section. First of all it is necessary to obtain initial estimates $\hat{a}_0$ and $\hat{b}_0$. We can take $\hat{a}_0$ and $\hat{b}_0$ by plotting t_i and x_i and estimating them visually or computing the regression equation

$$E(t_i) = a_0 + b_0 x_i, \qquad i = 1, \ldots, n, \qquad (3.5.6)$$

where the least squares estimates $\hat{a}_0$ and $\hat{b}_0$, obtained from (3.5.6), become the initial estimates in the iterative procedure.

If $\hat{a}_k$, $\hat{b}_k$ denote the estimates at the kth iteration, the values for the $(k+1)$th iteration are found by solving the two simultaneous equations

(3.5.7) and (3.5.8) in the two unknowns $\delta \hat{a}_k$ and $\delta \hat{b}_k$, where

$$A_k \delta \hat{a}_k + B_k \delta \hat{b}_k = D_k, \qquad (3.5.7)$$

$$B_k \delta \hat{a}_k + C_k \delta \hat{b}_k = E_k, \qquad (3.5.8)$$

and

$$A_k = \sum_{i=1}^{n} \left(\hat{a}_k + \hat{b}_k x_i \right)^{-2} - 2 \sum_{i=1}^{n} t_i \left(\hat{a}_k + \hat{b}_k x_i \right)^{-3} = \frac{\partial^2 \log_e L}{\partial a_k^2}, \qquad (3.5.9)$$

$$B_k = \sum_{i=1}^{n} x_i \left(\hat{a}_k + \hat{b}_k x_i \right)^{-2} - 2 \sum_{i=1}^{n} t_i x_i \left(\hat{a}_k + \hat{b}_k x_i \right)^{-3} = \frac{\partial^2 \log_e L}{\partial a_k \, \partial b_k}, \qquad (3.5.10)$$

$$C_k = \sum_{i=1}^{n} x_i^2 \left(\hat{a}_k + \hat{b}_k x_i \right)^{-2} - 2 \sum_{i=1}^{n} t_i x_i^2 \left(\hat{a}_k + \hat{b}_k x_i \right)^{-3} = \frac{\partial^2 \log_e L}{\partial b_k^2}, \qquad (3.5.11)$$

$$D_k = \sum_{i=1}^{n} \left(\hat{a}_k + \hat{b}_k x_i \right)^{-1} - \sum_{i=1}^{n} t_i \left(\hat{a}_k + \hat{b}_k x_i \right)^{-2} = - \frac{\partial \log_e L}{\partial a_k}, \qquad (3.5.12)$$

$$E_k = \sum_{i=1}^{n} x_i \left(\hat{a}_k + \hat{b}_k x_i \right)^{-1} - \sum_{i=1}^{n} t_i x_i \left(\hat{a}_k + \hat{b}_k x_i \right)^{-2} = - \frac{\partial \log_e L}{\partial b_k}, \qquad (3.5.13)$$

and $\delta \hat{a}_k = \hat{a}_{k+1} - \hat{a}_k$, $\delta \hat{b}_k = \hat{b}_{k+1} - \hat{b}_k$.

The asymptotic variance–covariance matrix for $(\hat{a}, \hat{b})$ is obtained from

$$\begin{bmatrix} \operatorname{Var} \hat{a} & \operatorname{Cov}(\hat{a}, \hat{b}) \\ \operatorname{Cov}(\hat{a}, \hat{b}) & \operatorname{Var} \hat{b} \end{bmatrix} = - \begin{bmatrix} E\left(\dfrac{\partial^2 \log_e L}{\partial a^2} \right) & E\left(\dfrac{\partial^2 \log_e L}{\partial a \, \partial b} \right) \\ E\left(\dfrac{\partial^2 \log_e L}{\partial a \, \partial b} \right) & E\left(\dfrac{\partial^2 \log_e L}{\partial b^2} \right) \end{bmatrix}^{-1}$$

$$(3.5.14)$$

following (3.1.4).

It is not hard to show that

$$\operatorname{Var} \hat{a} = \Delta^{-1} \sum_{i=1}^{n} x_i^2 (a + bx_i)^{-2}, \qquad (3.5.15)$$

$$\operatorname{Var} \hat{b} = \Delta^{-1} \sum_{i=1}^{n} (a + bx_i)^{-2}, \qquad (3.5.16)$$

$$\text{Cov}(\hat{a},\hat{b}) = \Delta^{-1} \sum_{i=1}^{n} x_i(a+bx_i)^{-2}, \tag{3.5.17}$$

and

$$\Delta = \left[\sum_{i=1}^{n} x_i^2(a+bx_i)^{-2} \right]\left[\sum_{i=1}^{n} (a+bx_i)^{-2} \right] - \left[\sum_{i=1}^{n} x_i(a+bx_i)^{-2} \right]^2. \tag{3.5.18}$$

Now we have

$$\text{Var}(\hat{a}+\hat{b}x_i) = \text{Var}\,\hat{a} + x_i^2\text{Var}\,\hat{b} + 2x_i\,\text{Cov}(\hat{a},\hat{b}). \tag{3.5.19}$$

Using (3.5.19), a $100(1-\alpha)$ percent *simultaneous* confidence interval for $(a+bx_i)$, $i=1,2,\ldots,n$ is approximately

$$\hat{a}+\hat{b}x_i - Z\left(1-\frac{\alpha}{2}\right)\hat{\sigma} < a+bx_i < \hat{a}+\hat{b}x_i + Z\left(1-\frac{\alpha}{2}\right)\hat{\sigma}, \tag{3.5.20}$$

where

$$\hat{\sigma} = \sqrt{\text{Vâr}(\hat{a}+\hat{b}x_i)}\;. \tag{3.5.21}$$

$\text{Vâr}(\hat{a}+\hat{b}x_i)$ is the estimated variance of $\hat{a}+\hat{b}x_i$ with $\hat{a}$ and $\hat{b}$ replacing a and b, respectively, in (3.5.19). The value $Z(1-\alpha/2)$ is the upper $100(1-\alpha/2)$ percentage point of the standard normal density. Thus a $100(1-\alpha)$ percent simultaneous confidence interval for $S_i(t) = \exp[-t(a+bx_i)^{-1}]$, $i=1,2,\ldots,n$, the probabilities that each of the n patients survives to at least time t, is

$$\exp(-tc_{1i}^{-1}) \leqslant S_i(t) \leqslant \exp(-tc_{2i}^{-1}), \tag{3.5.22}$$

where

$$c_{1i} = \hat{a}+\hat{b}x_i - Z\left(1-\frac{\alpha}{2}\right)\hat{\sigma} \tag{3.5.23}$$

and

$$c_{2i} = \hat{a}+\hat{b}x_i + Z\left(1-\frac{\alpha}{2}\right)\hat{\sigma}, \qquad i=1,2,\ldots,n. \tag{3.5.24}$$

To test whether the fitted model is appropriate with respect to the observed data, we compare the observed and expected numbers of patients dying in certain predetermined time intervals, using a chi-square goodness of fit test. If the time axis is divided into k subintervals, these determine for the ith patient the k quantile intervals

$$0 \leqslant t_i < t_i(1/k), \qquad t_i(1/k) \leqslant t_i < t_i(2/k), \ldots, t_i\left(\frac{k-1}{k}\right) \leqslant t_i < \infty.$$

The probability is $1/k$ the ith patient dies in any one interval, $i = 1, 2, \ldots, k$. If the model is a good fit, we would expect the survival times to be equally distributed in the k intervals.

Example 3.5.1. We use the same example presented by Feigl and Zelen. Data appear in Table 3.3 for two groups of patients who died of acute myelogenous leukemia. Patients were classified into two groups according to the absence or presence of a morphologic characteristic of white blood cells. Patients termed AG positive were identified by the presence of Auer rods and/or significant granulature of the leukemic cells in the bone marrow at diagnosis. For the AG negative patients these factors were absent.

In Table 3.3 the survival times t_i are given in weeks from date of diagnosis. The concomitant variate x_i is the logarithm to the base 10 of the white blood count at the time of diagnosis. The white blood count has long been recognized as a predictor of survival times for acute leukemia—the higher the initial white blood count, the lower the probability of surviving a specified length of time.

Table 3.4 shows the Feigl and Zelen results for the data from Table 3.3. Mantel and Myers [1971], utilizing the same data and the iterative approach described, have obtained slightly different numerical results, which are given in parentheses in Table 3.4. Apparently the Mantel-Myers estimates are correct; the reader should note, however, that in using the Mantel-Myers values he will find no appreciable change in the fit of this model to the survival data. Graphical examination of the data reveals that there is some question concerning the fit of the survival data versus $\log_{10}$ white blood count data as a straight line. The reader should bear in mind (1) that no white blood counts greater than 1000×10^2 were recorded, and (2) that there are possible outliers in these data (e.g., the last observation in the AG positive group). Moreover, Mantel and Myers correctly point out the need for reporting survival data accurately: when it is assumed a patient dies after 150 weeks on study (say), it is correct to state his survival time as 149.5 weeks, since death presumably occurred sometime during, rather than at the end of, these reported weekly periods.

Table 3.3 Observed survival times and white blood counts (WBC)

(AG positive) $n = 17$		(AG negative) $n = 16$	
WBC $\times 10^2$	Survival Time (weeks)	WBC $\times 10^2$	Survival Time (weeks)
23	65	44	56
7.5	156	30	65
43	100	40	17
26	134	15	7
60	16	90	16
105	108	53	22
100	121	100	3
170	4	190	4
54	39	270	2
70	143	280	3
94	56	310	8
320	26	260	4
350	22	210	3
1000	1	790	30
1000	1	1000	4
520	5	1000	43
1000	65		
Median values 100	56	200	7.5

Table 3.4

Patient Group	$\hat{a}$	$\hat{b}$	$\sqrt{\hat{\text{Var}}\,\hat{a}}$	$\sqrt{\hat{\text{Var}}\,\hat{b}}$	$\hat{\text{Cov}}(\hat{a}, \hat{b})$
AG positive	240. (248.7)	$-44.$ (-46.1)	95.5	20.1	-1914
AG negative	30. (36.9)	$-3.$ (-4.6)	35.1	8.2	-284

74

Table 3.5 provides a goodness of fit test of the adequacy of the model after we have divided the time axis into quantiles: χ^2 (AG positive) $= 0.317$, and χ^2 (AG negative) $= 3.50$. The goodness of fit chi-square values have 1 degree of freedom each, since there are four categories and two parameters estimated and these are clearly not significant at the 5 percent level. However, since the power of the chi-square is not high, and since there is some question concerning the straight line fit, the values of $\hat{a}$ and $\hat{b}$ for the two groups should be viewed with caution.*

Table 3.5

Group	Number of Patients	$0 < t < t(0.25)$	$t(0.25) < t$ $< t(0.50)$	$t(0.50) < t$ $< t(0.75)$	$t(0.75) < t < \infty$	Total
AG positive	Observed	5	4	3	5	17
	Expected	4.25	4.25	4.25	4.25	17
AG negative	Observed	4	3	2	7	16
	Expected	4	4	4	4	16

Zippin and Armitage [1966] modify the Feigl-Zelen model when the survival data are censored. Specifically, suppose n patients are on study in a clinical trial setting, for example, where (3.5.1) is the death density function for the ith patient (i.e., the exponential with concomitant information). However, as in Section 3.2, the ith patient's survival time t_i is known only if $t_i \leqslant T_i$, where T_i is the maximum observation time of the ith patient, $i = 1, 2, \ldots, n$. It then follows from (3.2.4) that

$$L(a,b) = \prod_{i=1}^{n} \left[(a+bx_i)^{-1} \exp -(a+bx_i)^{-1} t_i \right]^{\delta_i} \left[\exp -(a+bx_i)^{-1} T_i \right]^{1-\delta_i}$$

and that the $\log_e$ likelihood of the sample of observations is

$$\log_e L = - \sum_{i=1}^{n} \delta_i \log_e(a+bx_i) - \sum_{i=1}^{n} \delta_i t_i (a+bx_i)^{-1}$$
$$- \sum_{i=1}^{n} (1-\delta_i) T_i (a+bx_i)^{-1}, (3.5.25)$$

*The test of goodness of fit by a chi-square distribution is not strictly correct. See the discussion prior to Example 3.6.3.

where

$$
\delta_i = \left[\begin{array}{ll} 1 & \text{if the } i\text{th patient dies in the interval } 0 \leqslant t_i \leqslant T_i \\ 0 & \text{if the } i\text{th patient does not die in the interval } 0 \leqslant t_i \leqslant T_i. \end{array} \right.
$$

$$(3.5.26)$$

The maximum likelihood estimators $\hat{a}$ and $\hat{b}$ are obtained as solutions to

$$
\frac{\partial \log_e L}{\partial a} = - \sum_{i=1}^{n} \delta_i \left(\hat{a} + \hat{b} x_i \right)^{-1} + \sum_{i=1}^{n} \delta_i t_i \left(\hat{a} + \hat{b} x_i \right)^{-2}
$$

$$
+ \sum_{i=1}^{n} (1 - \delta_i) T_i \left(\hat{a} + \hat{b} x_i \right)^{-2} = 0 \qquad (3.5.27)
$$

and

$$
\frac{\partial \log_e L}{\partial b} = - \sum_{i=1}^{n} \delta_i x_i \left(\hat{a} + \hat{b} x_i \right)^{-1} + \sum_{i=1}^{n} \delta_i t_i x_i \left(\hat{a} + \hat{b} x_i \right)^{-2}
$$

$$
+ \sum_{i=1}^{n} (1 - \delta_i) T_i x_i \left(\hat{a} + \hat{b} x_i \right)^{-2} = 0. \quad (3.5.28)
$$

The solution of (3.5.27) and (3.5.28) is analogous to the solution of (3.5.4) and (3.5.5). Initial estimates $\hat{a}_0$ and $\hat{b}_0$ are obtained as the simple regression estimates for the linear regression model (3.5.6); however, if $t_i > T_i$, then T_i replaces t_i as the dependent variable for the ith patient or that observation is ignored. The estimates $\hat{a}_{k+1}$, $\hat{b}_{k+1}$ at the $(k+1)$th iteration are obtained by solving the two simultaneous equations (3.5.7) and (3.5.8) in the unknowns $\hat{a}_{k+1}$, $\hat{b}_{k+1}$ with the appropriate modifications to account for the censoring. Details of the modification are left as an exercise to the reader, but the second derivatives are given in (3.5.29) and (3.5.30).

The asymptotic variance–covariance matrix for $(\hat{a}, \hat{b})$ is obtained from (3.5.14), but the expected values of the second partial derivations are obtained with censoring. Thus we have

$$
A_0 = \frac{\partial^2 \log_e L}{\partial a^2} = \sum_{i=1}^{n} \delta_i (a + b x_i)^{-2} - 2 \sum_{i=1}^{n} \delta_i t_i (a + b x_i)^{-3}
$$

$$
- 2 \sum_{i=1}^{n} (1 - \delta_i) T_i (a + b x_i)^{-3}, \quad (3.5.29)
$$

$$A_1 = \frac{\partial^2 \log_e L}{\partial a \, \partial b} = \sum_{i=1}^n \delta_i x_i (a + bx_i)^{-2} - 2 \sum_{i=1}^n \delta_i t_i x_i (a + bx_i)^{-3}$$

$$- 2 \sum_{i=1}^n (1 - \delta_i) T_i x_i (a + bx_i)^{-3},$$

$$(3.5.30)$$

and

$$A_2 = \frac{\partial^2 \log_e L}{\partial b^2} = \sum_{i=1}^n \delta_i x_i^2 (a + bx_i)^{-2} - 2 \sum_{i=1}^n \delta_i t_i x_i^2 (a + bx_i)^{-3}$$

$$- 2 \sum_{i=1}^n (1 - \delta_i) T_i x_i^2 (a + bx_i)^{-3}.$$

$$(3.5.31)$$

For the ith patient the probability of death before T_i is $(1 - \exp(-\lambda_i T_i))$, where $\lambda_i = (a + bx_i)^{-1}$. It then follows that

$$E(A_j) = \sum_{i=1}^n \left[1 - \exp(-\lambda_i T_i) \lambda_i^2 x_i^j \right] - 2 \sum_{i=1}^n \int_0^{T_i} \lambda_i \exp(-\lambda_i t_i) t_i \lambda_i^3 x_i^j \, dt_i$$

$$- 2 \sum_{i=1}^n \exp(-\lambda_i T_i) T_i \lambda_i^3 x_i^j, \qquad j = 0, 1, 2.$$

$$(3.5.32)$$

Since

$$\int_0^{T_i} \exp(-\lambda_i t_i) t_i \, dt_i = \lambda_i^{-2} \left[(1 - \exp(-\lambda_i T_i)) - \lambda_i T_i \exp(-\lambda_i T_i) \right], \quad (3.5.33)$$

we then find

$$E(A_j) = - \sum_{i=1}^n x_i^j \lambda_i^2 (1 - \exp(-\lambda_i T_i)). \qquad (3.5.34)$$

Analogous to (3.5.15) through (3.5.17), it follows that

$$\text{Var} \, \hat{a} = \frac{- E(A_2)}{E(A_2) E(A_0) - [E(A_1)]^2}, \qquad (3.5.35)$$

$$\text{Var}\,\hat{b} = \frac{-E(A_0)}{E(A_2)E(A_0)-[E(A_1)]^2}, \tag{3.5.36}$$

and

$$\text{Cov}(\hat{a},\hat{b}) = \frac{-E(A_1)}{E(A_2)E(A_0)-[E(A_1)]^2}. \tag{3.5.37}$$

If patients enter the clinical trial uniformly during the time interval $(0, T)$, the density function for T_i is

$$g(T_i) = \begin{pmatrix} \dfrac{1}{T}, & 0 < T_i < T, \\ \\ 0, & \text{elsewhere.} \end{pmatrix} \tag{3.5.38}$$

Defining $E^*(A_j)$ as $E_{T_i}E_{t_i}(A_j|T_i)$, we find that

$$E^*(A_j) = -\sum_{i=1}^{n} x_i^j \lambda_i^2 \left\{ 1 - (\lambda_i T)^{-1}[1 - \exp(-\lambda_i T)] \right\} \tag{3.5.39}$$

The formulas for $\text{Var}\,\hat{a}$, $\text{Var}\,\hat{b}$, and $\text{Cov}(\hat{a},\hat{b})$ given by (3.5.35) through (3.5.37) change accordingly, with $E^*(A_j)$ replacing $E(A_j)$, $j = 0, 1, 2$.

Example 3.5.2. Consider the data in Table 3.3 on observed survival times and white blood counts for AG positive patients with acute leukemia. To obtain a comparable censored group, Zippin and Armitage randomly selected five of these patients to be "survivors." The surviving patients were followed for randomly determined proportions of their actual survival time. The results are given in Table 3.6. For these data the maximum likelihood estimates are $\hat{a} = 240$ and $\hat{b} = -42$, compared with $\hat{a} = 240$ and $\hat{b} = -44$, which Feigl and Zelen obtained for the complete sample (Table 3.4). Their corresponding standard errors, which are found using (3.5.29) through (3.5.31) directly (not taking expected values), are $s_{\hat{a}} = 104.2$ and $s_{\hat{b}} = 23.5$. This compares favorably with the results of Feigl and Zelen, who obtained for the complete sample $s_{\hat{a}} = 95.5$ and $s_{\hat{b}} = 20.1$. Note that $\log_{10}$ of the white blood count is used as before, as the concomitant variable.

Table 3.6 Follow-up time, white blood count (WBC), and status of leukemia patients

	AG positive, $n = 17$	
WBC $\times 10^2$	Length of Follow-up (weeks)	Status at End of Follow-up: Alive (A) or Dead (D)
23	65	D
7.5	156	D
43	100	D
26	13	A
60	16	D
105	108	D
100	121	D
170	4	D
54	39	D
70	143	D
94	34	A
320	5	A
350	13	A
1000	1	A
1000	1	D
520	5	D
1000	65	D

Further discussion on this topic can be found in Miller [1974], or a model by Cox [1972], which is discussed briefly in Section 7.3, can be used.

EXERCISES

1. Verify the results of Table 3.4 using the data of Table 3.3 and plot the values of $\lambda(t)$, $f(t)$, and $S(t)$ obtained from the computed $\hat{a}$ and $\hat{b}$. Compare these plots with histograms of the original data. Also plot $\log_{10}$ WBC against length of survival. Examine the data for outliers and lack of fit.

2. Suppose that $\lambda_i = a + bx_i$, $i = 1, 2, \ldots, n$, in the Feigl-Zelen model. That is, the hazard rate instead of the mean time to death is related linearly to the concomitant variable. Obtain equations for finding the maximum likelihood estimators $\hat{a}$ and $\hat{b}$. Also obtain the large sample variance–covariance matrix for $(\hat{a}, \hat{b})$.

3. Assume for the data in Table 3.6 that all patients enter the study together at a time T units before the close of follow-up (i.e., $T_i = T$ for all 17 patients). Using (3.5.35) through (3.5.37), compute the values of the variance–covariance matrix of $(\hat{a}, \hat{b}) = (240, -42)$ for $T = 25, 50, 100, 200, \infty$.

4. Assume that T_i has the density (3.5.38). Use (3.5.39) in conjunction with (3.5.35) through (3.5.37) to compute the variance–covariance matrix in Exercise 3 for the same values of T (i.e., $T = 25, 50, 100, 200, \infty$).

5. Compare the results in Exercises 3 and 4, demonstrating that the precision of the estimates based on uniform entry of patients is about the same as a study of half the duration in which all patients enter together.

3.6 DETERMINATION OF WHETHER AN EXPONENTIAL DISTRIBUTION FITS SURVIVAL DATA

In the preceding sections it has been assumed that the survival data follow an exponential distribution. Radioactive decay times are a classic example of data that follow an exponential distribution. In practice, it is often not obvious which distribution to assume. Because the exponential is one of the easiest distributions to use, it is logical to try to fit this distribution to the data. As pointed out at the end of Chapter 1, the hazard rate frequently presents the best means of identification. In the case of the exponential distribution, this is a constant λ. However, the effect of individual data values sometimes makes the plotted hazard rate so irregular that it is difficult to distinguish. One technique for overcoming this problem is to look at the cumulation of the hazard rate (see Nelson [1972]). This is directly analogous to the use of normal probability paper (see Dixon and Massey [1967]), so scaled that the cumulative normal curve $F(x)$ is a straight line if the data are normally distributed. In the case of the exponential distribution, the cumulative hazard rate is simply

$$\Lambda(t) = \int_0^t \lambda\, dx = \lambda t, \qquad t \geqslant 0, \qquad (3.6.1)$$

a linear function passing through zero. Thus exponential hazard paper is merely regular graph paper with values of t on the vertical axis and $\Lambda(t)$ on the horizontal axis. Yet the question remains, how do we choose the plotting position for $\Lambda(t)$? Since $\Lambda(t)$ is the theoretical cumulative hazard function, it must be estimated in applications. Suppose the times to failure are ordered $t_{(1)} \leqslant t_{(2)} \leqslant \cdots \leqslant t_{(n)}$. In Section 4.6 we observe that the expected value of $\Lambda(t_{(i)})$ [i.e., $E(\Lambda(t_{(i)}))$], is $1/n + 1/(n-1) + \cdots + 1/(n-i+1)$. Thus the surrogate plotting position for $\Lambda(t_{(i)})$, which is unknown, is taken to be $E(\Lambda(t_{(i)}))$, which is known. We note at this point that when the data are censored, the plotting position $E(\Lambda(t_{(i)}))$ may not be exact. Nelson points out, however, that the procedure is still satisfactory. The values $t_{(i)}$ are plotted on the vertical axis and the values $E(\Lambda(t_{(i)}))$ (which is the sum of the reverse ranks at the ith order failure) are plotted on the horizontal axis. The procedure for displaying the data follows:

1. Order the n observations from smallest to largest, regardless of whether they are t_i or T_i (time to death or censored time, respectively) marking the T_i times with check marks.
2. Calculate the hazard value for each time as $100/k$, where k is its reverse rank. The hazard value is the observed conditional probability of survival. Cumulate the hazard values downward. The cumulative can be larger than 100 percent.
3. Mark the vertical scale from 0 to the longest length of survival and the horizontal scale from 0 to the largest value of the cumulative hazard values.
4. Plot each survival time vertically against its corresponding cumulative hazard value.
5. By eye, fit a straight line through 0 and the points in the center of the distribution.

If the points tend not to cluster around this straight line (particularly the central 80 percent), it is not reasonable to assume an exponential distribution. When the assumption of exponentiality is not satisfied, it is possible at times to transform the data, permitting an exponential distribution to be fitted to the transformed data. When a transformation is not possible, another distribution should be considered. If the points fit the straight line, μ can be estimated by reading t from the vertical axis corresponding to 100 percent.

Example 3.6.1. We consider the data given in Exercise 1 in Section 3.4, assuming that the first of the two times, 23, is $T_{(i)}$ (censored) not $t_{(i)}$ (a time to death).

$t_{(i)}$ or $T_{(i)}$	Reverse Rank	Hazard Value	Cumulative Hazard Value
0	26	100/26 = 3.846	3.846
0	25	4.000	7.846
0	24	4.167	12.013
1	23	4.348	16.361
2	22	4.545	20.906
2	21	4.762	25.668
2	20	5.000	30.668
3	19	5.263	35.931
4	18	5.556	41.487
5	17	5.882	47.369
5	16	6.250	53.619
5	15	6.667	60.286
5	14	7.143	67.429
8	13	7.692	75.121
8	12	8.333	83.454
10	11	9.091	92.545
11	10	10.000	102.545
12	9	11.111	113.656
16	8	12.500	126.156
21	7	14.286	140.442
21	6	16.667	157.109
21	5	20.000	177.109
23+	4	—	—
23	3	33.333	210.442
25	2	50.000	260.442
41	1	100.000	360.442

In Example 3.6.1 the straight line fit (see Figure 3.1) does seem reasonable, although there exists some deviation of the points from a straight line. In this example the data were deliberately generated from an exponential distribution using the fact that if $F(x)$ is the cumulative distribution of any continuous random variable X, the distribution of Y, where $Y = F(X)$, is uniformly distributed on the interval $0 \leqslant y \leqslant 1$. (See Mood and Graybill [1963].)

Another simple graphical procedure for a test of exponentiality of nongrouped data utilizes the density function. Suppose the survival times of n individuals are ordered, that is, $t_{(1)} \leqslant t_{(2)} \leqslant \cdots \leqslant t_{(n)}$. The empirical survival probability for the value $t_{(i)}$ is $(n-i+1)/(n+1)$, which is used instead of $(n-i)/n$ to avoid a zero value. If the hypothesis of exponentiality is true, plotting $t_{(i)}$ against $-\log_e(n-i+1)/(n+1)$ should approximate

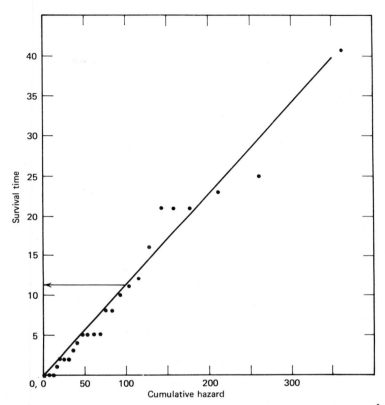

Figure 3.1   Cumulative hazard plot for the data on cancer survival times: $\hat{\mu} \doteq 11.3$, $\hat{\lambda} \doteq 0.09$.

a straight line; the slope of this line would estimate the mean of the distribution. Such plots are easiest to make using semilogarithmic graph paper; they should be nearly linear when the survival times are exponential.

Example 3.6.2.   Patients receiving treatment for an acute disease experienced toxicity from the drug they received. Seven patients experienced toxic side effects from the drug as follows: 5, 9, 12, 14, 21, 26, and 35 days. These toxicity times are plotted on semilogarithmic graph paper in Figure 3.2, which indicates that an exponential survival distribution would fit these data quite well. The mean toxicity time is 17.6 days for these patients. Note that in Figure 3.2 the points do cluster around the line.

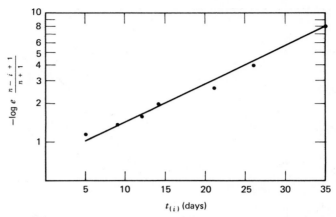

Figure 3.2   Toxicity times (days) of patients receiving a toxic drug for an acute disease.

Occasionally individual times to death are not available to the statistician or research scientist for analysis. Instead, as for example in a population life table, only the number of deaths in each of a set of mutually exclusive time intervals can be obtained. The set of time intervals taken together covers the entire nonnegative time axis. As an example, consider a group of 118 arthritic patients, each of whom receives an analgesic to relieve his discomfort. Table 3.7 shows number of patients receiving relief (in minutes) for 16 mutually exclusive time intervals. These data are a fictitious adaptation of data appearing in Table 3 of Scheuer [1968].

We treat the problem of testing grouped data for exponentiality. That is, we discuss the problem of fitting an exponential death density function to grouped data.

Suppose survival data on n individuals occur as in Table 3.7. More generally, suppose the nonnegative time axis to be divided into the mutually exclusive and exhaustive intervals $I_j : T_{j-1} \leqslant t < T_j$, $j = 1, 2, \ldots, k$, with $T_0 = 0$ and $T_k = \infty$. The data are the number of deaths observed in each interval: f_j deaths occurring in interval I_j, $j = 1, 2, \ldots, k$, $\Sigma_{j=1}^{k} f_j = n$.

We wish to test the hypothesis that these data have an exponential death density function

$$f(t; \lambda) = \lambda \exp(-\lambda t), \qquad t \geqslant 0, \lambda > 0, \tag{3.6.2}$$

and to estimate λ. When the hypothesis of exponentiality is true, the

Table 3.7 Relief times of arthritic patients receiving an analgesic

Interval (minutes)	Frequency
$0 \leqslant t < 20$	19
$20 \leqslant t < 40$	19
$40 \leqslant t < 60$	21
$60 \leqslant t < 80$	10
$80 \leqslant t < 100$	13
$100 \leqslant t < 120$	6
$120 \leqslant t < 140$	7
$140 \leqslant t < 160$	5
$160 \leqslant t < 180$	4
$180 \leqslant t < 200$	2
$200 \leqslant t < 220$	3
$220 \leqslant t < 240$	1
$240 \leqslant t < 260$	2
$260 \leqslant t < 280$	1
$280 \leqslant t < 300$	1
$300 \leqslant t < 320$	1
$320 \leqslant t < 340$	2
$t \geqslant 340$	1
Total	118

expected number of deaths in I_j is $np_j(\lambda)$, where $p_j(\lambda)$ is obtained by

$$p_j(\lambda) = \int_{T_{j-1}}^{T_j} \lambda \exp(-\lambda t)\, dt$$

$$= \exp(-\lambda T_{j-1}) - \exp(-\lambda T_j). \tag{3.6.3}$$

Clearly $\sum_{j=0}^{k} p_j(\lambda) = 1$. Thus to test whether (3.6.2) is a good fit to the data, we employ the chi-square statistic

$$\chi^2 = \sum_{j=1}^{k} \frac{(f_j - np_j(\lambda))^2}{np_j(\lambda)}, \tag{3.6.4}$$

which is asymptotically distributed as a chi-square variable with $k-1$ degrees of freedom when the hypothesis of exponentiality is true.

If λ were known, the test described by (3.6.4) would be appropriate for testing the grouped data for exponentiality. Ordinarily, however, λ is unknown and must be estimated from the data. Following Scheuer's development, we let χ^2 be defined as in (3.6.4), and we try to find that value of λ, λ' (say), which minimizes χ^2 with respect to the data. This is known as the chi-square minimum procedure, and λ' is the minimum chi-square estimator of λ. Formally, then,

$$\frac{\partial \chi^2}{\partial \lambda} = -\sum_{j=1}^{k} \left[\frac{2(f_j - np_j(\lambda))}{p_j(\lambda)} + \frac{(f_j - np_j(\lambda))^2}{np_j^2(\lambda)} \right] \frac{\partial p_j(\lambda)}{\partial \lambda}, \qquad (3.6.5)$$

where $p_j(\lambda)$ is given by (3.6.3) and

$$\frac{\partial p_j(\lambda)}{\partial \lambda} = T_j \exp(-\lambda T_j) - T_{j-1} \exp(-\lambda T_{j-1}), \qquad j = 1, 2, \ldots, k. \quad (3.6.6)$$

Since $T_k = +\infty$, $\exp(-\lambda T_k) = T_k \exp(-\lambda T_k) = 0$.

Thus λ', the minimum chi-square estimator of λ, is the solution to

$$\sum_{j=1}^{k} \left[\frac{(f_j - np_j(\lambda'))}{p_j(\lambda')} + \frac{(f_j - np_j(\lambda'))^2}{2np_j(\lambda')} \right] \frac{\partial p_j(\lambda')}{\partial \lambda'} = 0, \qquad (3.6.7)$$

which minimizes (3.6.4). Substituting λ' for λ in (3.6.4), it can be shown that χ^2 will have an asymptotic chi-square distribution with $k - 2$ degrees of freedom, when the null hypothesis of exponentiality is true.

To find λ' from (3.6.7), the Newton-Raphson method (discussed in Chapter 6) is recommended. Scheuer describes a computer program written in the JOSS* programming language for solution of (3.6.7) in terms of λ'. The appropriate starting value λ'_0 is

$$\lambda'_0 = n \left[\frac{\sum_{j=1}^{k-1} f_j(T_{j-1} + T_j)}{2} + f_k T_{k-1} \right]^{-1}. \qquad (3.6.8)$$

This value is chosen because it is the analog to the maximum likelihood estimator $\hat{\lambda}$ for the grouped data situation. The values $(T_{j-1} + T_j)/2$

---

*JOSS is the time-sharing computing system at the Rand Corporation, Santa Monica, Calif.

represent the midpoints of all the intervals except the last interval, for which we must use the lower limit T_{k-1} because the upper limit of the last interval is infinite.

Goodness of fit procedures can also be used when the survival data occur ungrouped and must be grouped for analysis. In this case intervals must also be determined. One method of determining the intervals is to divide the entire positive axis into k intervals, where the expected number of deaths in each interval is the same, given all the data. If we then compare the observed and expected numbers of deaths in each interval, the distribution of the test statistic is no longer a chi-square with $k-2$ degrees of freedom. However, using a chi-square with $k-3$ degrees of freedom will give us a conservative test; that is, at level α we will not reject as often as we should the null hypothesis that the fit is adequate. Dahiya and Gurland [1972] discuss this problem. They provide a procedure for testing the goodness of fit of a normal distribution that is exact.

Table 3.8 Observed and expected time interval frequencies with respect to relief times of arthritic patients receiving an analgesic

Interval (minutes)	Observed Frequency	Expected Frequency
$0 \leqslant t < 20$	19	23.10
$20 \leqslant t < 40$	19	18.58
$40 \leqslant t < 60$	21	14.94
$60 \leqslant t < 80$	10	12.02
$80 \leqslant t < 100$	13	9.66
$100 \leqslant t < 120$	6	7.77
$120 \leqslant t < 140$	7	6.25
$140 \leqslant t < 160$	5	5.03
$160 \leqslant t < 180$	4	4.04
$180 \leqslant t < 200$	2	3.25
$200 \leqslant t < 220$	3	2.62
$220 \leqslant t < 240$	1	2.10
$240 \leqslant t < 260$	2	1.69
$260 \leqslant t < 280$	1	1.36
$280 \leqslant t < 300$	1	1.09
$300 \leqslant t < 320$	1	0.88
$320 \leqslant t < 340$	2	0.71
$t \geqslant 340$	1	2.91
Total	118	118.00

Example 3.6.3. Consider the data in Table 3.7. Applying (3.6.8), we find $\lambda_0' = 0.0115$. Using 0.0115 as a starting value and applying the Newton-Raphson procedure to solve (3.6.7), we have $\lambda_0' = 0.0109$. That is, the minimum chi-square estimate of λ is 0.0109. Table 3.8 (Table 3 in Scheuer's report) compares the observed and expected frequencies in each interval based on the minimum chi-square estimate λ' for the data in Table 3.7. Using (3.6.4) we find that the computed value of chi-square, $\chi^2\text{cal} = 10.08 < \chi^2(16, .9) = 23.54$. Thus the hypothesis of exponentiality is accepted. However, note that Table 3.8 has numerous cells in which the expected frequency is less than 5.

We caution the reader that the chi-square test discussed here is often insensitive and often does not indicate significant results when the hypothesis of exponentiality is false. As Scheuer points out, a mere regrouping of the data from a large number into a smaller number of intervals can change the decision from rejection to acceptance of exponentiality. Nevertheless, this procedure is useful in many cases for detecting discrepancies from exponentiality, and it has the following two advantages: it estimates the unknown parameter λ, and it does not require the exact failure time of each individual.

EXERCISES

1. Plot the data given in Example 3.6.2 using cumulative hazard method and estimate λ.

2. Plot the data given in Example 3.6.1 using the second graphical method, where $t_{(i)}$ is plotted against $-\log_e(n - i + 1)/(n + 1)$.

3. Suppose the data in Table 3.7 are regrouped so that the last interval is now $t \geqslant 200$, which has an observed frequency 12. Using the chi-square minimum procedure, obtain λ', and show that the hypothesis of exponentiality is accepted at the 0.10 level of significance.

4. The chi-square minimum procedure is applicable in testing grouped data fitted to survival distributions other than the exponential. Suppose

$$f(t; \lambda, \gamma) = \lambda \gamma t^{\gamma - 1} \exp(-\lambda t^{\gamma}), \qquad \lambda > 0, \gamma > 0, t \geqslant 0, \qquad (3.6.9)$$

is proposed as the failure density function for grouped data (f_j, I_j), where f_j is the frequency of the jth interval $I_j : T_{j-1} \leqslant t < T_j, j = 1, 2, \ldots, k$, with $T_0 = 0$, and $T_k = +\infty$. Assuming that γ is known, obtain the equation whose solution yields the minimum chi-square estimator λ' of λ. If γ is unknown, find the two equations whose simultaneous solution yields the minimum chi-square estimators λ' and γ' of λ and γ, respectively.

5. The modified minimum chi-square estimator λ'', whose distribution is asymtotically equivalent to the minimum chi-square estimator λ', is obtained by omitting the second term in (3.6.7), since this term approaches zero as $n \to \infty$. If

there are no observations in the last interval, and if each interval but the last is short, show that

$$\lambda'' = \frac{2n}{\sum\limits_{j=1}^{k-1} f_j(T_{j-1}+T_j)}$$

is the approximate modified minimum chi-square estimator of λ.

6. Using the data in Table 3.8, find new intervals with more reasonable expected frequencies. Find the resulting estimate of λ and compare observed and expected frequencies.

3.7 COMPETING RISKS

In clinical trials the death of a patient may be due to a cause other than the disease for which he is under study. For example, a patient who is under study for prostate cancer may be involved in a fatal accident or he may succumb to a heart attack. These other risks, as well as the risk of death due to the disease under study, are called competing risks; since it is assumed that the death of the patient comes from only a single underlying cause, these risks thus compete for the patient's life.

The survival time for patients who have had surgery for invasive cancer of the bladder is their length of survival from surgery. This is a serious disease, and most patients will die from the cancer; however, some patients may die of other causes (e.g., heart failure or accidents). There are alternate ways to analyze such survival data. First, we can look at the patient survival times separately for each cause of death. Second, we can ignore the fact that several risks are competing for each patient's life. In this case we analyze all survival times together. Finally, and most logically, we can develop a method (or model) that allows for the analysis of competing risk data.

To this point we have assumed that the hazard rate is a constant with respect to time. In the development of the general competing risk model, the assumption of a constant hazard rate is abandoned to allow greater generality. However, the reader can easily apply the results of this section to the case when all the risks are constant. For estimation of parameters, the constant hazard rate is assumed.

The term "risk" refers to the probability of dying from a given cause *prior* to death. After death, the risk is the cause that was responsible for death. Suppose there are k independent competing risks such that the hazard rate of the individual due to the ith risk at time t is $\lambda_i(t)$, $i=1,2,\dots,k$. If $S(t)$ is the overall survival probability of an individual at time t, the k risks are assumed to be independent. Since from (1.2.8) the survival probability at time t for the ith risk is

$$S_i(t) = \exp\left(-\int_0^t \lambda_i(u)\,du\right), \tag{3.7.1}$$

it follows that

$$S(t) = \prod_{i=1}^k S_i(t) = \prod_{i=1}^k \exp\left(-\int_0^t \lambda_i(u)\,du\right) = \exp\left(-\sum_{i=1}^k \int_0^t \lambda_i(u)\,du\right). \tag{3.7.2}$$

If we define $\lambda(t)$ as the overall hazard rate,—that is, $\lambda(t)$ is defined by the integral equation

$$S(t) = \exp\left(-\int_0^t \lambda(u)\,du\right), \tag{3.7.3}$$

then from (3.7.2) and (3.7.3) and certain elementary rules of calculus, we obtain

$$\lambda(t) = \sum_{i=1}^k \lambda_i(t). \tag{3.7.4}$$

The competing risk model defined by (3.7.4) is called the additive, noninteractive model of cumulative risk. It has been discussed rather extensively in the literature by Berkson and Elveback [1960], Moeschberger and David [1971], and Chiang [1968].

Since (3.7.2) describes the survival probability for the additive, noninteractive competing risk model, we can by elementary considerations derive both $f(t)$ and $F(t)$, the death density and cumulative death distribution functions, respectively. These derivations are left as an exercise to the reader. Hence moments of $f(t)$ and the maximum likelihood estimators of the parameters can be obtained.

It may be important to isolate the probabilities an individual dies from the ith risk, $i = 1, 2, \ldots, k$. We thus consider the following. The probability an individual dies in the interval $t < x < t + \Delta t$ of risk i is the joint probability he survives to time t and dies of risk i in the interval $t < x < t + \Delta t$, given he has survived to time t. Mathematically, this probability is $S(t)\lambda_i(t)\Delta t$. If we add up these probabilities over the entire time axis and let $\Delta t \to 0$, we have Q_i, the probability of death due to the ith cause, given by

$$Q_i = \int_0^\infty (S(t)) \lambda_i(t) \, dt$$

$$= \int_0^\infty \exp\left(-\sum_{j=1}^k \int_0^t \lambda_j(u) \, du \right) \lambda_i(t) \, dt, \qquad i = 1, 2, \ldots, k. \quad (3.7.5)$$

The probability an individual dies in the interval $t < x < t + \Delta t$, given risk i, is the joint probability he lives to time t and dies in $t < x < t + \Delta t$ of risk i, divided by the probability he dies of risk i. Mathematically this probability is

$$f(t|\text{risk } i) \Delta t = \frac{S(t) \lambda_i(t) \Delta t}{\int_0^\infty S(t) \lambda_i(t) \, dt}. \qquad (3.7.6)$$

Again letting $\Delta t \to 0$, the death density function of those persons dying, given risk i, is

$$f(t|\text{risk } i) = \frac{S(t) \lambda_i(t)}{\int_0^\infty S(t) \lambda_i(t) \, dt}, \qquad i = 1, 2, \ldots, k. \qquad (3.7.7)$$

Consider now a situation for which the hazard rates for the k risks are proportional. That is, $\lambda_i(t) = c_i \lambda(t)$, where $c_i > 0$ is independent of t, $i = 1, 2, \ldots, k$. It can be demonstrated that Q_i, the probability of death due to risk i, is

$$Q_i = \frac{c_i}{\sum_{j=1}^k c_j}. \qquad (3.7.8)$$

To show this, let

$$y = \sum_{j=1}^k \int_0^t \lambda_j(u) \, du = \sum_{j=1}^k c_j \int_0^t \lambda(u) \, du.$$

Hence $dy = \sum_{j=1}^{k} c_j \lambda(t) \, dt$. Thus

$$Q_i = \int_0^\infty \exp\left[\left(-\sum_{j=1}^{k} c_j\right)\int_0^t \lambda(u)\,du\right] c_i\lambda(t)\,dt$$

$$= \frac{c_i}{\sum\limits_{j=1}^{k} c_j} \int_0^\infty e^{-y}\,dy = \frac{c_i}{\sum\limits_{j=1}^{k} c_j}. \qquad (3.7.9)$$

The death density function given the ith risk is then

$$f(t|\text{risk } i) = cS(t)\lambda(t), \qquad (3.7.10)$$

where $c = \sum_{i=1}^{k} c_i$. Hence the death density function, given the ith risk, does not depend on the risk.

Let us now consider the problem of estimating $c_i, i = 1, 2, \ldots, k$, when all hazard rates are constant with respect to time. The assumption of constant hazard rates for all k risks implies that they are proportional (but not conversely; see Exercise 4 at the end of this section). When the hazard rates are all constant there is no loss of generality in assuming $\lambda_i(t) = c_i$; $i = 1, 2, \ldots, k$. If t_{ij} is the survival time of the jth individual who dies from the ith risk, $j = 1, 2, \ldots, n_i$; $i = 1, 2, \ldots, k$, the likelihood of the sample is

$$L(c_1, \ldots, c_k) = \prod_{i=1}^{k} \prod_{j=1}^{n_i} S(t_{ij}) c_i$$

$$= \exp\left[\sum_{i=1}^{k} \sum_{j=1}^{n_i} c_i t_{ij}\right] \prod_{i=1}^{k} c_i^{n_i}. \qquad (3.7.11)$$

Thus

$$\log_e L(c_1, \ldots, c_k) = -\sum_{i=1}^{k} \sum_{j=1}^{n_i} c_i t_{ij} + \sum_{i=1}^{k} n_i \log_e c_i. \qquad (3.7.12)$$

It then follows that $\hat{c}_i$, the maximum likelihood estimator of c_i, for the ith risk is

$$\hat{c}_i = \frac{n_i}{\sum\limits_{j=1}^{n_i} t_{ij}}, \qquad i = 1, 2, \ldots, k. \qquad (3.7.13)$$

For large samples, the vector $\hat{\mathbf{c}} = (\hat{c}_1, \ldots, \hat{c}_k)$ has an approximate normal distribution with mean $\mathbf{c} = (c_1, \ldots, c_k)$ and variance–covariance matrix V, where

$$
V^{-1} = \begin{bmatrix} \sigma^{11} & \cdots & \sigma^{1k} \\ \cdot & & \cdot \\ \cdot & & \cdot \\ \cdot & & \cdot \\ \sigma^{k1} & \cdots & \sigma^{kk} \end{bmatrix} \qquad (3.7.14)
$$

and

$$
\sigma^{ij} = -E\left[\frac{\partial^2 \log_e L}{\partial c_i \, \partial c_j}\right], \qquad i,j = 1, 2, \ldots, k. \qquad (3.7.15)
$$

If we differentiate (3.7.12) with respect to i and then j we find

$$
\frac{-\partial^2 \log_e L}{\partial c_i^2} = \frac{n_i}{c_i^2} \qquad (3.7.16)
$$

and

$$
\frac{-\partial^2 \log_e L}{\partial c_i \, \partial c_j} = 0, \qquad i \neq j; i,j = 1, 2, \ldots, k. \qquad (3.7.17)
$$

The $n_1, n_2, \ldots, n_k$ have a joint multinomial probability distribution, where $E(n_i) = NQ_i$, $N = \Sigma_{j=1}^{k} n_j$, $Q_i = c_i / \Sigma_{j=1}^{k} c_j$, $i = 1, 2, \ldots, k$. Thus

$$
\sigma^{ii} = C\frac{n_i}{c_i} \qquad (3.7.18)
$$

and

$$
\sigma^{ij} = 0, \qquad i \neq j; i,j = 1, \ldots, k, \qquad (3.7.19)
$$

where

$$
C = \sum_{i=1}^{k} c_i.
$$

We have thus shown that for large samples $\hat{c}_1, \ldots, \hat{c}_k$ are approximately independently normally distributed variables with means $c_1, \ldots, c_k$, and variances $c_1/Cn_1, \ldots, c_k/Cn_k$, respectively.

It can be shown that an approximate $(1 - \alpha)$ 100 percent confidence interval for c_i is given by the interval

$$\hat{c}_i - \sqrt{\frac{\hat{c}_i}{\hat{C}n_i}} \; Z\left(1 - \frac{\alpha}{2}\right) < c_i < \hat{c}_i + \sqrt{\frac{\hat{c}_i}{\hat{C}n_i}} \; Z\left(1 - \frac{\alpha}{2}\right), \quad (3.7.20)$$

where $C = \sum_{i=1}^{k} c_i$ and $\hat{C} = \sum_{i=1}^{k} \hat{c}_i$ and $Z\left(1 - \frac{\alpha}{2}\right)$ is the upper $1 - \alpha/2$ percentile of the standard normal distribution.

In concluding this section it should be mentioned that competing risk models may also be developed when the survival times are censored. We refer the interested reader to Moeschberger and David [1971].

Example 3.7.1.   Suppose 20 patients are simultaneously exposed to the risk of heart disease and cancer. Suppose, further, that the survival times for both these risks are exponentially distributed. Eleven of the patients are assumed to die of cancer, and the other nine are assumed to die of heart disease. The survival times (in months) for the cancer deaths are: 8, 9, 11, 11, 13, 15, 17, 17, 17, 19, 20; the survival times (in months) for the heart disease deaths are: 7, 7, 9, 10, 10, 12, 14, 16, 17. The survival times for each of the two competing risks are $\sum_{j=1}^{11} t_{1j} = 157$ and $\sum_{j=1}^{9} t_{2j} = 102$ months each for patients dying of cancer and heart disease, respectively. Thus the estimated hazard rates for each disease are $\hat{c}_1 = 11/157 = 0.070$ death/month and $\hat{c}_2 = 9/102 = 0.088$ death/month for cancer and heart disease, respectively. Approximate 95 percent confidence intervals for c_1 and c_2 are,* respectively,

$$\left[\begin{array}{l} 0 < c_1 < 0.463 \text{ death/month} \\ 0 < c_2 < 0.576 \text{ death/month.} \end{array} \right. \quad (3.7.21)$$

In this chapter the reader is provided with a variety of estimation procedures for survival data that are exponentially distributed, including procedures to test survival data for exponentiality.

---

*The confidence interval for each of c_1 and c_2 includes negative values; however, hazard rates have a physical lower limit of zero.

We briefly review the various estimating procedures, to allow the reader to refer quickly and easily to the appropriate section.

Section 3.1 deals with noncensored survival data. Maximum likelihood estimators, obtained for the constant hazard λ rate and the mean time to death, μ, are given by (3.1.11) and (3.1.12), respectively. Furthermore, we demonstrate that the maximum likelihood estimator for μ is the best linear unbiased estimator (BLUE), the method of moments estimator, and the sufficient statistic. [See (3.1.13) through (3.1.15).]

Section 3.2 deals with the censored exponential case. The maximum likelihood estimators for μ and λ are again obtained and are given by (3.2.8) and (3.2.9), respectively.

In Section 3.3 we obtain both approximate and exact confidence intervals for λ in the noncensored case. These confidence intervals are given by (3.3.2) and (3.3.2'). In the censored case again approximate confidence intervals are obtained for both progressively and singly censored data, and if the number of deaths d is fixed prior to the analysis of the survival data, an exact confidence interval is obtained for the singly censored case. The requisite confidence intervals are given by (3.3.8), (3.3.14), and (3.3.15), respectively.

Section 3.4 is concerned with estimating the survival probability for an exponential survival distribution and concomitantly obtaining a confidence interval for the theoretical survival probability. It is assumed the survival data are censored. Equation 3.4.3 is the estimated survival probability. Approximate confidence intervals are obtained for the theoretical survival probability under the following assumptions: (1) the total time a surviving patient is on study is fixed, and (2) the total time a surviving patient is on study is a random variable, uniformly distributed over the length of the study. The confidence intervals are given by (3.4.7') in both cases; for fixed censored times, however, the estimate of the variance is given by (3.4.8) and for random censored times the estimated variance is given by (3.4.16).

In Section 3.5 we discuss the problem of analyzing survival times in the presence of concomitant information, under the assumption that the survival times follow an exponential distribution. It is further assumed that only one concomitant variable is present. Equation 3.5.2 describes the basic linear relationship between the mean survival time of the ith patient and his observed concomitant variable. Recall that $E(t_i) = a + bx_i$, where a and b are estimated by maximum likelihood both for the uncensored and censored cases. Equations 3.5.4 and 3.5.5 are the estimating equations in the uncensored case. For the censored case the estimating equations are (3.5.27) and (3.5.28). For the uncensored case, a $100(1 - \alpha)$ percent simultaneous confidence interval is obtained for the n survival probabilities, each of which must survive to at least time t. The interval is given by (3.5.22), (3.5.23), and (3.5.24).

Section 3.6 deals with the problem of fitting an exponential distribution to survival data. Two graphical procedures are discussed for fitting an exponential distribution to ungrouped survival data. Fitting an exponential distribution to grouped survival data is also discussed. Estimating the hazard rate, λ, is accomplished by the minimum chi-square method. The minimum chi-square estimator λ' is obtained by solving for λ' using (3.6.7).

In Section 3.7 we introduce the topic of competing risks and provide an example utilizing the exponential model.

EXERCISES

1. Suppose that 100 patients are simultaneously exposed to the risk of heart disease and cancer and that survival times for both risks are exponentially distributed. Let us further suppose that 40 patients die of cancer with a mean survival time of 14.5 months, and 60 patients die of heart disease with a mean survival time of 13.5 months. Find $\hat{c}_1$ and $\hat{c}_2$, the observed hazard rates for those patients dying of cancer and heart disease, respectively, and obtain individual 95 percent confidence intervals for these risks.

2. Starting with (3.7.2) derive $F(t)$ and $f(t)$, the cumulative death distribution and the death density function, respectively, for the additive noninteractive competing risk model.

3. Show that an approximate $(1 - \alpha)$ 100 percent confidence interval for c_i is given by (3.7.20).

4. Show that if the hazard rates for all k risks are constant with respect to time [i.e., $\lambda_i(t) = \lambda_i$; $i = 1, 2, \ldots, k$], all k risks are proportional to one another. Show, however, by means of a counterexample, that we can obtain k hazard rates all proportional to one another that are not constant with respect to time. *Hint*: Let $\lambda_i(t) = \lambda_i \gamma_i t^{\gamma_i - 1}$, $i = 1, 2, \ldots, k$. What restrictions on γ_i will then provide the counterexample?

CHAPTER 4

Estimating Parameters of the Weibull, Gamma, and Rayleigh Survival Distributions

4.1 INTRODUCTORY REMARKS

We discussed in Chapter 3 a variety of estimation procedures for exponentially distributed survival data. The principal reasons for devoting an entire chapter to the exponential are as follows: its estimators are simple to derive and use, and exponentially distributed survival data often arise in practice. On the other hand, a plot of the hazard rate may reveal that the data are not exponentially distributed. When this occurs a first step is to consider somewhat more complex distributions, three of which—the Weibull, the gamma, and the Rayleigh—comprise this chapter.

In Section 4.1 the Weibull and gamma death densities and their parameters are discussed. In Sections 4.2 and 4.3 maximum likelihood estimators for noncensored and censored samples are given. The large sample properties of these estimators are discussed in Section 4.4. We consider the problem of estimating parameters of truncated death density functions where the total sample size is unknown in Section 4.5. Methods of estimation of parameters other than maximum likelihood for the Weibull and gamma distribution are given in Section 4.6. These recent nonmaximum likelihood estimators are particularly important for the Weibull distribution because simulation studies indicate that they are less biased than the maximum likelihood estimators in small samples. Mann, Schafer, and Singpurwalla [1974] discuss these nonmaximum likelihood estimators of the Weibull in considerable detail. Section 4.7 presents graphical methods of estimation of parameters for the Weibull distribution.

Finally, in Section 4.8 we take up the problem of estimating parameters for the Rayleigh and generalized Rayleigh death density functions, that is, death density functions whose hazard rates are linear and polynomial, respectively. The Rayleigh and generalized Rayleigh distributions are being used with increasing frequency by many biostatisticians engaged in survival studies.

In Chapter 6, following a discussion of numerical methods, one additional hazard rate is considered, namely, the Gompertz hazard rate. Thus in Chapters 3, 4, and 6 we cover five distributions that are likely to be used in describing survival distributions.

In Chapter 1 we introduced the Weibull and gamma death density functions. For completeness we reintroduce these density functions here. The form of the two-parameter gamma death density function is

$$f(t;\lambda,\gamma) = \frac{\lambda^\gamma t^{\gamma-1}\exp(-\lambda t)}{\Gamma(\gamma)}, \qquad t \geqslant 0, \tag{4.1.1}$$

where $\lambda > 0$, $\gamma > 0$ are real parameters and

$$\Gamma(\gamma) = \int_0^\infty x^{\gamma-1}e^{-x}dx. \tag{4.1.2}$$

Equations 4.1.1 and 4.1.2 are a restatement of (1.3.8) and (1.3.9), respectively. The form of the two-parameter Weibull death density function is

$$f(t;\lambda,\gamma) = \lambda\gamma t^{\gamma-1}\exp(-\lambda t^\gamma), \qquad t \geqslant 0, \tag{4.1.3}$$

where $\lambda > 0$, $\gamma > 0$ are real parameters.

Recall that the hazard rate function for a given death density function and corresponding survival function is given by (1.2.6). That is, the hazard function $\lambda(t)$ is given by

$$\lambda(t) = \frac{f(t)}{S(t)}, \tag{4.1.4}$$

where $S(t)$ is the survival distribution corresponding to the death density function $f(t)$. Thus the hazard functions for the gamma and Weibull death density functions are, respectively,

$$\lambda(t) = \frac{t^{\gamma-1}\exp(-\lambda t)}{\displaystyle\int_t^\infty x^{\gamma-1}\exp(-\lambda x)\,dx}, \qquad t \geqslant 0, \tag{4.1.5}$$

and

$$\lambda(t) = \lambda\gamma t^{\gamma-1}. \tag{4.1.6}$$

The gamma hazard function is particularly cumbersome to graph unless γ is an integer, in which case $\lambda(t)$ in (4.1.5) reduces to (1.3.11); that is,

$$\lambda(t) = \frac{\lambda^n t^{n-1}/(n-1)!}{\displaystyle\sum_{\nu=0}^{n-1} (\lambda t)^{\nu}/\nu!} \tag{4.1.5'}$$

where $\gamma = n$, a positive integer.

Figures 4.1 and 4.2 are graphs of gamma and Weibull hazard rates for selected values of λ and γ, the scale and shape parameters, respectively. In contrasting the gamma and Weibull hazard rate, note that the gamma hazard rate approaches λ for $\gamma \geqslant 1$ and 0 for $\gamma < 1$ as $t \to \infty$, whereas the Weibull hazard rate approaches ∞ for $\gamma > 1$ and 0 for $\gamma < 1$ as $t \to \infty$. In Section 4.7 we discuss the use of cumulative hazard rate paper for the Weibull distribution, to determine whether the data appear to follow a Weibull distribution.

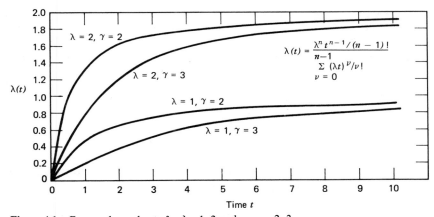

Figure 4.1 Gamma hazard rate for $\lambda = 1, 2$ and $n = \gamma = 2, 3$.

Corresponding to Figures 4.1 and 4.2 are graphs of the gamma and Weibull death density functions for the same choices of the parameters λ and γ. Figure 4.3 plots the gamma death density function for $\lambda = 2$, $\gamma = 2$; $\lambda = 2$, $\gamma = 3$; $\lambda = 1$, $\gamma = 2$; and $\lambda = 1$, $\gamma = 3$. Figure 4.4 plots the Weibull death density function for $\lambda = 2$, $\gamma = \frac{3}{2}$; $\gamma = 2$, $\gamma = \frac{2}{3}$; $\lambda = 1$, $\gamma = \frac{3}{2}$, and $\lambda = 1$, $\gamma = \frac{2}{3}$.

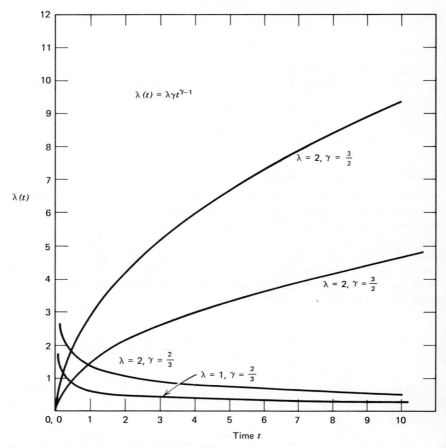

Figure 4.2 Weibull hazard rate for $\lambda = 1, 2$ and $\gamma = \frac{3}{2}$ or $\frac{2}{3}$.

It is not difficult to obtain the moment generating function m_θ for the gamma density function. By direct integration we find

$$m_\theta(t) = \left(1 - \frac{\theta}{\lambda}\right)^{-\gamma}. \qquad (4.1.7)$$

It then follows μ and σ^2 for the gamma death density function are, respectively,

$$\mu = \frac{\gamma}{\lambda} \qquad (4.1.8)$$

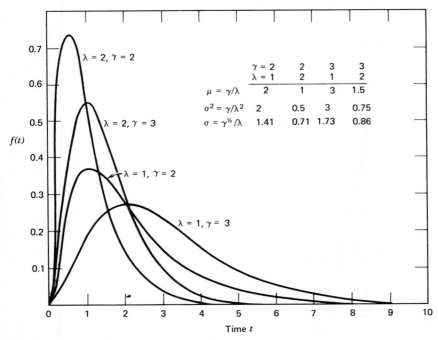

Figure 4.3 Gamma density function.

and

$$\sigma^2 = \frac{\gamma}{\lambda^2}. \tag{4.1.9}$$

The moment generating function for the Weibull death density function cannot be obtained in closed form. However, it is still possible to obtain its mean and variance directly. First we write

$$\mu = \int_0^\infty \lambda\gamma t^\gamma \exp(-\lambda t^\gamma)\, dt. \tag{4.1.10}$$

Making the substitution $u = t^\gamma$, it follows that

$$\mu = \int_0^\infty \lambda u^{1/\gamma} e^{-\lambda u}\, du = \frac{\Gamma(1+1/\gamma)}{\lambda^{1/\gamma}}. \tag{4.1.11}$$

Similarly

$$\sigma^2 = \frac{\Gamma(1+2/\gamma) - \{\Gamma(1+1/\gamma)\}^2}{\lambda^{2/\gamma}}. \tag{4.1.12}$$

The details of these derivations are left as an exercise to the reader.

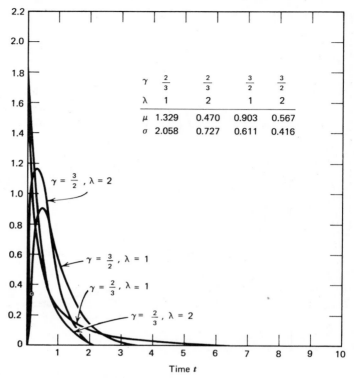

Figure 4.4   Weibull density function.

4.2 ESTIMATION OF THE GAMMA AND WEIBULL PARAMETERS BY MAXIMUM LIKELIHOOD WHEN ALL SURVIVAL TIMES ARE KNOWN

In a particular clinical trial, suppose that all n patients are followed until death. Their recorded survival times are $t_1, \ldots, t_n$, and it is assumed that the death density function for the ith patient is given by the gamma density function (4.1.1). The likelihood function $L(\lambda, \gamma)$ is given by

$$L(\lambda, \gamma) = \prod_{i=1}^{n} \frac{\lambda^{\gamma}}{\Gamma(\gamma)} t_i^{\gamma-1} \exp(-\lambda t_i)$$

$$= \frac{\lambda^{n\gamma}}{\Gamma^n(\gamma)} \prod_{i=1}^{n} t_i^{\gamma-1} \exp\left(-\lambda \sum_{i=1}^{n} t_i\right). \tag{4.2.1}$$

It then follows that

$$\log_e L(\lambda, \gamma) = n\gamma\log_e\lambda - n\log_e\Gamma(\gamma) + (\gamma - 1)\sum_{i=1}^{n}\log_e t_i - \lambda\sum_{i=1}^{n}t_i. \quad (4.2.2)$$

Thus

$$\frac{\partial\log_e L(\lambda, \gamma)}{\partial\lambda} = \frac{n\gamma}{\lambda} - \sum_{i=1}^{n}t_i \quad (4.2.3)$$

and

$$\frac{\partial\log_e L(\lambda, \gamma)}{\partial\gamma} = n\log_e\lambda - \frac{n\Gamma'(\gamma)}{\Gamma(\gamma)} + \sum_{i=1}^{n}\log_e t_i, \quad (4.2.4)$$

where

$$\Gamma'(\gamma) = \frac{d\Gamma(\gamma)}{d\gamma} = \int_0^{\infty}x^{\gamma-1}\log_e x e^{-x}dx. \quad (4.2.5)$$

Thus to find the maximum likelihood estimators $\hat{\lambda}$ and $\hat{\gamma}$ we must solve simultaneously

$$\hat{\lambda} = \hat{\gamma}\bar{t}^{-1} \quad (4.2.6)$$

and

$$n\log_e\hat{\lambda} + \sum_{i=1}^{n}\log_e t_i = \frac{n\Gamma'(\hat{\gamma})}{\Gamma(\hat{\gamma})} \quad (4.2.7)$$

for $\hat{\lambda}$ and $\hat{\gamma}$. Using (4.2.6) we can reduce (4.2.6) and (4.2.7) to a single equation in $\hat{\gamma}$:

$$n\log_e\hat{\gamma} - n\log_e\bar{t} + \sum_{i=1}^{n}\log_e t_i = \frac{n\Gamma'(\hat{\gamma})}{\Gamma(\hat{\gamma})}. \quad (4.2.8)$$

Dividing (4.2.8) by n and rearranging terms, we have

$$\frac{\Gamma'(\hat{\gamma})}{\Gamma(\hat{\gamma})} - \log_e\hat{\gamma} = \log_e R, \quad (4.2.9)$$

where

$$R = \frac{\prod_{i=1}^{n}t_i^{1/n}}{\bar{t}}. \quad (4.2.10)$$

Wilk, Gnanadesikan, and Huyett [1962a] demonstrate graphically the functional relationship between $\hat{\gamma}$, the maximum likelihood estimator of γ, and $(1-R)^{-1}$; Table 1 in the Appendix is a reproduction of Table 1 in their article. Linear interpolation for values of $(1-R)^{-1}$ between two values in the table yields values of $\hat{\gamma}$ accurate to four decimal places everywhere except between 1.000 and 1.001. Appendix Table 1 gives the reader adequate selection of values of $\hat{\gamma}$ for most applications he will encounter.

After $\hat{\gamma}$ has been determined, (4.2.6) is used to obtain $\hat{\lambda}$. Greenwood and Durand [1960] developed a method of solving (4.2.6) and (4.2.7) with corresponding tables that are similar to but less complete than those of Table 1.

Example 4.2.1. From a study by Furth, Upton, and Kimball [1959], ages at death of 208 male mice exposed to 240 rads of gamma radiation were recorded. We selected a random sample of 20 survival times (in weeks): 152, 152, 115, 109, 137, 88, 94, 77, 160, 165, 125, 40, 128, 123, 136, 101, 62, 153, 83, and 69. For these data, the arithmetic mean and geometric mean survival times are $\bar{t} = 113.450$ weeks and $\prod_{i=1}^{20} t_i^{1/20} = 107.068$ weeks, respectively. Thus $R = 107.068/113.450 = 0.9437$ and $(1-R)^{-1} = 17.78$, and by linear interpolation in Table 1 of the Appendix $\hat{\gamma} = 8.53$. We find $\hat{\lambda}$, the maximum likelihood estimate of λ, given by (4.2.6), to be 0.075; the estimated mean time to death $\hat{\mu}$, is $\hat{\mu} = \hat{\gamma}/\hat{\lambda} = 113.450$ weeks.

If the death density for the ith patient is given by the Weibull death density (4.1.3), the likelihood function $L(\lambda, \gamma)$ is

$$L(\lambda, \gamma) = \prod_{i=1}^{n} \lambda \gamma t_i^{\gamma-1} \exp(-\lambda t_i^{\gamma}). \tag{4.2.11}$$

Thus

$$\log_e L(\lambda, \gamma) = n\log_e\lambda + n\log_e\gamma + (\gamma-1)\sum_{i=1}^{n} \log_e t_i - \lambda \sum_{i=1}^{n} t_i^{\gamma}. \tag{4.2.12}$$

Differentiating (4.2.12) with respect to λ and γ, yield

$$\frac{\partial \log_e L(\lambda, \gamma)}{\partial \lambda} = \frac{n}{\lambda} - \sum_{i=1}^{n} t_i^{\gamma} \tag{4.2.13}$$

and

$$\frac{\partial \log_e L(\lambda, \gamma)}{\partial \gamma} = \frac{n}{\gamma} + \sum_{i=1}^{n} \log_e t_i - \lambda \sum_{i=1}^{n} t_i^{\gamma}\log_e t_i. \tag{4.2.14}$$

The maximum likelihood estimators $\hat{\lambda}$ and $\hat{\gamma}$ are found by solving simultaneously

$$\hat{\lambda} = \left(\frac{\sum_{i=1}^{n} t_i^{\hat{\gamma}}}{n} \right)^{-1}$$ (4.2.15)

and

$$\hat{\lambda} = \frac{n/\hat{\gamma} + \sum_{i=1}^{n} \log_e t_i}{\sum_{i=1}^{n} t_i^{\hat{\gamma}} \log_e t_i} .$$ (4.2.16)

Eliminating $\hat{\lambda}$ from (4.2.15) and (4.2.16), we obtain a single equation for $\hat{\gamma}$

$$\left[\frac{n}{\sum_{i=1}^{n} t_i^{\hat{\gamma}}} \right] \sum_{i=1}^{n} t_i^{\hat{\gamma}} \log_e t_i = \frac{n}{\hat{\gamma}} + \sum_{i=1}^{n} \log_e t_i .$$ (4.2.17)

Solving (4.2.17) in terms of $\hat{\gamma}$ can be accomplished by trial and error. However, we recommend the Newton-Raphson method, which is discussed in Chapter 6. The Newton-Raphson method is used to estimate γ, hence λ for the data in Example 4.2.2.

Example 4.2.2. In a study, 20 patients receiving an analgesic to relieve headache pain had the following recorded relief times (in hours): 1.1, 1.4, 1.3, 1.7, 1.9, 1.8, 1.6, 2.2, 1.7, 2.7, 4.1, 1.8, 1.5, 1.2, 1.4, 3.0, 1.7, 2.3, 1.6, and 2.0. Using a starting value $\hat{\gamma}_0 = 0.5$, the Newton-Raphson method applied to (4.2.17) yields $\hat{\gamma} = 2.79$. Applying (4.2.15), we see that $\hat{\lambda}$, the maximum likelihood estimate of λ, is 0.12, and $\hat{\mu}$, the estimated mean relief time, is $\hat{\mu} = \hat{\lambda}^{-1/\hat{\gamma}} \Gamma(1 + \hat{\gamma}) = 1.89$ hours. A table of $\Gamma(1 + p)$ for $p = .00$ to 1.00 by units of 0.01 is available in Pearson and Hartley [1966] as well as in many standard mathematical tables.

At this point we describe a simple graphical method for estimating the parameters of a Weibull death density function. The Weibull survival distribution is

$$S(t) = \exp(-\lambda t^{\gamma}).$$ (4.2.18)

If we take natural logarithms twice of both sides of (4.2.18), we obtain

$$\log_e \log_e \left[\frac{1}{S(t)} \right] = \log_e \lambda + \gamma \log_e t. \qquad (4.2.19)$$

Now let us suppose that in a clinical trial we observe n patients until they have all died and that we record sequentially their times to death as $t_{(1)} \leqslant t_{(2)} \leqslant \cdots \leqslant t_{(n)}$. That is, $t_{(1)} \leqslant t_{(2)} \leqslant \cdots \leqslant t_{(n)}$ are the n ordered observations of the patients' survival times. If these survival times have approximately a Weibull distribution, the graph of $\log_e t_{(i)}$ plotted on the horizontal axis, versus $\log_e \log_e [1/\hat{S}(t_{(i)})]$ plotted on the vertical axis, should be approximately a straight line, where $\hat{S}(t_{(i)}) = (n+1-i)/(n+1)$ is the empirical survival function.* The slope of the plotted line is then the estimator of γ and the exponential of the intercept is the estimator of λ. This follows from (4.2.19).

Example 4.2.3.   A Weibull plot of the data in Example 4.2.2 is obtained from Table 4.1, which gives the logarithms of the ordered relief times of patients and the logarithms of the logarithms of the inverse of the cumulative function. This plot appears in Figure 4.5. Although the points in Figure 4.5 do not lie precisely on a straight line, the straight line fit would be adequate except for the last three points.

The least square estimators $\tilde{\gamma}$ and $\tilde{\lambda}$ (say) of γ and λ, respectively, are obtained by the usual procedure. That is, if we let $y_i = \log_e \log_e [\hat{S}(t_{(i)})]^{-1}$ and $x_i = \log_e t_{(i)}$, then

$$\tilde{\gamma} = \frac{\displaystyle\sum_{j=1}^{20} (x_i - \bar{x})(y_i - \bar{y})}{\displaystyle\sum_{i=1}^{20} (x_i - \bar{x})^2} \qquad (4.2.20)$$

and

$$\log_e \tilde{\lambda} = \bar{y} - \tilde{\gamma}\bar{x}. \qquad (4.2.21)$$

Numerically, then, $\tilde{\gamma} = 3.15$, $\tilde{\lambda} = 0.09$, and the estimated mean relief time is $\tilde{\mu} = \tilde{\lambda}^{-1/\tilde{\gamma}} \Gamma(1 + 1/\tilde{\gamma}) = 1.92$ hours. This agrees quite favorably with results of Example 4.2.2.

---

*Note: Since $(n+1)$ is used instead of n, the last point can be plotted. Otherwise, $\log_e \log_e [1/\hat{S}(t_{(n)})] = +\infty$.

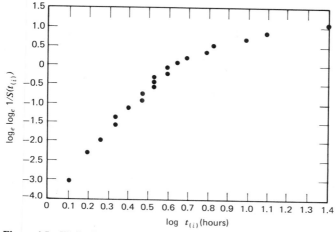

Figure 4.5 Weibull plot of headache pain relief times (data from Example 4.2.2).

Table 4.1

$\log_e t_{(i)}$	$\log_e \log_e [\hat{S}(t_{(i)})]^{-1}$	$\log_e t_{(i)}$	$\log_e \log_e [\hat{S}(t_{(i)})]^{-1}$
0.10	−3.02	0.53	−0.30
0.18	−2.30	0.59	−0.17
0.26	−1.87	0.59	−0.04
0.34	−1.55	0.64	0.09
0.34	−1.30	0.69	0.23
0.40	−1.09	0.79	0.36
0.47	−0.90	0.83	0.51
0.47	−0.73	0.99	0.67
0.53	−0.58	1.10	0.86
0.53	−0.44	1.41	1.11

Often the least squares estimator $\tilde{\gamma}$ is used as an initial value in obtaining the maximum likelihood estimator $\hat{\gamma}$ by the Newton-Raphson procedure. In addition, the least squares procedure can be used when the data are censored. Finally, it should be noted that in this example the expected straight line fit is not perfect. We could consider prior transformations on the survival time or the use of a different death density function. One of the advantages of this graphical technique is that we do obtain a rough idea of the correctness of our choice of distributions as well as estimates of the population parameters. With only 20 observations, of course, comparison of a histogram of relief times with the density function with the estimated $\hat{\gamma}$ and $\hat{\lambda}$ would be difficult to interpret.

EXERCISES

1. Graph the gamma hazard rate and death density functions for $\lambda = 3$, $\gamma = 2$; $\lambda = 3$, $\gamma = 3$; and $\lambda = 3$, $\gamma = 4$.

2. Graph the Weibull hazard rate and death density functions for the same choice of the parameters λ and γ in Exercise 1. Can you detect a difference in the death densities and/or the hazard rates? Explain your answer.

3. Verify in detail that $\mu = \Gamma(1 + 1/\gamma)/\lambda^{1/\gamma}$ and

$$\sigma^2 = [\Gamma(1 + 2/\gamma) - \{\Gamma(1 + 1/\gamma)\}^2]/\lambda^{2/\gamma},$$

where μ and σ^2 are the mean and the variance, respectively, of the Weibull density function.

4. Suppose that t has the Weibull death density function (4.1.3). Find $\mu'_n = E(t^n)$, the nth moment about the origin.

5. Suppose the data in Example 4.2.2 come from a gamma distributed population. Find the maximum likelihood estimates $\hat{\lambda}$ and $\hat{\gamma}$ of λ and γ, respectively, based on these data.

6. Suppose the data in Example 4.2.1 come from a Weibull distributed population. Find the maximum likelihood estimates $\hat{\lambda}$ and $\hat{\gamma}$ of λ and γ, respectively, based on these data.

7. Suppose t has the gamma death density function (4.1.1). Let $y = \log_e t$. Show that

$$E(y) = \frac{\Gamma'(\gamma)}{\Gamma(\gamma)} - \log_e \lambda. \tag{4.2.22}$$

Hint: It is not necessary to make a transformation of variables.

Table 4.2[a]

γ	c.v. (Weibull)	c.v. (gamma)
$\frac{1}{3}$	4.36	1.73
$\frac{1}{2}$	2.24	1.41
$\frac{4}{5}$	1.26	1.12
1	1.00	1.00
$\frac{5}{4}$	0.80	0.89
$\frac{5}{3}$	0.62	0.77
3	0.36	0.58
4	0.28	0.50

[a] See Cohen [1965] to find the Weibull coefficients of variation.

8. Make a Weibull plot of the data in Example 4.2.1. Find the least squares estimates $\tilde{\lambda}$ and $\tilde{\gamma}$ of λ and γ, respectively.

9. Define the coefficient of variation as c.v. $= \sigma / \mu$. Find the c.v. for both the gamma and Weibull density functions. Hence construct Table 4.2 as a function of γ. Using Table 4.2, the sample mean and standard deviation, $\bar{x}$ and s, respectively, we can obtain, using linear interpolation, an estimate of γ for both the Weibull and the gamma.

4.3 MAXIMUM LIKELIHOOD ESTIMATORS OF PARAMETERS FOR SINGLY AND PROGRESSIVELY CENSORED WEIBULL AND GAMMA DISTRIBUTIONS

First let us assume that n patients with an acute illness are all put on study at the same time. It is assumed that the survival times of the first $d \leqslant n$ patients are recorded and the survival times of the remaining $(n - d)$ patients are not recorded. Thus we assume the data are singly censored. If the death density for the ith patient is given by the gamma death density (4.1.1), the likelihood function for the two parameters λ and γ is

$$L(\lambda, \gamma) = \frac{n!}{(n-d)!} \frac{\lambda^{n\gamma}}{[\Gamma(\gamma)]^n} \left[\prod_{i=1}^{d} t_{(i)}^{\gamma-1} \right] \exp\left[-\lambda \sum_{i=1}^{d} t_{(i)} \right]$$
$$\times \left[\int_{t_{(d)}}^{\infty} t^{\gamma-1} \exp(-\lambda t)\, dt \right]^{n-d}.$$

$$(4.3.1)$$

Equation 4.3.1 follows from the general likelihood function for singly censored data, which is given by (3.2.12) and (3.2.13).

We now follow the approach of Wilk, Gnanadesikan, and Huyett [1962a] to obtain the maximum likelihood estimators $\hat{\lambda}$ and $\hat{\gamma}$. We rewrite (4.3.1) as

$$L(\tau, \gamma) = \frac{n!}{(n-d)!} \frac{\tau^{n\gamma}}{[\Gamma(\gamma)]^n} \frac{G^{d(\gamma-1)}}{t_{(d)}^d} \exp(-d\tau A) \left[\int_{1}^{\infty} t^{\gamma-1} \exp(-\tau t)\, dt \right]^{n-d},$$

$$(4.3.1')$$

where

$$G = \frac{\left(\prod_{i=1}^{d} t_{(i)} \right)^{1/d}}{t_{(d)}} \qquad A = \frac{\sum_{i=1}^{d} t_{(i)}}{d t_{(d)}} \qquad \tau = \lambda t_{(d)}.$$

The problem becomes maximizing the logarithm of (4.3.1′) jointly with respect to τ and γ. If $\hat{\tau}$ and $\hat{\gamma}$ are the respective maximum likelihood estimators of τ and γ, $\hat{\lambda}$ is simply $\hat{\tau}/t_{(d)}$. Differentiating $\log_e L(\tau,\gamma)$ with respect to τ and γ, those values $\hat{\tau}$ and $\hat{\gamma}$ which jointly maximize (4.3.1′) jointly satisfy

$$\log_e G = \frac{n}{d}\frac{\Gamma'(\hat{\gamma})}{\Gamma(\hat{\gamma})} - \frac{n}{d}\log_e \hat{\tau} - \left(\frac{n}{d}-1\right)\frac{J'(\hat{\gamma},\hat{\tau})}{J(\hat{\gamma},\hat{\tau})} \qquad (4.3.2)$$

and

$$A = \frac{\hat{\gamma}}{\hat{\tau}} - \frac{1}{\hat{\tau}}\left(\frac{n}{d}-1\right)\frac{\exp(-\hat{\tau})}{J(\hat{\gamma},\hat{\tau})}, \qquad (4.3.3)$$

where

$$J(\gamma,\tau) = \int_1^\infty t^{\gamma-1}e^{-\tau t}\,dt \qquad (4.3.4)$$

and

$$J'(\gamma,\tau) = \frac{\partial J(\gamma,\tau)}{\partial \gamma} = \int_1^\infty t^{\gamma-1}(\log_e t)e^{-\tau t}\,dt. \qquad (4.3.5)$$

The procedure for estimating $\hat{\tau}$ (hence $\hat{\lambda}$) and $\hat{\gamma}$ is rather involved. First, Table 2 in the Appendix (which is Appendix B in Wilk, Gnanadesikan, and Huyett) yields for a grid of values G, A, and n/d, the estimates $\hat{\gamma}$ and $\hat{\mu} = \hat{\gamma}/\hat{\tau}$. These tables are obtained by choosing specific values for G, A, and n/d and solving (4.3.2) and (4.3.3) in terms of $\hat{\gamma}$ and $\hat{\tau}$. This is accomplished by numerical approximations, which the authors discuss in detail in their paper. In most practical situations observed values of G, A, and n/d will not be tabled values. Thus an interpolation procedure is necessary to determine $\hat{\gamma}$ and $\hat{\mu}$. This procedure is now described.

Stage 1 Procedure

1. For any observed values of G_0, A_0, and n/d, make a note of the nearest tabulated values. Let G_1, G_2, A_1, and A_2 be the nearest tabulated values.

2. To obtain $\hat{\mu}_{ij}$, interpolate linearly on $\hat{\mu}_{ij}$ between G_i and A_j, $i,j = 1,2$.

3. To obtain $\hat{\gamma}_{ij}$, note that there are three regions marked off in Table 2 of the Appendix. In region I interpolate linearly on $\hat{\gamma}_{ij}$ between G_i and A_j. In region II interpolate linearly on $\hat{\gamma}_{ij}(n/d-0.9)$ between G_i and A_j. In region III interpolate linearly on $\hat{\gamma}_{ij}(n/d-0.8)$ between G_i and A_j, $i,j = 1,2$.

Stage 2 Procedure

From stage 1 we obtain the following table:

	G_1	G_2
A_1	$(\hat{\gamma}_{11}, \hat{\mu}_{11})$	$(\hat{\gamma}_{21}, \hat{\mu}_{21})$
A_2	$(\hat{\gamma}_{12}, \hat{\mu}_{12})$	$(\hat{\gamma}_{22}, \hat{\mu}_{22})$

The requisite values $\hat{\gamma}$ and $\hat{\mu}$ are now found by methods of bilinear interpolation as follows.

1. Write $\hat{\mu} = \hat{\mu}_{11}(1 - \theta - \phi + \phi\theta) + \hat{\mu}_{21}(\theta - \theta\phi) + \hat{\mu}_{12}(\phi - \theta\phi) + \hat{\mu}_{22}\theta\phi$, where $\theta = (G_0 - G_1)/(G_2 - G_1)$, $\phi = (A_0 - A_1)/(A_2 - A_1)$, and G_0 and A_0 are the observed values.

2. If $A_2 - G_1 > 0.16$, we use the bilinear interpolation formula on $\hat{\gamma}$. That is, $\hat{\gamma} = \hat{\gamma}_{11}(1 - \theta - \phi + \theta\phi) + \hat{\gamma}_{21}(\theta - \theta\phi) + \hat{\gamma}_{12}(\phi - \theta\phi) + \hat{\gamma}_{22}\theta\phi$. If $A_2 - G_1 \leqslant 0.16$, we use the bilinear interpolation formula on $\hat{\gamma}(A_0 - G_0)^2$ and $\hat{\gamma}_{ij}(A_i - A_j)^2$, $i, j = 1, 2$.

To illustrate this procedure we consider the same example taken by Wilk, Gnanadesikan, and Huyett in a different context.

Example 4.3.1. In a study on a new drug that induces remission in leukemia, a total of 34 patients were entered for treatment. All achieved remission, but at the end of the study 31 patients had relapsed. The ordered remission times (in months) were 3, 4, 5, 6, 6, 7, 8, 8, 9, 9, 9, 10, 10, 11, 11, 11, 13, 13, 13, 13, 13, 17, 17, 19, 19, 25, 29, 33, 42, 42, 52, 52$^+$, 52$^+$, and 52$^+$. The plus above the last three ordered observations implies that the patients were still in remission at the conclusion of the study. We assume that each remission time of patients has a gamma distribution for which we wish to estimate γ and λ by the censored data maximum likelihood method. For $\hat{\mu}_{12}$ we use linear interpolation. Since $G_0 = 0.2386$, and $A_0 = 0.3021$, we have $G_1 = 0.20$, $G_2 = 0.24$, $A_1 = 0.28$, and $A_2 = 0.32$. The corresponding values of $\hat{\mu}_{12}$ for $n/d = 1.0$ and $n/d = 1.1$ are, respectively, 0.320 and 0.419. Thus by linear interpolation $\hat{\mu}_{12} = 0.320 + (0.968)(0.099) = 0.416$. For $\hat{\gamma}$, we note that the values $G_1 = 0.20$ and $A_2 = 0.32$ lie in region II of Table 2 of the Appendix. Thus we interpolate for $\hat{\gamma}(n/d - 0.9)$. The corresponding values of $\hat{\gamma}_{12}$ for $n/d = 1.0$ and $n/d = 1.1$ are, respectively, 1.203 and 1.018. Furthermore, $\hat{\gamma}(n/d - 0.9)$ yields the values 0.1203 and 0.2036. Since $\hat{\gamma}_{12}(n/d - 0.9) = (\hat{\gamma}_{12})(0.1968) = 0.1203 + 0.968(0.2036 - 0.1203)$, we have $\hat{\gamma}_{12} = 1.021$. The final stage 1

112 ESTIMATING PARAMETERS OF THE WEIBULL, GAMMA, RAYLEIGH

table can be computed from similar considerations. Thus we have

	$G_1 = 0.20$	$G_2 = 0.24$
$A_1 = 0.28$	$\hat{\gamma}_{11} = 1.238$	$\hat{\gamma}_{21} = 2.009$
	$\hat{\mu}_{11} = 0.371$	$\hat{\mu}_{21} = 0.362$
$A_2 = 0.32$	$\hat{\gamma}_{12} = 1.021$	$\hat{\gamma}_{22} = 1.456$
	$\hat{\mu}_{12} = 0.416$	$\hat{\mu}_{22} = 0.407$

To obtain $\hat{\mu}$ in final form for stage 2, we use the bilinear interpolation formula given in item 1. Thus $\hat{\mu} = (0.371)(0.0157) + (0.362)(0.4318) + (0.416)(0.0193) + (0.407)(0.5332) = 0.387$. Since $A_2 - G_1 = 0.32 - 0.20 = 0.12 < 0.16$, we use bilinear interpolation with respect to $\hat{\gamma}(A_0 - G_0)^2$, $\hat{\gamma}_{11}(A_1 - G_1)^2$, $\hat{\gamma}_{12}(A_2 - G_1)^2$, $\hat{\gamma}_{21}(A_1 - G_2)^2$, and $\hat{\gamma}_{22}(A_2 - G_2)^2$. Computationally then,

$$\hat{\gamma}(0.004032) = (0.007925)(0.0157) + (0.003214)(0.4318)$$
$$+ (0.014702)(0.0193) + (0.009319)(0.5332)$$
$$= 0.006765.$$

Therefore, $\hat{\gamma} = 1.678$. Finally, $\hat{\lambda} = \hat{\gamma}/\hat{\mu}t_{(d)} = 1.678/(0.387)(52) = 0.0833$. For this example, Wilk, Gnanadesikan, and Huyett point out there is some slight inaccuracy in the value of $\hat{\lambda}$. This is due to a small inaccuracy in the final bilinear interpolation for $\hat{\gamma}$. Actually, the accurate value is $\hat{\gamma} = 1.625$, which yields a value $\hat{\lambda} = 0.0807$.

Harter and Moore [1965] consider estimating the parameters for the three-parameter censored gamma distribution. Their methodology is similar to the methodology just described above, but they do not carry out the details of obtaining the estimates as Wilk, Gnanadesikan, and Huyett do.

Let us now estimate the parameters from the singly censored Weibull distribution. The likelihood function $L(\lambda, \gamma)$ is

$$L(\lambda, \gamma) = \frac{n!}{(n-d)!} (\lambda\gamma)^d \prod_{i=1}^{d} t_{(i)}^{\gamma-1} \exp\left[-\left\{ \lambda \sum_{i=1}^{d-1} t_{(i)}^{\gamma} + (n-d+1)\lambda t_{(d)}^{\gamma} \right\} \right].$$

$$(4.3.6)$$

Estimates of λ and γ are found by taking the derivative of (4.3.6) with respect to $\hat{\lambda}$ and $\hat{\gamma}$ and setting the resulting expressions equal to zero.

Cohen [1965] obtains the same likelihood function (4.3.6). The expression (4.3.6) combines all $t_{(d)}$ values, whereas Cohen separates the last time to death from the other $t_{(d)}$ values that are censored. Equation 4.3.6 is considerably more simple than (4.3.1) for the gamma distribution because the survival probability is a closed expression. The maximum likelihood estimators $\hat{\lambda}$ and $\hat{\gamma}$ are then found by solving simultaneously

$$\hat{\lambda} = \frac{d}{\sum_{i=1}^{d-1} t_{(i)}^{\hat{\gamma}} + (n-d+1)t_{(d)}^{\hat{\gamma}}} \tag{4.3.7}$$

and

$$\frac{d}{\hat{\gamma}} - \sum_{i=1}^{d} \log_e t_{(i)} = \hat{\lambda}\left[\sum_{i=1}^{d-1} t_{(i)}^{\hat{\gamma}} \log_e t_{(i)} + (n-d+1)t_{(d)}^{\hat{\gamma}} \log_e t_{(d)} \right]. \tag{4.3.8}$$

Combining (4.3.7) and (4.3.8) we have an equation in $\hat{\gamma}$ alone. Thus $\hat{\gamma}$ is the solution of

$$\left[\frac{d}{\hat{\gamma}} + \sum_{i=1}^{d} \log_e t_{(i)} \right]\left[\sum_{i=1}^{d-1} t_{(i)}^{\hat{\gamma}} + (n-d+1)t_{(d)}^{\hat{\gamma}} \right] = d\left[\sum_{i=1}^{d-1} t_{(i)}^{\hat{\gamma}} \log_e t_{(i)} \right.$$

$$\left. + (n-d+1)t_{(d)}^{\hat{\gamma}} \log_e t_{(d)} \right] \tag{4.3.9}$$

or

$$\frac{\sum_{i=1}^{d-1} t_{(i)}^{\hat{\gamma}} \log_e t_{(i)} + (n-d+1)t_{(d)}^{\hat{\gamma}} \log_e t_{(d)}}{\sum_{i=1}^{d-1} t_{(i)}^{\hat{\gamma}} + (n-d+1)t_{(d)}^{\hat{\gamma}}} - \frac{1}{\hat{\gamma}} = \frac{1}{d}\sum_{i=1}^{d} \log_e t_{(i)}. \tag{4.3.10}$$

Cohen [1965] introduces the notation

$$\sum {}^* t_{(i)}^{\hat{\gamma}} [\log_e t_{(i)}]^j = \sum_{i=1}^{d-1} t_{(i)}^{\hat{\gamma}} [\log_e t_{(i)}]^j + (n-d+1)t_{(d)}^{\hat{\gamma}} [\log_e t_{(d)}]^j, \tag{4.3.11}$$

where $j = 0, 1$, which simplifies (4.3.10) somewhat.

Obtaining $\hat{\gamma}$ from (4.3.10), hence $\hat{\lambda}$ from (4.3.7), presents no severe complications over the noncensored case. That is, the Newton-Raphson method can be used to find $\hat{\gamma}$. The starting value $\hat{\gamma}_0$ can be obtained by treating the sample as a complete sample. We estimate $\bar{x}$ and s, the sample mean and standard deviation from this sample, and c.v. $= s/\bar{x}$. Then $\hat{\gamma}_0$ is obtained by linear interpolation in Table 4.2. (Table 4.2 can be extended quite easily for values of $\gamma > 4.0$). Now $\hat{\gamma}_0$ is the starting value in (4.3.11) for the Newton-Raphson technique. Alternatively (4.3.10) can be solved by trial and error.

The estimation procedures for the case of progressively censored data are the same as those given in Section 3.2. The expressions for $f(t)$ and $S(T)$ for the Weibull or gamma distribution are used in (3.2.4'). Cohen [1965] presents the first and second derivatives of the likelihood function for the Weibull distribution in a form convenient for estimating the parameters. Taking the first derivatives of the likelihood with respect to $\hat{\gamma}$ and $\hat{\lambda}$ and eliminating $\hat{\lambda}$ between the two equations, $\hat{\gamma}$ is the solution to

$$\frac{\sum_{i=1}^{d} t_{(i)}^{\hat{\gamma}}\log_e t_{(i)} + \sum_{j=1}^{n-d} T_{(j)}^{\hat{\gamma}}\log_e T_{(j)}}{\sum_{i=1}^{d} t_{(i)}^{\hat{\gamma}} + \sum_{j=1}^{n-d} T_{(j)}^{\hat{\gamma}}} - \frac{1}{\hat{\gamma}} = \frac{1}{d}\sum_{i=1}^{d}\log_e t_{(i)}. \qquad (4.3.12)$$

Then $\hat{\lambda}$ is the solution to

$$\hat{\lambda} = \frac{d}{\sum_{i=1}^{d} t_{(i)}^{\hat{\gamma}} + \sum_{j=1}^{n-d} T_{(j)}^{\hat{\gamma}}}. \qquad (4.3.13)$$

Whereas progressively censored samples for the gamma distribution are conceptually not difficult, the ensuing estimating equations are very difficult to solve. Thus no further consideration is given to progressively censored gamma samples.

Example 4.3.2. The following data, representing survival times of patients with chronic leukemia, are a fictitious adaptation of the data analyzed by Cohen [1966]. The survival times were 500, 700+, 800, 900, and 1000+ weeks, respectively. The plus signs indicate patients who were still alive at the end of the study. This is an example of progressively censored data, since there are two different censoring times. We thus find $\frac{1}{3}\sum_{i=1}^{3}\log_e t_{(i)} = 6.567$, which means that $\hat{\gamma}$ is the solution to

$$\frac{\sum_{i=1}^{3} t_{(i)}^{\hat{\gamma}}\log_e t_{(i)} + \sum_{j=1}^{2} T_{(j)}^{\hat{\gamma}}\log_e T_{(j)}}{\sum_{i=1}^{3} t_{(i)}^{\hat{\gamma}} + \sum_{j=1}^{2} T_{(j)}^{\hat{\gamma}}} - \frac{1}{\hat{\gamma}} = 6.567, \qquad (4.3.14)$$

where $t_{(i)}$ is the ith ordered time to death, $i = 1, 2, 3$, and $T_{(j)}$ is the jth ordered censored time, $j = 1, 2$. Using the Newton-Raphson procedure, or, following Cohen [1966] and selecting a trial-and-error approach, we find, $\hat{\gamma} = 4.41$: $\hat{\lambda}$ is then the solution of

$$\hat{\lambda} = \frac{3}{\sum_{i=1}^{3} t_{(i)}^{\hat{\gamma}} + \sum_{j=1}^{2} T_{(j)}^{\hat{\gamma}}}. \qquad (4.3.15)$$

Thus $\hat{\lambda} = 7.828 \times 10^{-14}$.

EXERCISES

1. Prove that $\gamma J(\gamma, \tau) + [\partial J(\gamma, \tau) / \partial \tau] \tau = e^{-\tau}$, hence obtain (4.3.3), where $J(\gamma, \tau)$ is given by (4.3.4).

2. In the analgesic study of Example 4.2.2, suppose the data are censored at 2 hours, giving the following recorded relief times and censored times (denoted with a plus): 1.1, 1.4, 1.3, 1.7, 1.9, 1.8, 1.6, 2.0+, 1.7, 2.0+, 2.0+, 1.8, 1.5, 1.2, 1.4, 2.0+, 1.7, 1.6 and 2.0, hours. Assuming these times to have the singly censored Weibull distribution, estimate $\hat{\lambda}, \hat{\gamma}$.

3. Let

$$G(\gamma) = \frac{\sum_{i=1}^{n} t_i^{\gamma} \log_e t_i}{\sum_{i=1}^{n} t_i^{\gamma}} - \frac{1}{\gamma} - \frac{1}{n} \sum_{i=1}^{n} \log_e t_i.$$

   Prove that $G(\gamma)$ is an increasing function in $\gamma, \gamma > 0$. Hence show that there is but a single value γ_0 such that $G(\gamma_0) = 0$. Can this be extended to the case of observations that are singly or progressively censored? Explain.

4. Assume that the relief times in Exercise 2 obey the censored gamma death density distribution. Using the Wilk, Gnanadesikan, and Huyett [1962a] procedure, estimate $\hat{\lambda}, \hat{\gamma}$.

4.4 LARGE SAMPLE PROPERTIES OF THE WEIBULL AND GAMMA MAXIMUM LIKELIHOOD ESTIMATORS

In this section we examine the large sample properties of the Weibull and gamma maximum likelihood estimators. For small samples (e.g., sample sizes smaller than 20), maximum likelihood estimators are likely to be quite biased. In such cases the experimenter may choose to estimate the parameters by alternate procedures. Such procedures are discussed in Section 4.6 for both these distributions. Furthermore, Mann, Schafer, and Singpurwalla [1974] have an excellent discussion on Weibull parameter estimation procedures.

Suppose $\hat{\lambda}$ and $\hat{\gamma}$ are the maximum likelihood estimators of λ and γ, respectively, where λ and γ are the parameters for either the gamma or Weibull death density function. The large sample variance–covariance matrix for $\hat{\lambda}$ and $\hat{\gamma}$ is given by

$$
\begin{bmatrix}
\operatorname{Var}\hat{\lambda} & \operatorname{Cov}(\hat{\lambda},\hat{\gamma}) \\
\operatorname{Cov}(\hat{\lambda},\hat{\gamma}) & \operatorname{Var}\hat{\gamma}
\end{bmatrix}
=
\begin{bmatrix}
-E\left(\dfrac{\partial^2 \log_e L}{\partial \lambda^2}\right) & -E\left(\dfrac{\partial^2 \log_e L}{\partial \lambda\,\partial \gamma}\right) \\[2ex]
-E\left(\dfrac{\partial^2 \log_e L}{\partial \lambda\,\partial \gamma}\right) & -E\left(\dfrac{\partial^2 \log_e L}{\partial \gamma^2}\right)
\end{bmatrix}^{-1},
$$

$$(4.4.1)$$

where the expectation is with respect to the distribution of the observations. When the observations are censored, the expected values in (4.4.1) often are not readily obtained. Frequently in such cases the second partials alone are used, yielding approximate large sample values of $\operatorname{Var}\hat{\lambda}$, $\operatorname{Cov}(\hat{\lambda},\hat{\gamma})$, and $\operatorname{Var}\hat{\gamma}$.

We consider now the four cases—gamma uncensored and censored and Weibull uncensored and censored.

Case 1: Gamma Uncensored

Suppose $t_1, t_2, \ldots, t_n$ are the survival times of n patients on a clinical study who are followed until death. The death density of the ith patient is the gamma. The likelihood function for the n patients is (4.2.1). It follows that

$$
\frac{\partial^2 \log_e L}{\partial \lambda^2} = -\frac{n\gamma}{\lambda^2},
\tag{4.4.2}
$$

$$
\frac{\partial^2 \log_e L}{\partial \lambda\,\partial \gamma} = \frac{n}{\lambda},
\tag{4.4.3}
$$

and

$$
\frac{\partial^2 \log_e L}{\partial \gamma^2} = -n\Psi'(\gamma),
\tag{4.4.4}
$$

where

$$
\Psi(\gamma) = \frac{d\log_e \Gamma(\gamma)}{d\gamma}
\tag{4.4.5}
$$

and

$$\Psi'(\gamma) = \frac{d^2 \log_e \Gamma(\gamma)}{d\gamma^2} , \text{ a trigamma function} \qquad (4.4.6)$$

Since (4.4.2) through (4.4.4) are devoid of terms involving the sample, it follows that

$$\begin{bmatrix} \operatorname{Var}\hat{\lambda} & \operatorname{Cov}(\hat{\lambda},\hat{\gamma}) \\ \operatorname{Cov}(\hat{\lambda},\hat{\gamma}) & \operatorname{Var}\hat{\gamma} \end{bmatrix} = \begin{bmatrix} \dfrac{n\gamma}{\lambda^2} & \dfrac{-n}{\lambda} \\ \dfrac{-n}{\lambda} & n\Psi'(\gamma) \end{bmatrix}^{-1}$$

$$= (n\Delta)^{-1} \begin{bmatrix} \Psi'(\gamma) & \dfrac{1}{\lambda} \\ \dfrac{1}{\lambda} & \dfrac{\gamma}{\lambda^2} \end{bmatrix}, \qquad (4.4.7)$$

where

$$\Delta = \left(\frac{\gamma}{\lambda^2} \Psi'(\gamma) - \frac{1}{\lambda^2} \right). \qquad (4.4.8)$$

According to the large sample theory of maximum likelihood estimators (see, e.g., Mood and Graybill [1963]), the joint distribution of the vector estimator $\begin{pmatrix} \hat{\lambda} \\ \hat{\gamma} \end{pmatrix}$ is approximately the bivariate normal distribution whose mean is the vector $\begin{pmatrix} \lambda \\ \gamma \end{pmatrix}$ and whose variance–covariance matrix is (4.4.7).

Before proceeding with developing confidence intervals for λ and γ, a brief discussion of elliptical confidence regions is given. (See Mood and Graybill [1963] or Draper and Smith [1966] for further results.)

Some Notes on Elliptical Confidence Regions

If θ_1, and θ_2 are two parameters whose corresponding (death) density function is $f(x; \theta_1, \theta_2)$, we denote $\hat{\theta}_1$ and $\hat{\theta}_2$ to be the maximum likelihood estimators of θ_1 and θ_2, respectively. Let $L(\theta_1, \theta_2) = \Pi_{i=1}^{n} f(x_i; \theta_1, \theta_2)$ be the likelihood function of θ_1 and θ_2 with respect to the data set $(x_1, x_2, \ldots, x_n)$. Then an approximate (large sample) $(1 - \alpha)100$ percent elliptical confidence region for θ_1 and θ_2, jointly, is given by

$$\sigma^{11}(\hat{\theta}_1,\hat{\theta}_2)(\theta_1-\hat{\theta}_1)^2 + \sigma^{22}(\hat{\theta}_1,\hat{\theta}_2)(\theta_2-\hat{\theta}_2)^2 + 2\sigma^{12}(\hat{\theta}_1,\hat{\theta}_2)(\theta_1-\hat{\theta}_1)(\theta_2-\hat{\theta}_2)$$

$$\doteq \chi^2(1-\alpha;2), \qquad (4.4.9)$$

where

$$\sigma^{ij}(\hat{\theta}_1,\hat{\theta}_2) = -E\left[\frac{\partial^2 \log_e L(\theta_1,\theta_2)}{\partial\theta_i\,\partial\theta_j}\right]_{\substack{\theta_1=\hat{\theta}_1 \\ \theta_2=\hat{\theta}_2}} \qquad \text{for } i,j=1,2, \quad (4.4.10)$$

and $\chi^2(1-\alpha;2)$ is the $(1-\alpha)100$ percentage point of the chi-square distribution with 2 degrees of freedom. If $\sigma^{12}(\hat{\theta}_1,\hat{\theta}_2)=0$, the elliptical region defined by (4.4.9) has its major and minor axes parallel to the θ_1 and θ_2 axes. Furthermore, if $\sigma^{11}(\hat{\theta}_1,\hat{\theta}_2)=\sigma^{22}(\hat{\theta}_1,\hat{\theta}_2)$, the region is circular. In the general case—that is, when $\sigma^{12}(\hat{\theta}_1,\hat{\theta}_2)\neq 0$ and $\sigma^{11}(\hat{\theta}_1,\hat{\theta}_2)\neq\sigma^{22}(\hat{\theta}_1,\hat{\theta}_2)$—the elliptical confidence region for (θ_1,θ_2) does not have its major and minor axes parallel to the θ_1 and θ_2 axes. In this case the major and minor axes are determined by rotation. Specifically, if ϕ is the angle that the major and minor axes make with the θ_1 and θ_2 axes, ϕ is determined by

$$\tan 2\phi = \frac{2\sigma^{12}(\hat{\theta}_1,\hat{\theta}_2)}{\left\{\sigma^{11}(\hat{\theta}_1,\hat{\theta}_2)-\sigma^{22}(\hat{\theta}_1,\hat{\theta}_2)\right\}}, \qquad (4.4.11)$$

which follows from the analytic geometry considerations of the rotation of axes to eliminate the cross product $(\hat{\theta}_1\hat{\theta}_2)$ term. Figure 4.6 shows the general case.

There are two points to note concerning these elliptical confidence regions: (1) a value of $(\hat{\theta}_1,\hat{\theta}_2)$ may seem to be reasonable based on the individual confidence intervals for θ_1 and θ_2, but it may lie outside the elliptical confidence region, hence may not be a reasonable value; and (2) the joint confidence region that is the correct region to use yields results different from the rectangular or square region obtained by considering the two confidence intervals separately.

Example 4.4.1. The survival times of 20 male mice in Example 4.2.1 yield as the maximum likelihood estimates for γ and λ, $\hat{\gamma}=8.53$ and $\hat{\lambda}=0.075$. An approximate 95 percent elliptical confidence region for γ and λ is

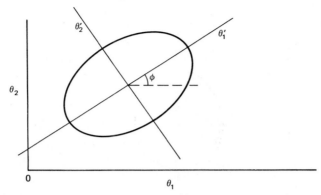

Figure 4.6 Elliptical confidence region for θ_1 and θ_2 general case.

then constructed using (4.4.2) through (4.4.4):

$$\text{(i)} \quad \sigma^{11}(\hat{\lambda},\hat{\gamma}) = \frac{n\hat{\gamma}}{\hat{\lambda}^2} = \frac{(20)(8.53)}{(0.075)^2} = 30{,}328.8889,$$

$$\text{(ii)} \quad 2\sigma^{12}(\hat{\lambda},\hat{\gamma}) = \frac{-2n}{\hat{\lambda}} = \frac{-40}{0.075} = -533.3333,$$

$$\text{(iii)} \quad \sigma^{22}(\hat{\lambda},\hat{\gamma}) = n\Psi'(\hat{\gamma}) = 20\Psi'(8.53) = 2.4800.$$

The angle of rotation ϕ is given by

$$\text{(iv)} \quad \tan 2\phi = \frac{-533.3333}{30{,}326.4089} = -0.0176.$$

This corresponds to a value of ϕ of less than 1°, which implies very little rotation at all.
The elliptical confidence region (Figure 4.7) is given by

$$\text{(v)} \quad 30{,}329.89(\lambda - 0.075)^2 + 2.48(\gamma - 8.53)^2$$

$$- 533.33(\gamma - 8.53)(\lambda - 0.075) = 5.99$$

The separate approximate 95 percent confidence intervals for λ and γ in Figure 4.7 are $\hat{\lambda} \pm 1.96\sqrt{\text{Var}\,\hat{\lambda}}$ and $\hat{\gamma} \pm 1.96\sqrt{\text{Var}\,\hat{\gamma}}$, respectively; where $\text{Var}\,\hat{\lambda}$ and $\text{Var}\,\hat{\gamma}$ are the diagonal elements of the variance-covariance matrix (4.4.7). Numerical values are obtained by replacing the parameter values

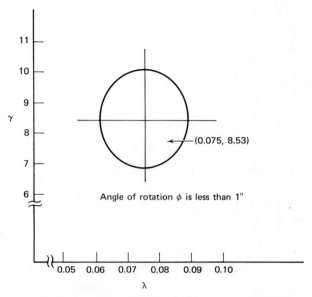

Figure 4.7 An approximate 95 percent elliptical confidence region for the two parameters λ and γ of the gamma survival distribution for the data in Example 4.4.1: $2.48\,(\gamma-8.53)^2 + 30{,}329.89(\lambda-0.075)^2 - 533.33(\lambda-0.075)\,(\gamma-8.53)=5.99$.

with their maximum likelihood estimators. We thus find $0.027<\lambda<0.123$, and $3.20<\gamma<13.86$.

For both the Weibull and gamma survival distributions, as well as other two-parameter survival distributions, we can approximate $\mathrm{Var}(\hat{\mu})$, where $\hat{\mu}$ is the estimated time to death. Thus

$$\mathrm{Var}\,\hat{\mu} \doteq \left(\frac{\partial\mu}{\partial\lambda}\right)^2\mathrm{Var}\,\hat{\lambda} + 2\left(\frac{\partial\mu}{\partial\lambda}\right)\left(\frac{\partial\mu}{\partial\gamma}\right)\mathrm{Cov}(\hat{\lambda},\hat{\gamma}) + \left(\frac{\partial\mu}{\partial\gamma}\right)^2\mathrm{Var}\,\hat{\gamma}. \quad (4.4.12)$$

For the uncensored gamma distribution, we write

$$\mathrm{Var}\,\hat{\mu} \doteq \left(\frac{\gamma}{\lambda^2}\right)^2\mathrm{Var}\,\hat{\lambda} - \frac{2\gamma}{\lambda^3}\mathrm{Cov}(\hat{\lambda},\hat{\gamma}) + \left(\frac{1}{\lambda}\right)^2\mathrm{Var}\,\hat{\gamma}, \quad (4.4.13)$$

where $\mathrm{Var}\,\hat{\lambda}$, $\mathrm{Cov}(\hat{\lambda},\hat{\gamma})$, and $\mathrm{Var}\,\hat{\gamma}$ are obtained from (4.4.7). An approximate $(1-\alpha)100$ percent confidence interval for μ is thus given by $\hat{\mu} \pm \sqrt{\mathrm{Var}\,\hat{\mu}}\ \ Z(1-\alpha/2)$, where $\hat{\lambda}$ and $\hat{\gamma}$ replace λ and γ in computing $\mathrm{Var}\,\hat{\mu}$, and $Z(1-\alpha/2)$ is the upper $(1-\alpha/2)\,100$ percentage point of the standard normal density.

An approximate 95 percent confidence interval for μ for the data in Example 4.4.1 is given by

$$96.67 < \mu < 130.79.$$

Case 2: Weibull Uncensored

If the death density function for the patients on study is given by Weibull distribution, the large sample variance–covariance matrix for λ and γ changes as $\partial^2 \log_e L / \partial \lambda^2$, $\partial^2 \log_e L / \partial \lambda \partial \gamma$, and $\partial^2 \log_e L / \partial \gamma^2$ change:

$$\frac{\partial^2 \log_e L}{\partial \lambda^2} = \frac{-n}{\lambda^2}, \tag{4.4.14}$$

$$\frac{\partial^2 \log_e L}{\partial \lambda \partial \gamma} = -\sum_{i=1}^{n} t_i^{\gamma} \log_e t_i, \tag{4.4.15}$$

$$\frac{\partial^2 \log_e L}{\partial \gamma^2} = -\left[\frac{n}{\gamma^2} + \lambda \sum_{i=1}^{n} t_i^{\gamma} (\log_e t_i)^2 \right]. \tag{4.4.16}$$

Unfortunately it is not so easy to obtain the expected values of the second derivatives (4.4.14) through (4.4.16), since they involve rather complicated functions of the sample. We now obtain $E(t^{\gamma} \log_e t)$.

$$E(t^{\gamma} \log_e t) = \lambda \gamma \int_0^{\infty} t^{2\gamma - 1} \log_e t \exp(-\lambda t^{\gamma}) \, dt. \tag{4.4.17}$$

Making the substitution $y = \lambda t^{\gamma}$, we have

$$E(t^{\gamma} \log_e t) = \frac{1}{\lambda \gamma} \int_0^{\infty} y (\log_e y - \log_e \lambda) e^{-y} \, dy. \tag{4.4.18}$$

It is not difficult to show that

$$\int_0^{\infty} y (\log_e y) e^{-y} \, dy = \Psi(2), \tag{4.4.19}$$

where $\Psi(\gamma)$ is given by (4.4.5). Thus

$$E(t^{\gamma} \log_e t) = \frac{1}{\lambda \gamma} (\Psi(2) - \log_e \lambda). \tag{4.4.20}$$

In a similar manner we can obtain $E(t^\gamma(\log_e t)^2)$, hence the large sample variance–covariance matrix for the maximum likelihood estimators of the Weibull parameters*

$$E\left(t^\gamma(\log_e t)^2\right) \doteq (\lambda\gamma^2)^{-1}\left[\Psi'(2) + [\Psi(2)]^2 - 2\log_e \lambda\Psi(2) + (\log_e \lambda)^2\right].$$

(4.4.21)

Thus the variance–covariance matrix of the Weibull maximum likelihood estimator vector $\begin{pmatrix} \hat{\lambda} \\ \hat{\gamma} \end{pmatrix}$ is

$$\begin{bmatrix} \text{Var}\,\hat{\lambda} & \text{Cov}(\hat{\lambda},\hat{\gamma}) \\ \text{Cov}(\hat{\lambda},\hat{\gamma}) & \text{Var}\,\hat{\gamma} \end{bmatrix}$$

$$= \begin{bmatrix} \dfrac{n}{\lambda^2} & n(\lambda\gamma)^{-1}(\Psi(2) - \log_e\lambda) \\ n(\lambda\gamma)^{-1}(\Psi(2) - \log_e\lambda) & n(\lambda\gamma^2)^{-1}\left[\Psi'(2) + [\Psi(2)]^2 - 2\log_e\lambda\Psi(2) + (\log_e\lambda)^2\right] \end{bmatrix}^{-1}.$$

(4.4.22)

It then follows that for large n, $\begin{pmatrix} \hat{\lambda} \\ \hat{\gamma} \end{pmatrix}$ is approximately bivariate normally distributed, with mean $\begin{pmatrix} \lambda \\ \gamma \end{pmatrix}$ and variance–covariance matrix given by (4.4.22). As for the gamma parameters, an elliptical confidence region for $\begin{pmatrix} \lambda \\ \gamma \end{pmatrix}$ can be constructed.

An approximate $100(1 - \alpha)$ percent confidence interval for μ is given as $\hat{\mu} \pm \sqrt{\text{Var}\,\hat{\mu}}\ Z(1 - \alpha/2)$, where

$$\text{Var}\,\hat{\mu} \doteq \left(\frac{\mu}{\lambda\gamma}\right)^2 \text{Var}\,\hat{\lambda} - \frac{2\mu^2}{\lambda\gamma^3}[\log_e\lambda + \Psi(1 + 1/\gamma)]\text{Cov}(\hat{\lambda},\hat{\gamma})$$

$$+ \frac{\mu^2}{\gamma^4}[\log_e\lambda + \Psi(1 + 1/\gamma)]^2\text{Var}\,\hat{\gamma}.$$

remembering that

$$\Psi(x) \equiv \frac{\Gamma'(x)}{\Gamma(x)}, \text{ for all } x > 0.$$

(4.4.23)

*Numerically, $\Psi(2) = 0.4228$, and $\Psi'(2) = 0.6449$.

The values $\text{Var}\,\hat{\lambda}$, $\text{Var}\,\hat{\gamma}$, and $\text{Cov}(\hat{\lambda},\hat{\gamma})$ are obtained by inverting the matrix in (4.4.22). In practice $\text{Var}\,\hat{\mu}$ is obtained numerically by substituting $\hat{\lambda}$ and $\hat{\gamma}$ for λ and γ respectively.

Example 4.4.2. For the data in Example 4.2.2 of relief times of headache sufferers, $\hat{\lambda}=0.121$, $\hat{\gamma}=2.79$, and $\hat{\mu}=1.89$ hours based on a sample size $n=20$. An approximate 95 percent confidence interval for μ is given by 1.59 hours $<\mu<2.19$ hours, where $\text{Var}\,\hat{\mu}=0.0230$.

Case 3: Gamma Censored

Let us assume as in Section 4.3 that n patients with an acute illness on study are singly censored with d deaths $t_{(1)}\leqslant t_{(2)}\leqslant\cdots\leqslant t_{(d)}$ and $n-d$ survivors, each of whom is observed for a time $t_{(d)}$. As we showed in Section 4.3 the likelihood function of the ordered sample is given by (4.3.1) when each patient has a gamma death density function. It then follows from (4.3.1′) through (4.3.3) that

$$\frac{\partial^2 \log_e L}{\partial \gamma^2} = -n\Psi'(\gamma)+(n-d)V'(\gamma,\tau), \tag{4.4.24}$$

$$\frac{\partial^2 \log_e L}{\partial \gamma\, \partial \tau} = \frac{n}{\tau}+\frac{(n-d)\exp(-\tau)}{\tau}\frac{V(\gamma,\tau)}{J(\gamma,\tau)}, \tag{4.4.25}$$

and

$$\frac{\partial^2 \log_e L}{\partial \tau^2} = \frac{-n\gamma}{\tau^2}+\frac{(n-d)e^{-\tau}\{(\gamma+\tau+1)J(\gamma,\tau)-e^{-\tau}\}}{\tau^2 J^2(\gamma,\tau)}, \tag{4.4.26}$$

where

$$V(\gamma,\tau)=\frac{\partial \log_e J(\gamma,\tau)}{\partial \gamma}, \tag{4.4.27}$$

and

$$V'(\gamma,\tau)=\frac{\partial^2 \log_e J(\gamma,\tau)}{\partial \gamma^2}. \tag{4.4.28}$$

For large samples the variance–covariance matrix of the estimates $\hat{\tau}$ and $\hat{\gamma}$ is approximately

$$\begin{bmatrix} \text{Var}\,\hat{\tau} & \text{Cov}(\hat{\gamma},\hat{\tau}) \\ \text{Cov}(\hat{\gamma},\hat{\tau}) & \text{Var}\,\hat{\gamma} \end{bmatrix} = \begin{bmatrix} -\dfrac{\partial^2 \log_e L}{\partial \tau^2} & -\dfrac{\partial^2 \log_e L}{\partial \gamma\, \partial \tau} \\ -\dfrac{\partial^2 \log_e L}{\partial \gamma\, \partial \tau} & -\dfrac{\partial^2 \log_e L}{\partial \gamma^2} \end{bmatrix}^{-1}_{\substack{\gamma=\hat{\gamma} \\ \tau=\hat{\tau}}}, \tag{4.4.29}$$

where the second partial derivatives are given by (4.4.24) through (4.4.26).

Theoretically it is possible to obtain an elliptical confidence region for the parameter vector $\begin{pmatrix} \tau \\ \gamma \end{pmatrix}$, hence $\begin{pmatrix} \lambda \\ \gamma \end{pmatrix}$, when the data are censored. As Harter and Moore [1967] showed, the large sample distribution of the joint maximum likelihood estimator vector $\begin{pmatrix} \hat{\lambda} \\ \hat{\gamma} \end{pmatrix}$ has an approximate bivariate normal distribution. Since $\hat{\tau} = t_{(d)}\hat{\lambda}$ where $t_{(d)}$ is fixed, it follows that the large sample distribution of $\begin{pmatrix} \hat{\tau} \\ \hat{\gamma} \end{pmatrix}$ is approximately a bivariate normal distribution with mean $\begin{pmatrix} \tau \\ \gamma \end{pmatrix}$ and variance–covariance matrix given by (4.4.29). Thus an approximate $(1-\alpha)100$ percent elliptical confidence region for $\begin{pmatrix} \tau \\ \gamma \end{pmatrix}$ is given by

$$
\left[\frac{n\hat{\gamma}}{\hat{\tau}^2} - (n-d)P(\hat{\gamma},\hat{\tau}) \right](\hat{\tau}-\tau)^2 + [n\Psi'(\hat{\gamma}) - (n-d)V'(\hat{\gamma},\hat{\tau})](\hat{\gamma}-\gamma)^2
$$

$$
- 2\left[\frac{n}{\hat{\tau}} + (n-d)L(\hat{\gamma},\hat{\tau}) \right](\hat{\tau}-\tau)(\hat{\gamma}-\gamma)
$$

$$
= \chi^2(1-\alpha;2), \qquad (4.4.30)
$$

where

$$
P(\hat{\gamma},\hat{\tau}) = \frac{(n-d)e^{-\hat{\tau}}\{(\hat{\gamma}+\hat{\tau}+1)J(\hat{\gamma},\hat{\tau}) - e^{-\hat{\tau}}\}}{\hat{\tau}^2 J^2(\hat{\gamma},\hat{\tau})}, \qquad (4.4.31)
$$

$$
L(\hat{\gamma},\hat{\tau}) = \frac{e^{-\hat{\tau}}V(\hat{\gamma},\hat{\tau})}{\hat{\tau}J(\hat{\gamma},\hat{\tau})}, \qquad (4.4.32)
$$

and where $\chi^2(1-\alpha;2)$ is the $(1-\alpha)100$ percentile of the chi-square distribution with two degrees of freedom.

Noting that $\hat{\tau} = t_{(d)}\hat{\lambda}$ and $\tau = t_{(d)}\lambda$, where $t_{(d)}$ is fixed, the joint confidence region for $\begin{pmatrix} \lambda \\ \gamma \end{pmatrix}$ is obtained by substituting $\hat{\lambda} = \hat{\tau}/t_{(d)}$ and $\lambda = \tau/t_{(d)}$ into

(4.4.30). Computationally the problem is quite difficult because computing the coefficients for the ellipse involves evaluating incomplete gamma, digamma, and trigamma functions. Conceptually, however, the reader should not find this procedure more difficult than the work required for a complete gamma sample.

Case 4: Weibull Censored

Let us again assume that the n patients on study during a clinical trial are singly censored; d of them die, and $n-d$ survive. In this case, the likelihood function of the ordered sample is given by (4.3.6). It then follows that

$$\frac{\partial^2 \log_e L}{\partial \lambda^2} = -\frac{d}{\lambda^2}, \tag{4.4.33}$$

$$\frac{\partial^2 \log_e L}{\partial \lambda \partial \gamma} = -\Sigma^* t_{(i)}^\gamma (\log_e t_{(i)}), \tag{4.4.34}$$

and

$$\frac{\partial^2 \log_e L}{\partial \gamma^2} = -\frac{d}{\gamma^2} - \lambda \Sigma^* t_{(i)}^\gamma (\log_e t_{(i)})^2, \tag{4.4.35}$$

using Cohen's [1965] star notation as defined in (4.3.11). Hence for large samples the variance–covariance matrix of the estimators $\hat{\lambda}$ and $\hat{\gamma}$ is given approximately by inverting the matrix of second partial derivatives. Thus we obtain the approximate values $\mathrm{Var}\,\hat{\lambda}$, $\mathrm{Var}\,\hat{\gamma}$, and $\mathrm{Cov}(\hat{\lambda},\hat{\gamma})$ as elements of the matrix

$$\begin{bmatrix} \mathrm{Var}\,\hat{\lambda} & \mathrm{Cov}(\hat{\lambda},\hat{\gamma}) \\ \mathrm{Cov}(\hat{\lambda},\hat{\gamma}) & \mathrm{Var}\,\hat{\gamma} \end{bmatrix} \doteq \begin{bmatrix} \dfrac{d}{\hat{\lambda}^2} & \Sigma^* t_{(i)}^{\hat{\gamma}} \log_e t_{(i)} \\ \Sigma^* t_{(i)}^{\hat{\gamma}} \log_e t_{(i)} & \dfrac{d}{\hat{\gamma}^2} + \hat{\lambda}\Sigma^* t_{(i)}^{\hat{\gamma}} (\log_e t_{(i)})^2 \end{bmatrix}^{-1}. \tag{4.4.36}$$

Again as for the censored gamma, Harter and Moore [1967] showed that the vector estimator $\begin{pmatrix} \hat{\lambda} \\ \hat{\gamma} \end{pmatrix}$ has an approximate bivariate normal distribution in large samples. Thus $\begin{pmatrix} \hat{\lambda} \\ \hat{\gamma} \end{pmatrix}$ has an approximate bivariate normal distribution with mean vector $\begin{pmatrix} \lambda \\ \gamma \end{pmatrix}$ and variance–covariance matrix given by (4.4.36). Thus an approximate $(1-\alpha)100$ percent elliptical confidence region for $\begin{pmatrix} \lambda \\ \gamma \end{pmatrix}$ is given by

$$\frac{d}{\hat{\lambda}^2}(\hat{\lambda}-\lambda)^2 + \left[\frac{d}{\hat{\gamma}^2} + \Sigma^* t_{(i)}^{\hat{\gamma}}(\log_e t_{(i)}^{\hat{\gamma}})^2\right](\hat{\gamma}-\gamma)^2$$

$$+ (2\Sigma^* t_{(i)}^{\hat{\gamma}}\log_e t_{(i)})(\hat{\lambda}-\lambda)(\hat{\gamma}-\gamma) = \chi^2(1-\alpha;2). \quad (4.4.37)$$

A similar development of a joint elliptical confidence region can be obtained for the progressively censored case.

Example 4.4.3.   Cohen [1965] analyzes the following ordered sample from a Weibull distribution: 0.001, 0.030, 0.071, 0.185, 0.345, 0.435, 0.469, 0.470, 0.505, 0.664, 0.806, 0.970, 1.033, 1.550, 1.550, 2.046$^+$, 3.532, 7.057, 9.098$^+$, 57.628. For an application to clinical trials, assume that these observations represent the length in months of patient remission times on patients treated for Hodgkin's disease. These survival times are thus progressively censored. It is further assumed that the patients with remission times 2.046 and 9.098 months, denoted by plus signs, were followed only to the completion of the study and were in remission when the study ended. First we obtain the maximum likelihood estimates $\hat{\lambda}$ and $\hat{\gamma}$ of the parameters λ and γ, respectively. Using (4.3.12) we find that $\hat{\gamma}$ is the solution of

$$\frac{\displaystyle\sum_{i=1}^{18} t_{(i)}^{\hat{\gamma}}\log_e t_{(i)} + \sum_{i=1}^{2} T_{(i)}^{\hat{\gamma}}\log_e T_{(i)}}{\displaystyle\sum_{i=1}^{18} t_{(i)}^{\hat{\gamma}} + \sum_{i=1}^{2} T_{(i)}^{\hat{\gamma}}} - \frac{1}{\hat{\gamma}} + 0.624 = 0, \quad (4.4.38)$$

where, numerically, $1/18\sum_{i=1}^{18}\log_e t_{(i)} = -0.624$. For the uncensored case Cohen found $\hat{\gamma}=0.506$. Setting $G(\gamma)$ defined in Exercise 3, Section 4.3, equal to the left-hand side of (4.4.38), it can be shown numerically that $G(0.4) = -0.703$ and $G(0.5)=0.166$. Thus $\hat{\gamma}$ must lie between 0.4 and 0.5. As Cohen indicates, linear interpolation can be used to find $\hat{\gamma}$. Table 4.3 demonstrates the procedure.

**Table 4.3 Linear
interpolation for $\hat{\gamma}$**

$\hat{\gamma}$	$G(\hat{\gamma})$
0.4	−0.703
$\hat{\gamma} = 0.481$	0.000
0.5	0.166

The maximum likelihood estimate $\hat{\lambda}$ is then the solution to

$$\hat{\lambda} = \frac{18}{\left(\sum_{i=1}^{18} t\hat{\gamma}_{(i)} + \sum_{i=1}^{2} T\hat{\gamma}_{(i)} \right)} = \frac{18}{26.245} = 0.686. \qquad (4.4.39)$$

Next we see that the asymptotic variance–covariance matrix for the estimator $\begin{pmatrix} \hat{\lambda} \\ \hat{\gamma} \end{pmatrix}$ is

$$\begin{bmatrix} \text{Var}\,\hat{\lambda} & \text{Cov}(\hat{\lambda},\hat{\gamma}) \\ \text{Cov}(\hat{\lambda},\hat{\gamma}) & \text{Var}\,\hat{\gamma} \end{bmatrix} = \begin{bmatrix} 38.249 & 38.715 \\ 38.715 & 182.538 \end{bmatrix}^{-1}$$

$$= \begin{bmatrix} 0.033 & -0.007 \\ -0.007 & 0.007 \end{bmatrix}, \qquad (4.4.40)$$

where

$$\frac{d}{\hat{\lambda}^2} = 38.249,$$

$$\sum_{i=1}^{18} t\hat{\gamma}_{(i)} \log_e t_{(i)} + \sum_{i=1}^{2} T\hat{\gamma}_{(i)} \log_e T_{(i)} = 38.715,$$

and

$$\frac{d}{\hat{\gamma}^2} + \hat{\lambda}\left[\sum_{i=1}^{18} t\hat{\gamma}_{(i)} \log_e t_{(i)} + \sum_{i=1}^{2} T\hat{\gamma}_{(i)} \log_e T_{(i)} \right] = 182.538.$$

The approximate distribution for the vector estimator $\begin{pmatrix} \hat{\lambda} \\ \hat{\gamma} \end{pmatrix}$ is the

bivariate normal distribution with mean $\begin{pmatrix} \lambda \\ \gamma \end{pmatrix}$ and variance–covariance

matrix(4.4.40). It follows that a 95 percent joint elliptical confidence

region for the parameter vector $\begin{pmatrix} \lambda \\ \gamma \end{pmatrix}$ is given by

$$38.249(\lambda - 0.686)^2 + 182.538(\gamma - 0.481)^2$$
$$+ 79.430(\lambda - 0.686)(\gamma - 0.481) = 5.99. \qquad (4.4.41)$$

An elliptical region can be drawn from (4.4.41) analogous to Figure 4.7.

EXERCISES

1. Derive a $(1 - \alpha)100$ percent elliptical confidence region for the Weibull parameter vector $\binom{\lambda}{\gamma}$. Using this region obtain a 95 percent elliptical confidence region for $\binom{\lambda}{\gamma}$ from the data in Example 4.2.2. Sketch the region on graph paper. What are the separate 95 percent confidence intervals for λ and γ?

2. Suppose the observations 2.046 and 9.098 in Example 4.4.3 are not censored. Using Cohen's method (see Exercise 3 in Section 4.3) for form $G(\gamma)$, show that $G(0.5) = -0.0432$ and $G(0.6) = 0.6501$. Hence by linear interpolation show that $\hat{\gamma} = 0.506$. Hence obtain $\hat{\lambda}$, as well as the approximate variance–covariance matrix, using (4.4.20). Obtain a 95 percent elliptical confidence region for $\binom{\lambda}{\gamma}$ and sketch this region on graph paper. What are the separate 95 percent confidence intervals for λ and γ?

3. For the data in Exercise 2 find, approximately, $\hat{\mu}$, $\text{Var}\,\hat{\mu}$, and an approximate 95 percent confidence interval for μ the mean remission time of patients with Hodgkin's disease.

4. Show that $E[t^\gamma(\log_e t)^2] = (\lambda\gamma^2)^{-1}[\Psi'(2) + [\Psi(2)]^2 - 2\log_e\lambda\Psi(2) + (\log_e\lambda)^2]$.

5. Suppose $\hat{\mu}$ is the estimated mean survival time based on the maximum likelihood estimators $\hat{\lambda}$ and $\hat{\gamma}$ of a two-parameter survival distribution. Show that approximately, $\text{Var}\,\hat{\mu} \doteq (\partial\mu/\partial\lambda)^2\text{Var}\,\hat{\lambda} + 2(\partial\mu/\partial\lambda)(\partial\mu/\partial\gamma)\text{Cov}(\hat{\lambda},\hat{\gamma}) + (\partial\mu/\partial\gamma)^2\text{Var}\,\hat{\gamma}$. *Hint*: $\mu = f(\gamma,\lambda)$ and $\hat{\mu} = f(\hat{\gamma},\hat{\lambda})$. Expand $\hat{\mu}$ in a two-dimensional Taylor series about μ, ignoring terms of higher than the first order.

4.5 TRUNCATED GAMMA AND WEIBULL DEATH DENSITY FUNCTIONS

In this section we again use the method of maximum likelihood. Here we are concerned with estimating the parameters of truncated gamma and

Weibull survival distributions. In the next two sections we discuss alternate procedures (alternate to maximum likelihood) for estimating parameters of gamma and Weibull survival distributions. The reader may omit Section 4.5 without any loss of continuity.

A truncated death density function differs from a censored one in that the total sample size N of patients on study is unknown, and if n deaths are recorded prior to some time T, these deaths represent the first n ordered observations from an unknown number of ordered observations. Thus if $f(t;\gamma,\lambda)$ is the death density of a patient who is followed until he dies, the corresponding truncated death density of this patient assuming he dies prior to time T is

$$g(t;\gamma,\lambda) = \frac{f(t;\gamma,\lambda)}{F(T;\gamma,\lambda)}, \tag{4.5.1}$$

where

$$F(T;\gamma,\lambda) = \int_0^T f(t;\gamma,\lambda)dt, \qquad 0 \leqslant t \leqslant T. \tag{4.5.2}$$

Suppose we record from hospital records the survival times of n individuals with a serious illness such as stroke. However, the total number of patients with the illness is not known to us, since some persons do not enter a hospital. We assume that the gamma death density function for the ith patient is

$$g(t_i;\gamma,\lambda) = \frac{t_i^{\gamma-1}\exp(-\lambda t_i)}{F(T;\gamma,\lambda)}, \qquad 0 \leqslant t_i \leqslant T, \tag{4.5.3}$$

where

$$F(T;\gamma,\lambda) = \int_0^T t_i^{\gamma-1}\exp(-\lambda t_i)dt_i, \qquad i=1,2,\ldots,n, \tag{4.5.4}$$

and T is at least as large as the maximum observed survival time.

Estimating the parameters γ and λ is accomplished by the method of maximum likelihood. Thus

$$L(\gamma,\lambda) = \frac{\left(\prod_{i=1}^{n} t_i^{\gamma-1}\right)\exp\left(-\lambda \sum_{i=1}^{n} t_i\right)}{F^n(T;\gamma,\lambda)}, \tag{4.5.5}$$

where

$$F^n(T;\gamma,\lambda) \equiv \{F(T;\gamma,\lambda)\}^n. \tag{4.5.6}$$

It then follows that the values $\hat{\gamma}$ and $\hat{\lambda}$ that are the solutions to

$$\left.\frac{\partial \log_e L(\gamma,\lambda)}{\partial \gamma}\right|_{\substack{\gamma=\hat{\gamma} \\ \lambda=\hat{\lambda}}} = 0$$

and

$$\left.\frac{\partial \log_e L(\gamma,\lambda)}{\partial \lambda}\right|_{\substack{\gamma=\hat{\gamma} \\ \lambda=\hat{\lambda}}} = 0$$

are the requisite maximum likelihood estimators of γ and λ, respectively. We find that

$$\frac{\partial \log_e L}{\partial \gamma} = \sum_{i=1}^{n} \log_e t_i - \frac{nJ'(\gamma,\lambda,T)}{J(\gamma,\lambda,T)} \tag{4.5.7}$$

and

$$\frac{\partial \log_e L}{\partial \lambda} = \sum_{i=1}^{n} t_i - \frac{nT^\gamma \exp(-\lambda T)}{\lambda J(\gamma,\lambda,T)} + \frac{n\gamma}{\lambda}, \tag{4.5.8}$$

where

$$J(\gamma,\lambda,T) = \int_0^T t^{\gamma-1} \exp(-\lambda t)\,dt$$

and

$$J'(\gamma,\lambda,T) = \frac{\partial J(\gamma,\lambda,T)}{\partial \gamma} = \int_0^T t^{\gamma-1} \log_e t \exp(-\lambda t)\,dt.$$

If the death density function for the ith patient is of the Weibull rather than the gamma form, we write

$$g(t_i;\gamma,\lambda) = \frac{\lambda \gamma t_i^{\gamma-1} \exp(-\lambda t_i^\gamma)}{1 - \exp(-\lambda T^\gamma)}, \qquad 0 \leqslant t_i \leqslant T; i = 1,2,\ldots,n. \tag{4.5.9}$$

The maximum likelihood estimators $\hat{\lambda}$ and $\hat{\gamma}$ of λ and γ, respectively, are the solutions to

$$\frac{n}{\hat{\lambda}} - \sum_{i=1}^{n} t_i^{\hat{\gamma}} - \frac{nT^{\hat{\gamma}} \exp(-\lambda T^{\hat{\gamma}})}{1 - \exp(-\hat{\lambda} T^{\hat{\gamma}})} = 0 \tag{4.5.10}$$

and

$$\frac{n}{\hat{\gamma}} + \sum_{i=1}^{n} \log_e t_i - \hat{\lambda} \sum_{i=1}^{n} t_i^{\hat{\gamma}} \log_e t_i - \frac{n\hat{\lambda} T^{\hat{\gamma}} \log_e T \exp(-\hat{\lambda} T^{\hat{\gamma}})}{1 - \exp(-\hat{\lambda} T^{\hat{\gamma}})} = 0. \quad (4.5.11)$$

An important special case of both the truncated gamma and Weibull death densities is the truncated exponential density; we have here

$$g(t;\lambda) = \begin{cases} \dfrac{\lambda \exp(-\lambda t)}{1 - \exp(-\lambda T)}, & 0 \leqslant t \leqslant T \\ 0, & \text{elsewhere} \end{cases} . \quad (4.5.12)$$

If $t_1, t_2, \ldots, t_n$ is a random sample of times to death from a population whose death density function is (4.5.12), the maximum likelihood estimator $\hat{\lambda}$ of λ is obtained by solving

$$\frac{1}{\hat{\lambda}} - \frac{T}{\exp(\hat{\lambda} T) - 1} - \bar{t} = 0 \quad (4.5.13)$$

in terms of $\hat{\lambda}$. It is not difficult to show that the function

$$\tau(\lambda) = \frac{1}{\lambda} - \frac{T}{\exp(\lambda T) - 1} - \bar{t} \quad (4.5.14)$$

is a decreasing function in λ, hence crosses the origin just once. Hence that value $\hat{\lambda}$ for which $\tau(\hat{\lambda}) = 0$ is the maximum likelihood estimator of λ.

Example 4.5.1. Suppose we are interested in recording the life span of a particular insect whose total population size in a given region is unknown to us. We decide to sample from this population for 1000 days, noting that an insect will appear in our sample only when it dies. Deaths are recorded on 100 insects and are grouped into 10 intervals as shown in Table 4.4. In Table 4.4 $\bar{t} = 251.5$ days. To obtain $\hat{\lambda}$, we use the Newton-Raphson procedure discussed in Chapter 6, noting that

$$\tau'(\lambda) = \frac{-1}{\lambda^2} + \frac{T^2 \exp(\lambda T)}{[\exp(\lambda T) - 1]^2}. \quad (4.5.15)$$

As a starting value $\hat{\lambda}_0$, we choose $\hat{\lambda}_0 = \bar{t}^{-1} = 0.00398$. We thus find, after three iterations, $\hat{\lambda} = 0.00358$. Since the approximate variance of $\hat{\lambda}$ is $-[\tau'(\hat{\lambda})]^{-1}$, where $\tau'(\lambda)$ is given by (4.5.15), an approximate 95 percent confidence interval for λ is $0.00354 < \lambda < 0.00362$.

Table 4.4

Class (days)	Frequency	Cumulative Frequency
$0 \leqslant t < 100$	30	30
$100 \leqslant t < 200$	21	51
$200 \leqslant t < 300$	18	69
$300 \leqslant t < 400$	10	79
$400 \leqslant t < 500$	7	86
$500 \leqslant t < 600$	5	91
$600 \leqslant t < 700$	5	96
$700 \leqslant t < 800$	2	98
$800 \leqslant t < 900$	1	99
$900 \leqslant t < 1000$	1	100

EXERCISES

1. Show that $\tau(\lambda)$ given by (4.5.14) is a decreasing function of λ.
2. Suppose that t is a random survival time whose death density function is a truncated Weibull given by (4.5.9). Show that

$$E(t^\gamma) = \frac{1}{\lambda} - \frac{T^\gamma}{\exp(\lambda T^\gamma) - 1}.$$

3. Let $g(t; \lambda)$ be the truncated exponential death density function (4.5.12). Show that

$$\lim_{\lambda \to 0} g(t; \lambda) = \begin{cases} \dfrac{1}{T}, & 0 \leqslant t \leqslant T \\ 0, & \text{elsewhere.} \end{cases},$$

4. Compute the mean and variance of the truncated exponential death density (4.5.12). Using these results, find an approximate 95 percent confidence interval for the mean time to death from the data in Example 4.5.1.

4.6 NONMAXIMUM LIKELIHOOD METHODS OF ESTIMATING PARAMETERS OF GAMMA AND WEIBULL DISTRIBUTIONS

Although the primary focus in this book has been estimation of the parameters of survival distributions by the method of maximum likelihood, in this section we examine a few other methods of parameter estimation with special reference to the gamma and Weibull distributions.

Mann, Schafer, and Singpurwalla [1974] discuss a number of techniques for estimating Weibull distribution parameters. Their material is particularly useful if one wishes to estimate the parameters from a three-parameter Weibull. The procedures they describe, however, are generally useful for estimating Weibull distribution parameters.

Let $t_1, t_2, \ldots, t_n$ be independent survival times of n individuals with a gamma death density function

$$f(t; \lambda, \gamma) = \frac{\lambda^\gamma t^{\gamma-1} \exp(-\lambda t)}{\Gamma(\gamma)}, \qquad (4.6.1)$$

with scale parameter $\lambda > 0$ and shape parameter $\gamma > 0$. The noncensored maximum likelihood estimators of λ and γ were obtained in Section 4.2. The censored case was treated in Section 4.4.

A simple alternative approach to the estimation of λ and γ for the noncensored case is the method of moments. Recalling that $E(t)$ and $\mathrm{Var}(t)$ for the gamma density are γ/λ and γ/λ^2, respectively [see (4.1.5) and (4.1.6)], the method of moments estimators $\tilde{\lambda}$ and $\tilde{\gamma}$ of λ and γ, respectively, are the unique solution to

$$\frac{\tilde{\gamma}}{\tilde{\lambda}} = \bar{t} \qquad (4.6.2)$$

and

$$\frac{\tilde{\gamma}}{\tilde{\lambda}^2} = s_t^2, \qquad (4.6.3)$$

where $\bar{t} = \sum_{i=1}^{n} t_i / n$ and $s_t^2 = (\sum_{i=1}^{n} t_i^2 - n\bar{t}^2)/(n-1)$.

Thus

$$\tilde{\lambda} = \frac{\bar{t}}{s_t^2} \qquad (4.6.4)$$

and

$$\tilde{\gamma} = \frac{\bar{t}^2}{s_t^2}. \qquad (4.6.5)$$

It is well known, however, that unlike the maximum likelihood estimators $\hat{\lambda}$ and $\hat{\gamma}$, $\tilde{\lambda}$ and $\tilde{\gamma}$ are not asymptotically efficient estimators of λ and γ. That is, the variances of $\tilde{\lambda}$ and $\tilde{\gamma}$ are not as small as possible (i.e., not as small as the variances of $\hat{\lambda}$ and $\hat{\gamma}$) in large samples. (See Cramér [1946] regarding this point.) Nevertheless, because of the simplicity in obtaining $\tilde{\lambda}$ and $\tilde{\gamma}$, the method is not entirely without merit.

Wilk, Gnanadesikan, and Huyett [1962b] describe a graphical method to estimate λ for the gamma death density function for a known value of γ. Although analytically the maximum likelihood estimator $\hat{\lambda}$ is easily obtainable for γ known, we discuss briefly this alternative graphical procedure, which allows for the estimation of an unknown location parameter as well. We can rewrite (4.6.1) to allow for a location parameter ξ. Thus we have

$$f(t;\lambda,\gamma,\xi) = \frac{\lambda^{\gamma}(t-\xi)^{\gamma-1}\exp[-\lambda(t-\xi)]}{\Gamma(\gamma)},$$

$$\gamma>0, \lambda>0, -\infty<\xi<\infty, t \geqslant \xi. \qquad (4.6.6)$$

In survival studies as well as other reliability studies, we usually have $\xi > 0$, referring to this as the guaranteed life of the individual.

The distribution function of t is

$$F(t;\lambda,\gamma,\xi) = \int_{\xi}^{t} f(t';\lambda,\gamma,\xi)dt', \qquad (4.6.7)$$

where $f(t;\lambda,\gamma,\xi)$ is given by (4.6.6). Let $t_{(1)} \leqslant t_{(2)} \leqslant \cdots \leqslant t_{(n)}$ be an ordered sample of n survival times whose death density function is (4.6.6). Following Wilk, Gnanadesikan, and Huyett, let $b_1, b_2, \ldots, b_n$ be n suitably chosen percentage points of the gamma distribution. Furthermore, let $\tilde{t}_{(i)}$, satisfy

$$f\left(\tilde{t}_{(i)};\lambda,\gamma,\xi\right) = b_i, \qquad i = 1,2,\ldots,n. \qquad (4.6.8)$$

It then follows that if $t_{(1)}, \ldots, t_{(n)}$ is an ordered sample of survival times whose death density is the gamma given by (4.6.6), the points $(\tilde{t}_{(i)}, t_{(i)})$, $i = 1,2,\ldots,n$, will tend to lie along a straight line with unit slope and zero intercept, provided $b_1, \ldots, b_n$ denote appropriately chosen percentage points corresponding to the observed order statistics. The choice is $b_i = (i - \frac{1}{2})/n$, $i = 1,2,\ldots,n$, which is precisely the midpoint of the interval $((i-1)/n, i/n)$. This choice avoids the cumbersome values 0 and 1.

To obtain estimates of λ and ξ, we make the linear transformation $x = \lambda(t - \xi)$. The distribution function of x is then

$$F(x;0,1,\gamma) = \frac{1}{\Gamma(\gamma)} \int_0^x Z^{\gamma-1}e^{-Z}dZ. \qquad (4.6.9)$$

Let $\tilde{x}_{(i)}$ satisfy

$$F(\tilde{x}_{(i)}; 0, 1, \gamma) = \frac{i - \frac{1}{2}}{n}, \qquad i = 1, 2, \ldots, n. \qquad (4.6.10)$$

Because of the linear transformation from t to x, the points $(\tilde{x}_{(i)}, t_{(i)})$ tend to lie along a straight line with slope λ^{-1} and intercept ξ. Hence the estimators λ' and ξ' of λ and ξ, respectively, are obtained by fitting a straight line through the points $(\tilde{x}_{(i)}, t_{(i)})$, $i = 1, 2, \ldots, n$.

Wilk, Gnanadesikan, and Huyett describe the numerical procedure for obtaining the quantile values $\tilde{x}_{(i)}$, $i = 1, 2, \ldots, n$, as functions of γ and n. That is, they describe the numerical procedure for solving (4.6.10). Tables of quantiles of a wide range of values of γ are provided. Once the quantile values have been determined, λ and ξ are obtained from the straight line plot through the points $(\tilde{x}_{(i)}, t_{(i)})$, $i = 1, 2, \ldots, n$.

Example 4.6.1. Consider the survival times of rats exposed to gamma radiation in Example 4.2.1. Ordering these times (in weeks) we have: 40, 62, 69, 77, 83, 88, 94, 101, 109, 115, 123, 125, 128, 136, 137, 152, 152, 153, 160, and 165. Since the maximum likelihood estimate for γ is $\hat{\gamma} = 8.53$, linear interpolation is used between the η columns $\eta = 8.5$ and $\eta = 9.0$ to obtain the quantile values corresponding to percentages in Table 3 in the Appendix, which are taken from Wilk, Gnanadesikan, and Huyett [1962b]. For percentages not tabled, we used bilinear interpolation to obtain the requisite quantile values as in Example 4.3.1. For example, for $n = 20$ we find, corresponding to $(1 - \frac{1}{2})/20 = 0.025$, the quantile value 3.78 under $\eta = 8.5$ and 4.12 under $\eta = 9.0$. Thus by linear interpolation the quantile value corresponding to $\eta = 8.53$ is 3.80. The quantile value corresponding to $(3 - \frac{1}{2})/20 = 0.125$ must be found using bilinear interpolation. That is, we must interpolate bilinearly between the values given in Table 4.5 to obtain the value corresponding to $(0.125, 8.53)$. Thus the requisite value q (say) is given by

$$q = (5.04)(0.47) + (5.56)(0.47) + (5.43)(0.03) + (5.97)(0.03) = 5.32.$$

The complete set of 20 quantile values is then: 3.80, 4.75, 5.32, 5.80, 6.22, 6.61, 6.96, 7.32, 7.66, 8.02, 8.37, 8.75, 9.14, 9.57, 10.02, 10.54, 11.17, 11.95, 13.02, 15.14. These values are graphed in Figure 4.8.

If we estimate λ and ξ by ordinary least squares methods,* we find as

---

*Since the observations are ordered, a more efficient least squares procedure would involve weighted least squares. However, the computational difficulties overshadow the usefulness of the procedure here.

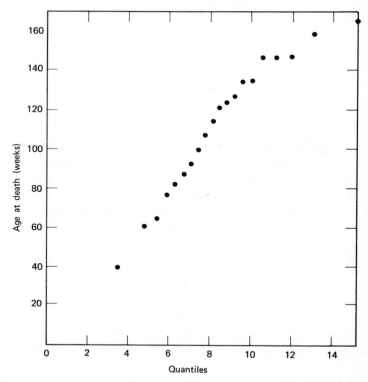

Figure 4.8 Gamma probability plot for ages at death of rats exposed to radiation (data from Example 4.2.1).

our least squares estimates $\lambda'^{-1} = 11.877$ and $\xi' = 12.420$. Thus $\lambda' = 0.084$ for these data. In Example 4.2.1 we assumed $\xi = 0$ and found $\hat{\lambda} = 0.075$. The mean death times, however, are almost identical: the mean time to death using the graphical method is 113.97 weeks, compared with 113.45 weeks using maximum likelihood. [Note: $\mu' = \xi' + \hat{\gamma}/\lambda'$ for the graphical procedure.]

Table 4.5

η	0.100	0.150
8.5	$5.04(q_{11})$	$5.56(q_{12})$
9.0	$5.43(q_{21})$	$5.97(q_{22})$

We assume now that $t_1, t_2, \ldots, t_n$ are independent survival times of n individuals for which the common death density function is the Weibull death density function. That is, we write

$$f(t; \lambda, \gamma) = \lambda \gamma t^{\gamma-1} \exp(-\lambda t^\gamma), \qquad \lambda > 0, \gamma > 0, t > 0. \qquad (4.6.11)$$

Let $u(t, \lambda, \gamma)$ be a function of t, λ, and γ, such that
  1. The density function of u is independent of λ and γ,
  2. The ordering of the transformed u values is determined from the ordering of the t values.
Bain and Antle [1967] then suggest estimators λ^* and γ^* of λ and γ, respectively, which jointly maximize the agreement between $u_{(i)}$ and $E(u_{(i)})$, where $u_{(i)}$ is the ith largest value of $u(t_i; \lambda, \gamma)$ in the sample. As the problem is now stated, it is not certain how we should maximize the agreement between $u_{(i)}$ and $E(u_{(i)})$, as well as the choice of the function $u(t; \lambda, \gamma)$. To ease the ambiguity, we use the least squares method to maximize this agreement. Thus λ^* and γ^* are chosen so that

$$\sum_{i=1}^{n} \left\{ \tau(u_{(i)}) - \tau[E(u_{(i)})] \right\}^2 \qquad \text{is a minimum,}$$

where τ is a monotone function of both $u_{(i)}$ and $E(u_{(i)})$. We consider two choices of $u(t; \lambda, \gamma)$ which yield simple estimators λ^* and γ^*.

Choice 1

$$u(t; \lambda, \gamma) = \lambda t^\gamma. \qquad (4.6.12)$$

This choice of u yields the density function

$$f(u) = e^{-u}. \qquad (4.6.13)$$

Ordering the transformed sample values $u_{(1)} \leqslant u_{(2)} \leqslant \cdots \leqslant u_{(n)}$, it is not difficult to show that

$$E(u_{(i)}) \equiv \xi_i = \sum_{j=1}^{i} \frac{1}{n-j+1}. \qquad (4.6.14)$$

To obtain closed form estimators λ^* and γ^*, we choose τ as the natural logarithm function. Thus we minimize the function

$$\tau_1(\lambda, \gamma) = \sum_{i=1}^{n} \left(\log_e \lambda t_{(i)}^\gamma - \log_e \xi_i \right)^2 \qquad (4.6.15)$$

with respect to λ and γ. The minimizing values are

$$\lambda^* = \left[\frac{\prod_{i=1}^{n} \xi_i}{\prod_{i=1}^{n} t_{(i)}^*} \right]^{1/n} \tag{4.6.16}$$

and

$$\gamma^* = \frac{\sum_{i=1}^{n} \log_e \xi_i \log_e t_{(i)} - \dfrac{\sum_{i=1}^{n} \log_e \xi_i \sum_{i=1}^{n} \log_e t_{(i)}}{n}}{\sum_{i=1}^{n} (\log_e t_{(i)})^2 - \dfrac{\left[\sum_{i=1}^{n} \log_e t_{(i)} \right]^2}{n}}. \tag{4.6.17}$$

Choice 2

$$u(t; \lambda, \gamma) = 1 - \exp(\lambda t^\gamma). \tag{4.6.18}$$

For this function u is uniformly distributed on the interval $0 \leqslant u \leqslant 1$. Ordering the values $u_{(1)} \leqslant u_{(2)} \leqslant \cdots \leqslant u_{(n)}$ it follows that

$$E(u_{(i)}) \equiv \eta_i = \frac{i}{n+1}, \qquad i = 1, 2, \ldots, n.$$

The closed form estimators λ^{**} and γ^{**} are obtained by choosing $\tau(x) = \log_e \log_e(1 - x)$ for $x = u$ and $E(u)$, and minimizing the function

$$\tau_2(\lambda, \gamma) = \sum_{i=1}^{n} \left(\log_e \log_e(1 - u_{(i)}) - \log_e \log_e(1 - \eta_i) \right)^2 \tag{4.6.19}$$

with respect to λ and γ. Hence

$$\lambda^{**} = \prod_{i=1}^{n} \left[\log_e \left(\frac{n+1}{n+1-i} \right) t_{(i)}^{**} \right]^{1/n}, \tag{4.6.20}$$

and

$$\gamma^{**} = \frac{\displaystyle\sum_{i=1}^{n} \log_e \log_e\left(\frac{n+1}{n+1-i}\right)\log_e t_{(i)} - \frac{\displaystyle\sum_{i=1}^{n} \log_e \log_e\left(\frac{n+1}{n+1-i}\right)\sum_{i=1}^{n} \log_e t_{(i)}}{n}}{\displaystyle\sum_{i=1}^{n} (\log_e t_{(i)})^2 - \frac{\left[\displaystyle\sum_{i=1}^{n} \log_e t_{(i)}\right]^2}{n}}.$$

$$(4.6.21)$$

It should be noted that the estimators λ^*, γ^*, λ^{**}, and γ^{**} are easily modified to the case of censored failure times. We leave the details of this modification to the reader.

Other estimators of λ and γ have been studied by Menon [1963], Miller and Freund [1965], and Gumbel [1958]. Bain and Antle [1967] perform Monte Carlo simulations to compare the biases and the variance among four of the five estimators that are potential substitutes for the maximum likelihood estimators of the Weibull distribution parameters. These estimators, which are compared in samples of sizes 5, 10, 20, 25, and 30, are the Bain-Antle choice 1 estimators, the Menon estimators, the Miller and Freund estimators, and the Gumbel estimators. The conclusion of the simulation was that all these estimators are quite good and the differences among them are small. Menon's estimators improved as the sample size increased. Menon's estimators, however, cannot be used in the presence of censoring.

Table 4.6, a comparison of the variances of the Menon estimator and the maximum likelihood estimator of γ, demonstrates that comparing the variances of the Menon and maximum likelihood estimators for a variety of sample sizes favors the maximum likelihood estimator. Furthermore, since the Menon estimator and the other three estimators appear to have similar properties, the maximum likelihood estimator of γ, hence λ, appears to be superior to the others, especially in larger samples.

Example 4.6.2   A Weibull death density function was fitted to the data on patients receiving an analgesic for headache relief (Example 4.2.2) for which $\hat{\lambda} = 0.121$ and $\hat{\gamma} = 2.79$ are the maximum likelihood estimates of λ and γ, respectively. As a comparison, we compute λ^*, γ^*, λ^{**}, and γ^{**}. All three sets of estimators are summarized in Table 4.7.

Table 4.6 Variance of Menon's and the maximum likelihood (ML) estimators of γ the Weibull shape parameter[1]

Sample Size	Menon Estimator	Maximum Likelihood Estimator
5	0.334	0.320
6	0.236	0.215
8	0.147	0.124
10	0.108	0.087
12	0.086	0.067
14	0.073	0.055
16	0.063	0.047
18	0.056	0.041
20	0.050	0.036
25	0.040	0.028
30	0.034	0.023
35	0.029	0.020
40	0.026	0.017
45	0.023	0.015
50	0.021	0.014
60	0.017	0.011
70	0.015	0.010
80	0.013	0.008
100	0.011	0.006
120	0.009	0.005

[1]Source: Thoman, Bain, and Antle [1969].

Table 4.7 Maximum likelihood and Bain-Antle estimates of Weibull parameters for data of Example 4.2.2

Estimates	λ	γ
$(\hat{\lambda}, \hat{\gamma})$	0.121	2.79
(λ^*, γ^*)	0.081	3.44
$(\lambda^{**}, \gamma^{**})$	0.083	3.34

Note that the Bain and Antle estimates differ to some degree from the maximum likelihood estimates. This is probably because the maximum likelihood estimates are biased in small samples, whereas a Monte Carlo simulation by Bain and Antle [1967] demonstrates that their estimators are not as biased in small samples. Finally, μ^* and μ^{**}, the estimated mean relief times (using λ^*, γ^*, and λ^{**}, γ^{**}, respectively, as estimates for λ and γ) are 1.87 and 1.89 hours. These values agree very well with $\hat{\mu} = 1.89$ hours calculated in Example 4.2.2.

Recently Gross et al. [1975] further demonstrated, by means of a Monte Carlo simulation, that the Bain-Antle estimates are less biased in small samples than the corresponding maximum likelihood estimates. However, the large sample properties of the maximum likelihood estimators still make them quite attractive for sample sizes of 50 or more. Table 4.8 shows one of the simulations by Gross and colleagues.

Table 4.8 Estimated biases and root mean squared errors (RMS) for the three estimators for 1000 random samples of sizes 10 and 100 generated from a Weibull density function where $\gamma = 5.0$ and $\lambda = 1.0$

n	$\hat{\lambda}$	$\hat{\gamma}$	λ^*	γ^*	λ^{**}	γ^{**}
$n = 10$						
Bias	0.110	0.883	0.107	−0.420	−0.003	−0.590
RMS	0.472	1.934	0.379	1.589	0.313	1.581
$n = 100$						
Bias	0.009	0.072	0.004	−0.219	−0.010	−0.247
RMS	0.110	0.409	0.105	0.546	0.100	0.554

White [1969] reports a graphical procedure for estimating λ and γ when the data have a Weibull density function. His method, which we describe, is a slight modification and generalization of the graphical methods in Section 4.2. It allows for singly censored observations.

Suppose $t_{(1)} \leqslant t_{(2)} \leqslant \cdots \leqslant t_{(d)}$ are the ordered survival times recorded for the d out of n individuals who die in a clinical trial; the remaining $n - d$ individuals are singly censored. Suppose further that the common death density function for all n patients on study is given by the Weibull death density (4.6.11). Thus $S(t)$, the survival probability at time t, is

$$S(t) = \exp(-\lambda t^{\gamma}), \qquad \lambda > 0, \gamma > 0, t > 0. \tag{4.6.22}$$

If we take the natural logarithm of (4.6.22) twice and rearrange terms, we have the regression equation

$$y = a + bx, \tag{4.6.23}$$

where $y = \log_e t$, $x = \log_e(-\log_e S(t))$, $a = -\log_e \lambda / \gamma$, and $b = \gamma^{-1}$. [Note that (4.6.23) differs from (4.2.19) in that the roles of x and y are interchanged.] Values of y are obtained directly from the data (e.g., $y_1 = \log_e t_{(1)}$, etc.). The choice of x is somewhat more complicated. In Section 4.2 x_i was obtained using an estimated value $\hat{S}(t_{(i)}) = (n+1-i)/(n+1)$, where $t_{(i)}$ is the value of the ith order statistic, $i = 1, 2, \ldots, d$. White chooses to estimate x_i by $E(X_{(i)})$, where $X_{(i)}$ is the ith order statistic, and X is distributed as a reduced log-Weibull variable. Tables of $E(X_{(i)})$ are provided for samples up to size $n = 20$. As White and other authors have pointed out, the choice of x_i, $i = 1, 2, \ldots, d$ is an unresolved problem.

Estimates of λ and γ are obtained by estimating a and b using regression techniques. It is important to indicate that the variables $y_1, \ldots, y_n$ do not have equal variance, nor is the covariance between pairs y_i and y_j $(i \neq j)$ zero. Thus the regression estimators $\hat{a}$ and $\hat{b}$ using usual least squares techniques are not the best linear unbiased estimators (BLUE) of a and b, respectively. They are, however, unbiased, although they do not have minimum variance with respect to all linear unbiased estimators of a and b, respectively.

The simple least squares estimators $\hat{a}$ and $\hat{b}$ are given by

$$\hat{a} = \bar{y} - \hat{b}\bar{x} \tag{4.6.24}$$

and

$$\hat{b} = \frac{\displaystyle\sum_{i=1}^{d} (x_i - \bar{x})(y_i - \bar{y})}{\displaystyle\sum_{i=1}^{d} (x_i - \bar{x})^2}, \tag{4.6.25}$$

where $\bar{y} = \displaystyle\sum_{i=1}^{d} y_i / d$ and $\bar{x} = \displaystyle\sum_{i=1}^{d} x_i / d$.

White [1969] also discusses weighted least squares estimators for a and b. Although the weighted least squares estimators produce an equal variance for the y values, they do not eliminate the nonzero covariance problem.

Example 4.6.3.   For the data in Example 4.2.2 a least squares plot is given in Figure 4.9. Table 1 in White [1969] for $n = 20$ provides the x values, and the natural logarithms of the ordered observations are the y values. Using (4.6.24) and (4.6.25), we find $\hat{a} = 0.727$ and $\hat{b} = 0.239$. Letting $\gamma^{***} = \hat{b}^{-1}$ and $\lambda^{***} = \exp(-\hat{a}/\hat{b})$, $\gamma^{***} = 4.189$ and $\lambda^{***} = 0.048$. γ^{***} and λ^{***} provide us with yet another set of estimates of γ and λ for the

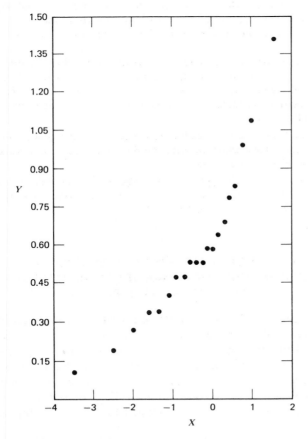

Figure 4.9   A least squares Weibull plot for data in Example 4.2.2: $Y = \log_e t$, $X = E(X_{(i)})$.

data in Example 4.2.2. Finally, we note that $\mu^{***} = 1.88$ hours, and μ^{***} is the estimated mean relief time using γ^{***} and λ^{***} as estimates for γ and λ, respectively. Thus μ^{***} compares very favorably with $\hat{\mu}$, μ^*, and μ^{**}.

EXERCISES

1. Modify the graphical procedure for estimating λ and ξ in Example 4.6.1, if the last three ordered observations are censored assuming $\gamma = 8.5$. For the censored case describe a general procedure for estimating λ and ξ of a three-parameter gamma death density function when γ is assumed known.

2. Modify the Bain and Antle methods for estimating parameters of the Weibull death density function when the survival data are singly censored. Apply this modification to the data in Example 4.3.2 to estimate λ and γ and compare these estimates with the maximum likelihood estimates $\hat{\lambda} = 7.828 \times 10^{-14}$, $\hat{\gamma} = 4.41$.

3. Suppose $t_1, \ldots, t_n$ are n independent survival times whose common death density is $f(t) = e^{-t}$, $t > 0$. Show that if $t_{(1)} \leqslant \cdots \leqslant t_{(n)}$ is the corresponding ordered sample, and if $y_i = (n - i + 1)(t_{(i)} - t_{(i-1)})$, then $y_1, \ldots, y_n$ are independent, each having the same death density as $t_1, \ldots, t_n$. Hence, or otherwise, derive (4.6.14).

4. Apply White's least squares procedure to the data in Example 4.3.2 to estimate λ and γ. Compare these estimates with the Bain and Antle estimates as well as the maximum likelihood estimates.

5. Suppose $x_1, x_2, \ldots, x_n$ is a sample of size n whose common density function is $f(x) = 1, 0 < x < 1$. Let $x_{(1)} \leqslant x_{(2)} \leqslant \cdots \leqslant x_{(n)}$ be the corresponding ordered sample. Prove that $E(x_{(i)}) = i/(n + 1)$, $i = 1, 2, \ldots, n$.

4.7 GRAPHICAL METHODS OF ESTIMATION OF SURVIVAL DISTRIBUTION PARAMETERS BY MEANS OF CUMULATIVE HAZARD PLOTS

For the special cases of the exponential and Weibull survival distributions, graphical methods of estimating the population parameters from the cumulative hazard function are practical (see Section 3.6 for the exponential distribution). In this section we treat the problem of graphical estimation in a slightly more general way for the Weibull survival distributions. The gamma distribution may be included as a special case, but its actual use is quite cumbersome. The procedure developed by Nelson [1972] allows for censoring.

Let $\Lambda(t)$ be the cumulative hazard function corresponding to a death density function $f(t)$ with survival distribution function $S(t)$. That is,

$\Lambda(t) = \int_0^t \lambda(x)\,dx$. With this definition and the use of (1.2.8), it follows that

$$\Lambda(t) = -\log_e S(t). \qquad (4.7.1)$$

For the exponential distribution with hazard rate $\lambda > 0$, it follows that

$$\Lambda(t) = \lambda t. \qquad (4.7.2)$$

This case was discussed in Section 3.6. For the Weibull distribution with scale parameter $\lambda > 0$ and shape parameter $\gamma > 0$,

$$\Lambda(t) = \lambda t^\gamma. \qquad (4.7.3)$$

If we take the logarithm of both sides of (4.7.3), we have

$$\log_e \Lambda(t) = \log_e \lambda + \gamma \log_e t. \qquad (4.7.3')$$

Hence we can use log-log paper to determine whether the data follow a Weibull distribution, by checking to see whether the cumulative hazard function lies on a straight line. For the gamma distribution, $\Lambda(t)$ is somewhat more complicated. First, for the gamma survival distribution with scale and shape parameters $\lambda > 0$ and $\gamma > 0$, respectively,

$$S(t) = \int_t^\infty \frac{\lambda^\gamma x^{\gamma-1} \exp(-\lambda x)\,dx}{\Gamma(\gamma)}, \qquad (4.7.4)$$

which cannot be simplified unless γ is an integer. Thus

$$\Lambda(t) = \log_e \left[\int_t^\infty \frac{\lambda^\gamma x^{\gamma-1} \exp(-\lambda x)\,dx}{\Gamma(\gamma)} \right]. \qquad (4.7.5)$$

Other specific values $\Lambda(t)$ can be obtained for survival distributions discussed in Chapter 1. Nelson obtains the relationships between the cumulative hazard functions and time for the normal and lognormal survival distributions. We do not discuss these here. The use of $\Lambda(t)$ provides one more technique to add to those presented in Chapter 2 in our attempt to identify the appropriate distribution. However, we must first discuss a method for determining $\Lambda(t)$ empirically, given a sample of times to death and/or times to censoring.

Suppose n individuals are placed on a clinical trial for which some patients may still be alive at the end of the trial. Let us denote the ordered observation times of the n patients as $t_{(1)} \leqslant t_{(2)} \leqslant \cdots \leqslant t_{(n)}$, where an observation time is the time to death if a patient dies while on study or his total time on study if he survives beyond the length of the clinical trial. Let d be the number of patients who die while on study and $t_{r(1)} \leqslant \cdots \leqslant t_{r(d)}$ their corresponding times to death. Nelson defines $r(1) > r(2) > \cdots > r(d)$ as the reverse ranks. As an illustration, from the survival data in Example 4.3.2 we see that $t_{(1)} = 500$, $t_{(2)} = 700^+$, $t_{(3)} = 800$, $t_{(4)} = 900$, and $t_{(5)} = 1000^+$, where 700^+ and 1000^+ are the times on study of the two surviving patients. Thus $n = 5$, $d = 3$, $t_{r(1)} = 500$, $t_{r(2)} = 800$, and $t_{r(3)} = 900$, and $r(1) = 5$, $r(2) = 3$, and $r(3) = 2$.

Nelson indicates that a satisfactory choice of plotting positions for the values $\Lambda(t_{r(m)})$ are the values $[r(1)]^{-1} + \cdots + [r(m)]^{-1}$, $m = 1, 2, \ldots, d$. These values are selected because if $t_{(1)} \leqslant t_{(2)} \leqslant \cdots \leqslant t_{(n)}$ is a complete sample with a continuous survival distribution function, the differences $\Lambda(t_{(i)}) - \Lambda(t_{(i-1)})$, $i = 1, 2, \ldots, n$, $\Lambda(t_{(0)}) \equiv 0$, are statistically independent and exponentially distributed, where $E(\Lambda(t_{(i)}) - \Lambda(t_{(i-1)})) = 1/(n - i + 1)$. Hence

$$E(\Lambda(t_{(i)})) = n^{-1} + (n-1)^{-1} + \cdots + (n-i+1)^{-1}, \qquad i = 1, 2, \ldots, n. \quad (4.7.6)$$

A similar result holds if the number of patients who die on study is fixed. In the case, however, where d (the number of patients who die during the clinical trial) is random because the length of time of the clinical trial is fixed, the choice of plotting positions using this criterion is not exact, but satisfactory. Finally, the graphical method of estimation entails plotting on the x axis of the requisite graph paper the value $[r(1)]^{-1} + \cdots + [r(m)]^{-1}$, and on the y axis the value $t_{r(m)}$, $m = 1, 2, \ldots, d$. That is, only those points are plotted for which a death has occurred; however, the censored observations are accounted for in the determination of the reverse ranks $r(1), r(2), \ldots, r(d)$. To apply the procedure outlined, it is necessary to specify the survival distribution. Summarizing the steps of this procedure, we stress the following points:

1. Choose the hazard paper for the requisite theoretical survival distribution.

2. On the chosen hazard paper, plot each time to death vertically against its corresponding cumulative hazard value—that is, $[r(1)]^{-1} + \cdots + [r(m)]^{-1}$—on the horizontal axis. This provides a plot of the sample cumulative hazard function.

3. If the plot of the sample times to failure reasonably lies on a straight line, the distribution fits the data adequately. Then by eye we can fit a straight line through these points.

4. Although it is helpful and indeed useful to have appropriate hazard paper, we can use ordinary arithmetic graph paper if we consider the appropriate transformation on the data.

Note the similarity between cumulative hazard plotting and the Kaplan-Meier product limit estimate, discussed in Section 2.3.

Example 4.7.1. Gehan [1971] considers the following survival data, which are ordered remission times of leukemia patients who have received 6 MP (a mercaptopurine used in the treatment of leukemia) to maintain their previously induced remission.* The data first appear in Freireich et al. [1963]. The ordered remission times in weeks are:6, 6, 6, 6$^+$, 7, 9$^+$, 10, 10$^+$, 11$^+$, 13, 16, 17$^+$, 19$^+$, 20$^+$, 22, 23, 25$^+$, 32$^+$, 32$^+$, 34$^+$, 35$^+$. A plus value indicates a censored survival time. Table 4.9 is a synopsis of the hazard method calculations.

**Table 4.9 Hazard rate method calculations
for leukemia patients**

Reverse Rank	Remission Time	Hazard (percent)	Cumulative Hazard (percent)
21	6	4.76	4.76
20	6	5.00	9.76
19	6	5.26	15.02
18	6$^+$	—	—
17	7	5.88	20.90
16	9$^+$	—	—
15	10	6.67	27.57
14	10$^+$	—	—
13	11$^+$	—	—
12	13	8.33	35.90
11	16	9.09	44.99
10	17$^+$	—	—
9	19$^+$	—	—
8	20$^+$	—	—
7	22	14.29	59.28
6	23	16.67	75.95
5	25$^+$	—	—
4	32$^+$	—	—
3	32$^+$	—	—
2	34$^+$	—	—
1	35$^+$	—	—

*Note that $100/21 = 4.76$, $100/20 = 5.00$, etc.

Gehan contends that these remission times are exponentially distributed. As a check, Figure 4.10 shows an exponential cumulative hazard plot of the data plotted on simple arithmetic graph paper with line of best fit through the origin. The result follows from (4.7.2). We find that the line intersects the point (35.9, 13); thus $\tilde{\mu} = 13/0.359 = 36.2$ weeks is the average remission time. If we compute $\hat{\mu}$, the maximum likelihood estimate, we find $\hat{\mu}$ given by (3.2.8) is 39.89 weeks. Thus Gehan's contention would seem to be justified. As with normal probability paper, principal attention should be paid to verify that the points in the central portion of the cumulative lie on a straight line.

The same data can be plotted on log-log cumulative hazard paper, if it is desired to see whether a Weibull distribution would be more appropriate. Since the fits are much the same (see Figure 4.11), it is simpler to work with the exponential distribution, as Gehan did.

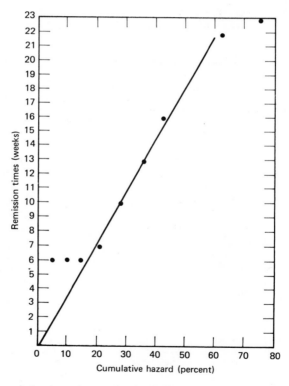

Figure 4.10 Cumulative hazard rate plot for leukemia patients assuming an exponential distribution.

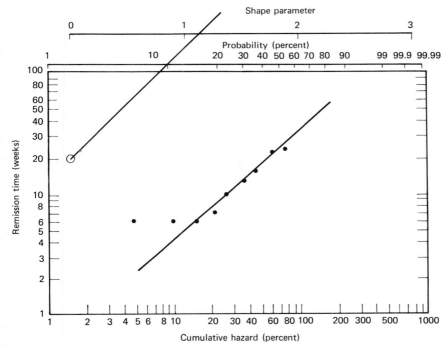

Figure 4.11 Cumulative hazard plot for leukemia patients assuming Weibull distribution.

Example 4.7.2. Consider the data on Hodgkin's disease remission times that we discussed in Example 4.4.3. These data were supposed to follow the Weibull distribution, and we use the cumulative hazard plot method to fit the Weibull survival distribution to them. Equation 4.7.3′ allows us two approaches in applying the cumulative hazard plot method. We can take logarithms of both the times to death and the observed cumulative hazard values, plotting these logarithms on ordinary arithmetic graph paper. Alternatively, we can plot the times to death and the cumulative hazard values on log-log graph paper, as we did for Gehan's data. The second approach is obviously easier if the reader has log-log graph paper (preferably two or three cycles). On the other hand, arithmetic graph paper is easy to obtain, and the student may perform a Weibull fit using ordinary graph paper.

Table 4.10 shows the computations necessary for making both arithmetic and log-log plots of the data. Figure 4.12 is a Weibull hazard plot of the

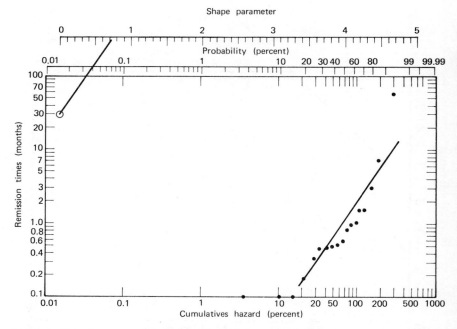

Figure 4.12   Hazard plot of Hodgkin's disease remission times, Weibull shape parameter.

data in Table 4.10. To estimate the shape parameter, we draw a straight line parallel to the fitted straight line so that it passes through the circled dot marked in the upper left-hand corner of the paper and through the shape parameter scale as in Figure 4.12. The graphical estimate of the shape parameter is 0.7, which is a bit larger than 0.481, the maximum likelihood estimate. Note that the fit of a Weibull distribution to these data is not good. In practice, the form of the distribution can be difficult to determine.

For examples of the cumulative hazard plot method involving the normal and lognormal survival distributions, we refer the reader to Nelson [1972].

EXERCISES

1. For the data in Example 4.7.2, plot the log cumulative hazard rates on the x axis and the log remission times on the y axis, using arithmetic graph paper. Hence obtain an estimate of γ the shape parameter.

2. For the data in Example 4.2.2, which is a complete sample of headache remission times, make a cumulative hazard plot to determine an appropriate distribution.

Table 4.10 Hazard rate method calculations for
Hodgkin's disease patients

Reverse Rank	Remission Time (months)	log Remission Time	Hazard (percent)	Cumulative* Hazard (percent)	log Cumulative Hazard
20	0.001	−6.91	5.00	5.00	1.61
19	0.030	−3.51	5.26	10.26	2.33
18	0.071	−2.65	5.56	15.82	2.76
17	0.185	−1.69	5.88	21.70	3.08
16	0.345	−1.06	6.25	27.95	3.33
15	0.435	−0.83	6.67	34.62	3.54
14	0.469	−0.76	7.14	41.76	3.73
13	0.470	−0.76	7.69	49.45	3.90
12	0.505	−0.68	8.33	57.78	4.06
11	0.664	−0.41	9.09	66.87	4.20
10	0.806	−0.22	10.00	76.87	4.34
9	0.970	−0.03	11.11	87.98	4.48
8	1.033	0.03	12.50	100.48	4.61
7	1.550	0.44	14.29	114.77	4.74
6	1.550	0.44	16.67	131.44	4.88
5	2.046+	0.72+			
4	3.532	1.26	25.00	156.44	5.05
3	7.057	1.95	33.33	189.77	5.25
2	9.098+	2.21+			
1	57.628	4.05	100.00	289.77	5.67

* The reader should note that the cumulative hazard may exceed 100 percent.

3. Suppose $t_{(1)} \leqslant t_{(2)} \leqslant \cdots \leqslant t_{(n)}$ is a complete ordered sample with a continuous survival distribution $S(t)$. Show that the differences $\Lambda(t_{(i)}) - \Lambda(t_{(i-1)})$ are statistically independent, $i = 1, 2, \ldots, n$, $\Lambda(t_{(0)}) \equiv 0$, with $E(\Lambda(t_{(i)}) - \Lambda(t_{(i-1)})) = 1/(n - i + 1)$.

4. For the data in Example 4.7.2, check whether an exponential distribution would be applicable by plotting the cumulative hazard rates on arithmetic paper.

5. Obtain the Kaplan-Meier product limit estimate $\hat{P}(t)$ for the data in Example 4.7.1 at the values $t = 6, 12, 18, 24, 36$.

4.8 ESTIMATING PARAMETERS FOR RAYLEIGH AND
GENERALIZED RAYLEIGH DISTRIBUTIONS

In many survival studies the experimenter plots his hazard rate as a function of time and discovers at least for some relatively long period that the plot is linear or can be transformed to a linear plot. That is, he observes $\lambda(t) = \lambda_0 + 2\lambda_1 t$ as his hazard rate. It turns out that a linear hazard rate with $\lambda_1 > 0$ is often a reasonable assumption; for example, the survival patterns of middle age groups demonstrate this phenomenon.

In this section we estimate the parameters of the linear hazard rate distribution, which is commonly called the Rayleigh distribution. We also treat the generalization of the linear hazard to a polynomial hazard rate. That is, we consider estimating the parameters of the death density function whose hazard rate is $\lambda(t) = \beta_1 + \beta_2 t + \cdots + \beta_m t^{m-1}$, where the degree of the polynomial m is less than the sample size n. In ordinary applications $m \leqslant 3$. We define death density functions with polynomial hazard rates as generalized Rayleigh distributions.

With the exception of the exponential distribution, the simplest form of the generalized Rayleigh death density function is defined as

$$f(t; \lambda_0, \lambda_1) = (\lambda_0 + 2\lambda_1 t)\exp[-(\lambda_0 t + \lambda_1 t^2)], \qquad t \geqslant 0, \text{ and } \lambda_0 \geqslant 0, \lambda_1 > 0.$$
(4.8.1)

The usual form of the Rayleigh density has $\lambda_0 = 0$. Using theory developed in Chapter 1, it is easy to show that $S(t)$ and $\lambda(t)$, the survival probability and hazard rate functions at time t, are given, respectively, as

$$S(t) = \exp[-(\lambda_0 t + \lambda_1 t^2)]$$
(4.8.2)

and

$$\lambda(t) = \lambda_0 + 2\lambda_1 t.$$
(4.8.3)

The hazard rate of the Rayleigh distribution is linear as a function of t. It is, therefore, known also as the linear hazard rate distribution, which was introduced by Kodlin [1967].

We note that since $f(t; \lambda_0, \lambda_1)$ is a density function,

$$\lambda_0 + 2\lambda_1 t > 0.$$
(4.8.4)

For (4.8.4) to hold, it is sufficient that $\lambda_0 > 0$ and $\lambda_1 \geqslant 0$. These conditions on λ_0 and λ_1 are noted in the definition of the Rayleigh death density function. We observe, further, that $\lambda_0 + 2\lambda_1 t$ is increasing if and only if $\lambda_1 > 0$.

Suppose we observe n patients on a study whose survival times are $t_1, t_2, \ldots, t_n$. That is, we follow all patients to death. The death density of the ith patient is given by (4.8.1). To estimate λ_0 and λ_1 we again use the method of maximum likelihood. Thus the likelihood function is

$$L(\lambda_0, \lambda_1) = \prod_{i=1}^{n} (\lambda_0 + 2\lambda_1 t_i)\exp\left[-\left(\sum_{i=1}^{n} \lambda_0 t_i + \sum_{i=1}^{n} \lambda_1 t_i^2\right)\right]. \quad (4.8.5)$$

It then follows that $\hat{\lambda}_0$ and $\hat{\lambda}_1$, the maximum likelihood estimators of λ_0 and λ_1, respectively, are the solutions to

$$\sum_{i=1}^{n} \left(\hat{\lambda}_0 + 2\hat{\lambda}_1 t_i \right)^{-1} - \sum_{i=1}^{n} t_i = 0 \qquad (4.8.6)$$

and

$$2 \sum_{i=1}^{n} t_i \left(\hat{\lambda}_0 + 2\hat{\lambda}_1 t_i \right)^{-1} - \sum_{i=1}^{n} t_i^2 = 0, \qquad (4.8.7)$$

which jointly maximize the likelihood function (4.8.5).

We can solve (4.8.6) and (4.8.7) simultaneously in terms of $\hat{\lambda}_0$ and $\hat{\lambda}_1$. Combining those equations by multiplying (4.8.6) by $\hat{\lambda}_0$ and (4.8.7) by $\hat{\lambda}_1$ and then adding these together, we obtain

$$\hat{\lambda}_0 \sum_{i=1}^{n} t_i + \hat{\lambda}_1 \sum_{i=1}^{n} t_i^2 = n. \qquad (4.8.8)$$

Substituting (4.8.8) for $\hat{\lambda}_0$ in (4.8.6), we have

$$\sum_{i=1}^{n} \frac{1}{\left(n - S^2 \hat{\lambda}_1 + m\hat{\lambda}_1 t_i \right)} = 1, \qquad (4.8.9)$$

where $S^2 = \sum_{i=1}^{n} t_i^2$ and $m = 2 \sum_{i=1}^{n} t_i$.

Finally, solution of (4.8.9) in terms of $\hat{\lambda}_1$ is obtained by trial and error or the Newton-Raphson method discussed in Chapter 6. At this point we caution the reader not to use $\hat{\lambda}_{10} = 0$ as a starting value because $\hat{\lambda}_1 = 0$ satisfies (4.8.9) trivially. Instead, a plot of the estimated hazard rate by life table methods discussed in Chapter 2 is recommended to estimate $\hat{\lambda}_{10}$.

Obtaining the theoretical large sample variance–covariance matrix is quite a difficult problem. Alternatively, for large samples

$$\begin{bmatrix} \text{Var } \hat{\lambda}_0 & \text{Cov}(\hat{\lambda}_0, \hat{\lambda}_1) \\ \text{Cov}(\hat{\lambda}_0, \hat{\lambda}_1) & \text{Var } \hat{\lambda}_1 \end{bmatrix} \doteq \begin{bmatrix} \dfrac{-\partial^2 \log_e L}{\partial \lambda_0^2} & \dfrac{-\partial^2 \log_e L}{\partial \lambda_0 \partial \lambda_1} \\ \dfrac{-\partial^2 \log_e L}{\partial \lambda_0 \partial \lambda_1} & \dfrac{-\partial^2 \log_e L}{\partial \lambda_1^2} \end{bmatrix}^{-1}_{\substack{\lambda_0 = \hat{\lambda}_0 \\ \lambda_1 = \hat{\lambda}_1}}$$

$$(4.8.10)$$

where

$$\frac{-\partial^2 \log_e L}{\partial \lambda_0^2} = \sum_{i=1}^{n} (\lambda_0 + 2\lambda_1 t_i)^{-2}, \qquad (4.8.11)$$

$$\frac{-\partial^2 \log_e L}{\partial \lambda_0 \partial \lambda_1} = 2 \sum_{i=1}^{n} t_i (\lambda_0 + 2\lambda_1 t_i)^{-2}, \qquad (4.8.12)$$

and

$$\frac{-\partial^2 \log_e L}{\partial \lambda_1^2} = 4 \sum_{i=1}^{n} t_i^2 (\lambda_0 + 2\lambda_1 t_i)^{-2}, \qquad (4.8.13)$$

let

$$a_\nu = 2^\nu \sum_{i=1}^{n} t_i^\nu (\hat{\lambda}_0 + 2\hat{\lambda}_1 t_i)^{-2}, \qquad \nu = 0, 1, 2. \qquad (4.8.14)$$

An approximate $(1 - \alpha)$ 100 percent elliptical confidence region for λ_0 and λ_1 is given by

$$a_0 (\hat{\lambda}_0 - \lambda_0)^2 + a_2 (\hat{\lambda}_1 - \lambda_1)^2 + 2a_1 (\hat{\lambda}_0 - \lambda_0)(\hat{\lambda}_1 - \lambda_1) = \chi^2(1 - \alpha; 2), \quad (4.8.15)$$

where $\chi^2(1 - \alpha; 2)$ is the $(1 - \alpha)$100th percentage point of the chi-square distribution with 2 degrees of freedom.

Example 4.8.1.   From a leukemia study described by Gehan [1970], the remission times of 21 patients who receive placebo (a simulated treatment corresponding to no medication) appear in Table 4.11.
To obtain the maximum likelihood estimates $\hat{\lambda}_0$ and $\hat{\lambda}_1$, we used (4.8.8) and (4.8.9). Equation 4.8.9 involved using the Newton-Raphson procedure, discussed in detail in Chapter 6. In this example we merely sketch the procedure. The initial estimate $\hat{\lambda}_{10}$ is obtained by setting $\hat{\lambda}_{10} = n/S^2$, where $n = 21$ and $S^2 = \sum_{i=1}^{21} t_i^2 = 2414$ for the data in Table 4.14. Thus $\hat{\lambda}_{10} = 0.0087$. The ith iteration $\hat{\lambda}_{1i}$, $i = 1, 2, \ldots$, is then given by

$$\hat{\lambda}_{1i} = \hat{\lambda}_{1,i-1} - \frac{L(\hat{\lambda}_{1,i-1})}{L'(\hat{\lambda}_{1,i-1})}, \qquad (4.8.16)$$

where

$$L(\hat{\lambda}_{1,i-1}) = \sum_{i=1}^{21} \frac{1}{(21 - 2414\hat{\lambda}_{1,i-1} + 364\hat{\lambda}_{1,i-1}t_i)} - 1 \qquad (4.8.17)$$

Table 4.11 Leukemia remission times (weeks)

Midpoint	Frequency
1	2
2	2
3	1
4	2
5	2 ·
8	4
11	2
12	2
15	1
17	1
22	1
23	1

and

$$L'(\hat{\lambda}_{1,i-1}) = \sum_{i=1}^{21} \frac{(2414 - 364t_i)}{(21 - 2414\hat{\lambda}_{1,i-1} + 364\hat{\lambda}_{1,i-1}t_i)^2}, \qquad (4.8.18)$$

noting that $2m = 2\sum_{i=1}^{21} t_i = 364$. After the fifth iteration this process converges to $\hat{\lambda}_1$, which equals 0.0039. The intermediate values were found to be $\hat{\lambda}_{11} = 0.0074$, $\hat{\lambda}_{12} = 0.0059$, $\hat{\lambda}_{13} = 0.0046$, $\hat{\lambda}_{14} = 0.0040$. Using (4.8.8) it then follows that $\hat{\lambda}_0 = 0.0637$. Thus the maximum likelihood estimate of the linear hazard rate for the remission times of these leukemia patients is

$$\lambda(t) = 0.0637 + 0.0078t. \qquad (4.8.19)$$

Finally, we obtain a 95 percent joint confidence region for the vector parameter $\begin{pmatrix} \lambda_0 \\ \lambda_1 \end{pmatrix}$. The values for a_0, a_1, and a_2 using (4.8.14) are 1,766.1965; 17,825.9241; and 327,175.1024; respectively. Thus an approximate 95 percent elliptical confidence region for $\begin{pmatrix} \lambda_0 \\ \lambda_1 \end{pmatrix}$ is given by

$$1.766.20(\lambda_0 - 0.0637)^2 + 323{,}175.10(\lambda_1 - 0.0039)^2$$
$$+ 35{,}651.85(\lambda_0 - 0.0637)(\lambda_1 - 0.0039) = 5.99. \qquad (4.8.20)$$

The separate 95 percent confidence intervals for λ_0 and λ_1 are $0 < \lambda_0 < 0.1332$, and $0 \leqslant \lambda_1 < 0.0090$. The lower bound of zero for both intervals follows by the way in which the Rayleigh death density is defined. The confidence interval for λ_1 includes zero, verifying Gehan's [1970, p. 24] observation that these remission times have an exponential death density.

If the observations are censored to ensure that only the first d deaths are recorded, $n - d$ patients surviving, the likelihood function (4.8.5) becomes

$$L(\lambda_0, \lambda_1) = \prod_d (\lambda_0 + 2\lambda_1 t_i) \exp[-(\lambda_0 t_i + \lambda_1 t_i^2)] \prod_s \exp[-(\lambda_0 T_i + \lambda_1 T_i^2)],$$

(4.8.21)

where $\prod_d$ is the product over the d patients who die on study, $\prod_s$ is the product over the $s = n - d$ patients surviving, t_i are the times to death, and T_i are times surviving patients are observed. The maximum likelihood estimators $\hat{\lambda}_0$ and $\hat{\lambda}_1$ are obtained by solving simultaneously

$$\sum_d (\hat{\lambda}_0 + 2\hat{\lambda}_1 t_i)^{-1} - \left(\sum_d t_i + \sum_s T_i\right) = 0$$

(4.8.22)

and

$$2\sum_d t_i (\hat{\lambda}_0 + 2\hat{\lambda}_1 t_i)^{-1} - \left(\sum_d t_i^2 + \sum_s T_i^2\right) = 0$$

(4.8.23)

in terms of $\hat{\lambda}_0$ and $\hat{\lambda}_1$. Note that (4.8.22) and (4.8.23) are obtained by finding $\hat{\lambda}_0$ and $\hat{\lambda}_1$ such that the derivatives of $\log_e L(\lambda_0, \lambda_1)$ with respect to λ_0 and λ_1 are equal to zero.

Using Kodlin's procedure, we again note, as in the complete sample case, that for large samples

$$
\begin{bmatrix} \text{Var}\,\hat{\lambda}_0 & \text{Cov}(\hat{\lambda}_0, \hat{\lambda}_1) \\ \text{Cov}(\hat{\lambda}_0, \hat{\lambda}_1) & \text{Var}\,\hat{\lambda}_1 \end{bmatrix} \doteq \begin{bmatrix} \dfrac{-\partial^2 \log_e L}{\partial \lambda_0^2} & \dfrac{-\partial^2 \log_e L}{\partial \lambda_0 \partial \lambda_1} \\ \dfrac{-\partial^2 \log_e L}{\partial \lambda_0 \partial \lambda_1} & \dfrac{-\partial^2 \log_e L}{\partial \lambda_1^2} \end{bmatrix}_{\substack{\lambda_0 = \hat{\lambda}_0 \\ \lambda_1 = \hat{\lambda}_1}}^{-1}
$$

(4.8.24)

where

$$\frac{-\partial^2 \log_e L}{\partial \lambda_0^2} = \sum_d (\lambda_0 + 2\lambda_1 t_i)^{-2}, \tag{4.8.25}$$

$$\frac{-\partial^2 \log_e L}{\partial \lambda_0 \partial \lambda_1} = 2 \sum_d t_i (\lambda_0 + 2\lambda_1 t_i)^{-2}, \tag{4.8.26}$$

and

$$\frac{-\partial^2 \log_e L}{\partial \lambda_1^2} = 4 \sum_d t_i^2 (\lambda_0 + 2\lambda_1 t_i)^{-2}. \tag{4.8.27}$$

To solve (4.8.22) and (4.8.23) simultaneously in terms of $\hat{\lambda}_0$ and $\hat{\lambda}_1$, we use the following iterative process, which is equivalent to a two-dimensional Newton-Raphson procedure. Let $\hat{\lambda}_{00}$ and $\hat{\lambda}_{10}$ be initial estimates of $\hat{\lambda}_0$ and $\hat{\lambda}_1$, respectively. Let $\delta \hat{\lambda}_{0i} = (\hat{\lambda}_{0i} - \hat{\lambda}_{0,i-1})$ and $\delta \hat{\lambda}_{1i} = (\hat{\lambda}_{1i} - \hat{\lambda}_{1,i-1})$, $i = 1, 2, \ldots,$. Obtain the adjustments $\delta \hat{\lambda}_{0i}$ and $\delta \hat{\lambda}_{1i}$ by solving simultaneously

$$a_{3,i-1} \delta \hat{\lambda}_0 + a_{4,i-1} \delta \hat{\lambda}_{1i} = -a_{1,i-1} \tag{4.8.28}$$

and

$$a_{4,i-1} \delta \hat{\lambda}_{0i} + a_{5,i-1} \delta \hat{\lambda}_{1i} = -a_{2,i-1}, \qquad i = 1, 2, \ldots, \tag{4.8.29}$$

where the partial derivatives are utilized in

$$a_{1,i-1} = \sum_d \left(\hat{\lambda}_{0,i-1} + 2 \hat{\lambda}_{1,i-1} t_j \right)^{-1} - \left(\sum_d t_j + \sum_s T_j \right), \tag{4.8.30}$$

$$a_{2,i-1} = 2 \sum_d t_j \left(\hat{\lambda}_{0,i-1} + 2 \hat{\lambda}_{1,i-1} t_j \right)^{-1} - \left(\sum_d t_j^2 + \sum_s T_j^2 \right), \tag{4.8.31}$$

$$a_{3,i-1} = -\sum_d \left(\hat{\lambda}_{0,i-1} + 2 \hat{\lambda}_{1,i-1} t_j \right)^{-2}, \tag{4.8.32}$$

$$a_{4,i-1} = -2 \sum_d t_j \left(\hat{\lambda}_{0,i-1} + 2 \hat{\lambda}_{1,i-1} t_j \right)^{-2}, \tag{4.8.33}$$

and

$$a_{5,i-1} = -4 \sum_d t_j^2 \left(\hat{\lambda}_{0,i-1} + 2 \hat{\lambda}_{1,i-1} t_j \right)^{-2}. \tag{4.8.34}$$

This process continues until the adjustments are less than some predetermined value—for example, $|\delta\hat{\lambda}_{ni}| < 10^{-4}$, $n = 0, 1$.

To obtain initial estimates $\hat{\lambda}_{00}$ and $\hat{\lambda}_{10}$, Kodlin suggests plotting the hazard rate $\lambda(t)$ against t. If the theoretical hazard rate is linear, the plot of $\lambda(t)$ versus t should be approximately a straight line, particularly if censoring is not extensive. The intercept and slope of this line can then be taken as $\hat{\lambda}_{00}$ and $2\hat{\lambda}_{10}$, respectively.

Example 4.8.2.   Boag [1949] presents survival time information on 121 patients with breast cancer. Kodlin [1967] considers the problem of estimating the parameters of the linear hazard rate survival distribution for those patients who die from causes *other* than breast cancer. The data are presented in Table 4.12. Using life table methods as discussed in Chapter 2, estimates of the hazard rate, $\lambda(t)$, are obtained for each interval representing a year's survival experience. Our estimates differ slightly from Kodlin's estimates in that the number at risk in the ith interval is given by $n_i = n_i' - \frac{1}{2}(w_i + l_i)$, whereas Kodlin uses n_i', the number under observation at the beginning of the ith interval.

**Table 4.12   Survival time information on 121
breast cancer patients**

Year	Number Under Observation at Beginning of Year n_i'	Number Dying from Other Causes d_i	Number Dying of Cancer + Number Lost Alive $w_i + l_i$	Estimate of $\lambda(t)$	Average
0–1	121	3	13+0	$3/114.5 = 0.03$	
1–2	105	2	25+0	$2/92.5 = 0.02$	0.02
2–3	78	0	9+0	$0/73.5 = 0$	
3–4	69	2	11+0	$2/63.5 = 0.03$	
4–5	56	1	6+0	$1/53.0 = 0.02$	
5–6	49	2	0+0	$2/49.0 = 0.04$	0.03
6–7	47	1	3+0	$1/45.5 = 0.02$	
7–8	43	2	4+0	$2/41.1 = 0.05$	
8–9	37	0	3+0	$0/35.5 = 0$	
9–10	34	3	1+6	$3/30.5 = 0.10$	
10–11	24	1	2+4	$1/21.0 = 0.05$	0.04
11–12	17	0	0+6	$0/14.0 = 0$	
12–13	11	1	0+1	$1/10.5 = 0.09$	

Inspection of the last two columns of Table 4.12 indicates that a linear hazard rate is reasonable. Using the average values in the last column of Table 4.12, the initial estimates of $\hat{\lambda}_0$ and $\hat{\lambda}_1$, the maximum likelihood estimates of λ_0 and λ_1, respectively, are $\hat{\lambda}_{00} = 0.02$ and $\hat{\lambda}_{10} = 0.002$. This follows, since the graphical estimate $\hat{\lambda}(t) = \hat{\lambda}_{00} + 2\hat{\lambda}_{10}t$ from Table 4.14 yields intercept value 0.02 and slope value 0.001. Using $\hat{\lambda}_{00} = 0.02$ and $\hat{\lambda}_{10} = 0.001$ as initial values in (4.5.7) and (4.5.8), we find after the first iteration, $\hat{\lambda}_{01} = 0.0163$ and $\hat{\lambda}_{11} = 0.00135$. After the second and final iteration we find $\hat{\lambda}_0 = 0.0166$ and $\hat{\lambda}_1 = 0.00135$ to be the requisite maximum likelihood estimates. Thus the maximum likelihood estimate of the linear hazard rate is

$$\hat{\lambda}(t) = 0.0166 + 0.0027t. \tag{4.8.35}$$

A confidence region for the parameter vector $\begin{pmatrix} \lambda_0 \\ \lambda_1 \end{pmatrix}$ can be constructed as was done in the complete sample case. This is left as an exercise for the reader in the censored sample case.

A generalization of Kodlin's linear hazard rate is described by Gehan and Siddiqui [1973], who provide regression procedures for estimating parameters of survival distributions whose hazard rates or logarithmic transformations are linear functions of the distribution parameters. The authors assume the availability of a large sample of survival times, which are grouped into intervals according to life table format. (See Table 2.2, Chapter 2.) Estimates of the hazard function for each interval (except the last) are obtained by methods described in Chapter 2. A linear model is then assumed, relating the estimated hazard rate to the theoretical hazard rate. Least squares (regression) techniques are then used to estimate the parameters. The survival times may or may not be censored. We refer the interested reader to Gehan and Siddiqui [1973] for the details of the procedure.

Krane [1963] also provides a general regression procedure for estimating the parameters of a survival distribution. In describing this method briefly, let us assume that $\Lambda(t)$, where $\lambda(t) \equiv d\Lambda(t)/dt$ is the hazard rate, can be expressed in the form of a polynomial of degree m; that is,

$$\Lambda(t) = \beta_1 t + \beta_2 t^2 + \cdots + \beta_m t^m. \tag{4.8.36}$$

Since $S(t) = \exp(-\Lambda(t))$ is the survival function, we can write $\Lambda(t) = -\log_e S(t)$. If we can group the survival data into $k + 1$ intervals in the form of a life table, letting s_i be the proportion of individuals who survive at the end of the ith interval, then $y_i = -\log_e s_i$ is the observed value at t_i corresponding to the theoretical value $\Lambda(t_i)$. Thus by (4.8.36) we have, for large samples,

$$y_i = \beta_1 t_i + \beta_2 t_i^2 + \cdots + \beta_m t_i^m + e_i, \qquad i = 1, 2, \ldots, k, \qquad (4.8.37)$$

where e_i is a random variable with zero mean and $k > m$.*

To obtain estimates $\tilde{\beta}_1, \tilde{\beta}_2, \ldots, \tilde{\beta}_m$ of $\beta_1, \beta_2, \ldots, \beta_m$, the method of least squares is used. However, weighted least squares must be applied because the random variables $e_1, e_2, \ldots, e_k$ do not have equal variances but are uncorrelated. Letting $\tilde{\beta}' = (\tilde{\beta}_1, \tilde{\beta}_2, \ldots, \tilde{\beta}_m)$ be the row vector of the requisite estimators, it can be shown that $\tilde{\beta}'$ has the form

$$\tilde{\beta}' = y' W T' (TWT')^{-1} \qquad (4.8.38)$$

where $y' = (y_1, y_2, \ldots, y_k)$ is the row vector of observations and W is a $k \times k$ matrix whose inverse W^{-1} is written

$$W^{-1} = \begin{bmatrix} a_1 & -b_1 & 0 & \cdots & 0 & 0 \\ -b_1 & a_2 & -b_2 & \cdots & 0 & 0 \\ 0 & -b_2 & a_3 & \cdots & 0 & 0 \\ \cdot & \cdot & \cdot & \cdots & \cdot & \cdot \\ \cdot & \cdot & \cdot & \cdots & \cdot & \cdot \\ \cdot & \cdot & \cdot & \cdots & \cdot & \cdot \\ 0 & 0 & 0 & \cdots & a_{k-1} & -b_{k-1} \\ 0 & 0 & 0 & \cdots & -b_{k-1} & a_k \end{bmatrix} . \qquad (4.8.39)$$

The elements of the matrix (4.8.39) are given by

$$b_i = \frac{s_i s_{i+1}}{s_i - s_{i+1}}, \qquad i = 0, \ldots, k, \qquad (4.8.40)$$

$$a_i = b_{i-1} + b_i, \qquad i = 1, \ldots, k,$$

*For (4.8.37), the interval $t_k \leqslant t < \infty$ is the $(k+1)$th interval in which $s_{k+1} = 0$.

and $s_0 = 1$ and $s_{k+1} = 0$. In addition, T' is the $k \times m$ matrix whose transpose T is given by

$$
T = \begin{bmatrix}
t_1 & t_2 & \cdots & t_k \\
t_1^2 & t_2^2 & \cdots & t_k^2 \\
\cdot & \cdot & \cdots & \cdot \\
\cdot & \cdot & \cdots & \cdot \\
\cdot & \cdot & \cdots & \cdot \\
t_1^m & t_2^m & \cdots & t_k^m
\end{bmatrix}, \tag{4.8.41}
$$

where t_i^j is the jth power of the ith end point, $i = 1, \ldots, k$; $j = 1, 2, \ldots, m$.

A simple test for the fit of the overall regression is a test of the hypothesis $H_0 : \beta_1 = \beta_2 = \cdots = \beta_m = 0$. First the total sum of squares can be separated into the regression sum of squares and error sum of squares, as in Table 4.13.

Table 4.13 Analysis of variance table for regression

Source of Variation	Degrees of Freedom	Sum of Squares	Mean Squares
Regression	m	$\mathbf{y}' W T' \tilde{\boldsymbol{\beta}}$	$\mathbf{y}' W T' \tilde{\boldsymbol{\beta}}/m$
Error	$k - m$	$(\mathbf{y}' - \tilde{\boldsymbol{\beta}}'T) W \mathbf{y}$	$(\mathbf{y}' - \tilde{\boldsymbol{\beta}}'T) W \mathbf{y}/(k - m)$
Total	k	$\mathbf{y}' W \mathbf{y}$	—

Second, from Table 4.13 the decision rule for testing H_0 at level α is to reject H_0 only if $[(k - m)/m][\mathbf{y}' W T' \tilde{\boldsymbol{\beta}}/(\mathbf{y}' - \tilde{\boldsymbol{\beta}}'T) W \mathbf{y}]$ exceeds $F(1 - \alpha; m, k - m)$, where $F(1 - \alpha; m, k - m)$ is the $(1 - \alpha)100$ percentage point of the F-ratio distribution for m degrees of freedom for the numerator and $k - m$ degrees of freedom for the denominator. Finally, as in an ordinary multiple regression, the regression sum of squares can be partitioned into its component parts, allowing a degree of freedom for each power of t from the values 1 to m.

In conclusion it should be emphasized that Krane's procedure for fitting a polynomial hazard rate to survival data differs from the method proposed by Gehan and Siddiqui [1973] in that Gehan and Siddiqui fit a linear hazard rate or the logarithm of a linear hazard rate to survival data. Actually, the two methods are quite similar in application. The difference lies in the choice of a model. Often a plot of the empirical hazard rate indicates the appropriate method.

Example 4.8.3.   We consider the following fictitious adaptation of the data, which were analyzed by Krane. Patients who have a history of chronic hypertension show in a particular way the survival pattern in Table 4.14.

It was decided to fit $\Lambda(t)$ to a polynomial of degree 3. That is, it was thought that an adequate description of survival would be attained if $\Lambda(t) \underset{\sim}{=} \beta_1 t + \beta_2 t^2 + \beta_3 t^3$. Now the weighted least squares estimators $\tilde{\beta}_1, \tilde{\beta}_2$, and $\tilde{\beta}_3$ are the solution to the vector equation (4.8.38), where the values y_i are given as $-\log_e s_i, i = 1, 2, \ldots, 8$ (as, e.g., $y_1 = -\log_e 0.990 = 0.0100$) and W and T are the matrices

$$
W = \begin{bmatrix}
152.46 & -53.46 & 0 & 0 & 0 & 0 & 0 & 0 \\
-53.46 & 86.23 & -32.77 & 0 & 0 & 0 & 0 & 0 \\
0 & -32.77 & 50.01 & -17.24 & 0 & 0 & 0 & 0 \\
0 & 0 & -17.24 & 23.56 & -6.321 & 0 & 0 & 0 \\
0 & 0 & 0 & -6.321 & 11.391 & -5.070 & 0 & 0 \\
0 & 0 & 0 & 0 & -5.070 & 9.752 & -4.682 & 0 \\
0 & 0 & 0 & 0 & 0 & -4.682 & 7.752 & -3.070 \\
0 & 0 & 0 & 0 & 0 & 0 & -3.070 & 3.070
\end{bmatrix}
$$

$$(4.8.42)$$

and

$$
T = \begin{bmatrix}
0.5 & 1.5 & 2.5 & 3.5 & 4.5 & 5.5 & 6.5 & 7.5 \\
0.25 & 2.25 & 6.25 & 12.25 & 20.25 & 30.25 & 42.25 & 56.25 \\
0.125 & 3.375 & 15.625 & 42.875 & 91.125 & 166.375 & 274.625 & 421.875
\end{bmatrix}.
$$

$$(4.8.43)$$

The elements of T are easy for the reader to verify. To compute the elements of W, the element in the first row, first column is $a_1 = b_0 - b_1$ $= (1 \times 0.990)/(1 - 0.990) + (0.990)(0.972)/(0.990 - 0.972) = 152.46$, as an example. The other elements in W are computed in a similar way. The vector $\mathbf{y}'$ is given as $\mathbf{y}' = (0.0100, 0.0284, 0.0576, 0.1109, 0.2433, 0.3871, 0.5226, 0.6992)$. Using (4.8.38) and performing the elementary calculations on a desk calculator, the reader can verify without much difficulty

Table 4.14 Survival probabilities of patients with chronic hypertension

Length of Time on Study t_i (years)	Proportion of Patients Surviving s_i
0.5	0.990
1.5	0.972
2.5	0.944
3.5	0.895
4.5	0.784
5.5	0.679
6.5	0.593
7.5	0.497
$t > 7.5$	0.000

that $\tilde{\beta}' = [0.0086587, 0.0047138, 0.00081858]$ is the weighted least squares vector estimate of $\beta = [\beta_1, \beta_2, \beta_3]$. The fitted survival form $\tilde{S}(t)$ is then

$$\tilde{S}(t) = \exp[-(0.008659t + 0.004714t^2 + 0.0008186t^3)]. \quad (4.8.44)$$

If we compare s_i and $\tilde{S}(t_i)$ at each measured time point, $i = 0.5, \ldots, 7.5$, we find remarkable agreement. The comparison is left as an exercise to the reader.

Finally, the analysis of variance table for testing the hypothesis $H_0: \beta_1 = \beta_2 = \cdots = \beta_m = 0$ is given in Table 4.15. The computed F value is significant beyond the 0.005 level, implying that the hypothesis of no regression effect is contradicted. A more detailed analysis of this regression is contained in the exercises.

In this chapter we cover in some detail the estimation of parameters for the gamma, Weibull, Rayleigh, and generalized Rayleigh survival distributions. In recent years these four survival distributions have been very useful in analyzing a variety of survival data. In addition, the Gompertz hazard rate is introduced in Section 6.5 and Cox's Model at the end of Section 7.3.1.

We reintroduce the gamma and Weibull distributions in Section 4.1: Section 4.2 is concerned with obtaining the maximum likelihood estimators of the gamma and Weibull distribution parameters in the uncensored or complete sample case. In Section 4.3 we obtain the maximum likelihood estimators of parameters of the gamma distribution and Weibull for the

Table 4.15 Analysis of variance table for regression fit of
$$\Lambda(t) = 0.008659t + 0.004714t^2 + 0.0008186t^3$$
for patients with chronic hypertension.

	Degrees of Freedom	Sum of Squares	Mean Squares	F Ratio
Regression	3	0.477226	0.159075	31.777
Error	5	0.025033	0.005006	—
Total	8	0.502259	—	—

singly censored case. Extension of the maximum likelihood estimators for the progressively censored Weibull distribution is relatively simple and straightforward. However, it is not a simple extension for the progressively censored gamma distribution. Section 4.4 deals with the large sample properties of the maximum likelihood estimators obtained in Sections 4.2 and 4.3, including a discussion of joint elliptical confidence regions for the parameters. In Section 4.5 truncated gamma and Weibull (as opposed to censored gamma and Weibull) distributions are introduced. The reader may choose to skip this section with no major loss of continuity. Section 4.6 presents methods alternative to maximum likelihood for estimating gamma and Weibull distribution parameters. These methods are especially useful if the sample size is small, since maximum likelihood estimators are biased in small samples. Section 4.7 covers a graphical method introduced by Nelson [1972]. This method, which is called cumulative hazard plotting, can be used by the experimenter to determine whether any of the distributions presented in this chapter provides an adequate fit to the data. Section 4.8 describes both the Rayleigh and generalized Rayleigh survival distributions. These distributions are also quite useful in practice.

At this point the experimenter probably wonders which distribution is applicable to his data. When faced with this problem, he should arrange the data in a life table format, calculate the empirical hazard rate, and obtain the cumulative hazard rate. With the aid of the cumulative hazard rate, the experimenter is likely to be able to choose the appropriate survival distribution that best fits the data. If, at this point, no adequate fit of a distribution exists, the experimenter may opt to analyze his data by way of the nonparametric life table or Kaplan and Meier techniques presented in Chapter 2.

EXERCISES

1. Compute the mean and variance of the death density function (4.8.1). *Hint*: Consider the transformation $y = \lambda_0^2 / 4\lambda_1 + \lambda_0 t + \lambda_1 t^2$.

2. Compute the theoretical median survival time t_{50} for an individual whose death density function is given by (4.8.1).

3. Compute the mean remission time for the data in Example 4.8.1, using the maximum likelihood estimates $\hat{\lambda}_0 = 0.0637$ and $\hat{\lambda}_1 = 0.0039$ as values for λ_0 and λ_1. Hence obtain an approximate 95 percent confidence interval for μ, the mean remission time.

4. Compute the median remission time for the data in Example 4.8.1, using $\hat{\lambda}_0$ and $\hat{\lambda}_1$ for λ_0 and λ_1, respectively.

5. Plot a 95 percent elliptical confidence region for the parameter vector $\begin{pmatrix} \lambda_0 \\ \lambda_1 \end{pmatrix}$

   using the data from Example 4.8.1.

6. Individuals having a certain type of cancer have been shown to have a hazard rate $\lambda_d(t) = \lambda_0 + 2\lambda_1 t$. Their hazard rate from all other causes is known to be constant and independent of the hazard rate due to cancer. That is, $\lambda_c(t) = \lambda$, where $\lambda_d(t)$ is the hazard rate due to cancer, and $\lambda_c(t)$ is the hazard rate due to all other causes. What is the overall hazard rate of these individuals? Is it the same form as either $\lambda_d(t)$ or $\lambda_c(t)$? Explain.

7. In Example 4.8.3, verify that $\hat{\beta}' = (0.0086587, 0.0047138, 0.00081858)$, hence construct Table 4.16.

Table 4.16 Estimated survival probabilities of patients with chronic hypertension

Length of Time on Study t_i (years)	Estimated Proportion of Patients Surviving $\tilde{S}(t_i)$
0.5	0.994
1.5	0.975
2.5	0.938
3.5	0.884
4.5	0.811
5.5	0.722
6.5	0.619
7.5	0.509
> 7.5	0

Noting the excellent agreement between Table 4.14 and Table 4.16, verify the regression analysis in Table 4.15 (Krane [1963]).

8. Suppose we denote the T matrix appropriate to an h-degree polynomial regression as $T_{(h)}$ and the estimated parameter vector as $\tilde{\beta}_{(h)}$. Show that the regression sum of squares can be partitioned into single degree of freedom

sums of squares as given in Table 4.17. Hence show that in Example 4.8.3 $\tilde{\beta}_{(1)}$ $= 0.0413846$, and $\tilde{\beta}_{(2)} = (-0.0042457, 0.0121321)$, and that the regression analysis for the partitioning is according to Table 4.18. Hence conclude that the hypothesis $H_0: \beta_3 = 0$ cannot be rejected but evidence exists to support $\beta_2 \neq 0$ and $\beta_1 \neq 0$ (Krane [1963]). (A plot of the data in Table 4.15 illustrates this conclusion.)

9. Noting that $\Lambda(t) = \int_0^t \lambda(x)\, dx$, $\lambda(x) \geqslant 0$ for all x, and assuming that $\Lambda(t)$ $= \beta_0 + \beta_1 t + \cdots + \beta_m t^m$, prove: (1) $\beta_0 = 0$, (2) $\beta_m > 0$, and (3) $\sum_{j=1}^m j \beta_j t^{j-1}$ $\geqslant 0$, $t \geqslant 0$ (Krane [1963]).

10. Show that if the quadratic model $\Lambda(t) = \beta_1 t + \beta_2 t^2$ is used in Example 4.8.3, which leads to the estimating vector $\tilde{\beta}_{(2)} = (-0.0042457, 0.0121321)$, then condition (3) in Exercise 9 is violated for small values of t. Tell how you would then adjust your regression (Krane [1963]).

11. For the data in Example 4.8.1 fit the quadratic model $\Lambda(t) = \beta_1 t + \beta_2 t^2$. Compare this with Kodlin's estimators.

Table 4.17 Partitioning of the regression sum of squares

Source of Variation	Degrees of Freedom	Sum of Squares
Linear regression	1	$y' W T'_{(1)} \tilde{\beta}_{(1)}$
Quadratic regression	1	$y' W (T'_{(2)} \tilde{\beta}_{(2)} - T'_{(1)} \tilde{\beta}_{(1)})$
$\vdots$		
mth degree regression	1	$y' W (T'_{(m)} \tilde{\beta}_{(m)} - T'_{(m-1)} \tilde{\beta}_{(m-1)})$
Error	$k - m$	$(y' - \tilde{\beta}' T) W y$
Total	k	$y' W y$

Table 4.18 Partitioning of sum of squares for regression in example 4.8.2

Source of Variation	Degrees of Freedom	Sum of Squares	Mean Squares	F Ratio
Linear regression	1	0.252392	—	50.418
Quadratic regression	1	0.216061	—	43.160
Cubic regression	1	0.008773	—	1.752
Error	5	0.025033	0.005006	—
Total	8	0.502259	—	—

CHAPTER 5

Growth in and Assessment of Survivability

5.1 INTRODUCTORY REMARKS

In Chapters 3 and 4 we examined censored distributions. These arise, for example, in clinical trial situations when there are survivors at the end of the trial. In this chapter we discuss the distribution of the number of survivors at the end of a clinical trial (which is binomial) and the effects of modifications in the clinical trial on new groups of patients with respect possibly to increasing the proportion of survivors.

To investigate these effects, we assume that the clinical trial is carried out in independent stages wherein the clinical procedure is modified between successive stages with the aim of increasing the proportion of survivors from stage to stage. It is also assumed that to ensure independent stages, each stage of the clinical trial contains a new group of patients.

To this end suppose a clinical trial is conducted in k stages such that at the ith stage n_i patients enter the study each with a probability p_i of surviving the ith stage. At each stage a new group of patients is entered; thus stages are assumed to be independent as well as identical in their time periods. After the ith stage the probability there are x_i survivors is

$$\text{pr}\{X = x_i\} = \binom{n_i}{x_i} p_i^{x_i}(1-p_i)^{n_i - x_i}, \qquad i = 1, 2, \ldots, k. \qquad (5.1.1)$$

In Section 5.2 we discuss in some detail the general parametric form that is assumed by p_i for measuring survivability growth from stage to stage. Section 5.3 deals with some specific examples of the general model. In Section 5.4 we look at the survivability assessment problem from the empirical standpoint.

5.2 GENERAL PARAMETRIC GROWTH MODELS AND THE HYPERBOLIC GROWTH MODEL

We assume that the proportion of survivors increases from stage to stage, subject of course to random fluctuations. On the basis of this assumption our task is to hypothesize a parametric form for the parameter p_i, $i = 1, 2, \ldots, k$ in (5.1.1). In this chapter we assume that p_i is a function of at most two parameters, as well as the stage number i. In this section and in Section 5.3 we discuss both in the text and exercises three specific parametric functions of p_i. Model 1 is defined as

$$p_i = p_\infty - \alpha G(i), \tag{5.2.1}$$

where $p_\infty, 0 < p_\infty \leqslant 1$, is the ultimate theoretical proportion of survivors achievable, $\alpha > 0$ is a second parameter that quantifies the amount of growth between stages 1 and k, and $G(i)$ is a known decreasing function of i [e.g., $G(i) = 1/i$, $i = 1, 2, \ldots, k$]. The model for which $G(i) = 1/i$ is known as a hyperbolic growth model. Model 2 is given by

$$p_i = 1 - \alpha_1 \exp(-\alpha_2 i), \tag{5.2.2}$$

where $0 < \alpha_1 < e^{\alpha_2}$, $\alpha_2 > 0$ are the parameters of the model, $i = 1, 2, \ldots, k$. This model is termed an exponential growth model. In some applications, estimators of p_∞ in Model 1 and α_1 and $\exp(\alpha_2)$ in Model 2 violate their physical constraints. The fairly straightforward remedy for this is discussed in Section 5.3. Model 3 has the defining equation

$$p_i = \{1 + \exp[-(\alpha_1 + \alpha_2 i)]\}^{-1}, \tag{5.2.3}$$

where α_1 and α_2 are the parameters of the model, $i = 1, 2, \ldots, k$. This model is a special case of the more general class of logistic models. Models 1 and 2 are discussed in Gross and Kamins [1968].

Generally, $p_i \equiv p_i(\alpha_1, \alpha_2)$ is a function of two parameters α_1 and α_2 and the stage number of the trial i, where $i = 1, 2, \ldots, k$. Then, given that a clinical trial has been conducted in k stages with x_1 survivors of n_1 patients in the first stage, x_2 survivors of n_2 patients in the second stage, $\ldots$, x_k survivors of n_k patients in the kth and last stage, the problem is to estimate the parameters α_1 and α_2.

Two methods commonly used to estimate α_1 and α_2 are least squares and maximum likelihood. We discuss both procedures in this chapter for the following reasons. Maximum likelihood estimators are generally preferred to least squares estimators because of the desirability of the large sample properties of maximum likelihood estimators. (See Chapters 3 and

4 regarding these properties). On the other hand, the least squares estimators are often obtainable in closed form and are a good first approximation to the maximum likelihood estimators.

Least Squares

Let us define $\psi(\alpha_1, \alpha_2)$ as

$$\psi(\alpha_1, \alpha_2) \equiv \sum_{i=1}^{k} \left[\frac{x_i}{n_i} - p_i(\alpha_1, \alpha_2) \right]^2. \tag{5.2.4}$$

The least squares estimators α_1^* and α_2^* are the values of the parameters α_1 and α_2, respectively, that simultaneously minimize $\psi(\alpha_1, \alpha_2)$. Assuming that $p_i(\alpha_1, \alpha_2)$ is differentiable in α_1 and α_2, $i = 1, 2, \ldots, k$, and that the minimum is obtained by differentiation, we find α_1^* and α_2^* as the solution of

$$\sum_{i=1}^{k} \left[\frac{x_i}{n_i} - p_i(\alpha_1^*, \alpha_2^*) \right] \frac{\partial p_i(\alpha_1^*, \alpha_2^*)}{\partial \alpha_1^*} = 0 \tag{5.2.5}$$

and

$$\sum_{i=1}^{k} \left[\frac{x_i}{n_i} - p_i(\alpha_1^*, \alpha_2^*) \right] \frac{\partial p_i(\alpha_1^*, \alpha_2^*)}{\partial \alpha_2^*} = 0, \tag{5.2.6}$$

where

$$\frac{\partial p_i(\alpha_1^*, \alpha_2^*)}{\partial \alpha_j^*} \equiv \frac{\partial p_i(\alpha \ \alpha_2)}{\partial \alpha_j} \bigg|_{\substack{\alpha_1 = \alpha_1^* \\ \alpha_2 = \alpha_2^*}} \qquad j = 1, 2.$$

That is, α_1^* and α_2^* are the least squares estimators of α_1 and α_2, respectively. Examples of the least squares procedure are given here and in Section 5.3 for the three principal models.

Maximum Likelihood

Equation 5.1.1 gives the probability that x_i patients of the n_i on trial at stage i survive, $i = 1, \ldots, k$. Thus assuming the k stages to be statistically independent, the likelihood function for all k stages is given by

$$L(\alpha_1, \alpha_2) = \prod_{i=1}^{k} \binom{n_i}{x_i} [p_i(\alpha_1, \alpha_2)]^{x_i} [1 - p_i(\alpha_1, \alpha_2)]^{n_i - x_i}. \tag{5.2.7}$$

Ordinarily the parameter vector (α_1, α_2) can be maximized with respect to the observations by the usual maximum likelihood procedures (cf. Chapters 3 and 4). However, it should be noted that $0 \leqslant p_i \leqslant 1$, $i = 1, 2, \ldots, k$, places an additional constraint on the maximum likelihood estimators, as well as the least squares estimators. These questions of physical constraints are addressed for each model as the problems arise.

Assuming that the parameter vector (α_1, α_2) can be maximized with respect to the observations by maximum likelihood procedures, the maximum likelihood estimators $\hat{\alpha}_1$ and $\hat{\alpha}_2$ are the values of the parameters α_1 and α_2, respectively, that simultaneously maximize $L(\alpha_1, \alpha_2)$ or, equivalently, $\log_e L(\alpha_1, \alpha_2)$. The vector $(\hat{\alpha}_1, \hat{\alpha}_2)$ is then the simultaneous solution of

$$\sum_{i=1}^{k} \frac{x_i}{p_i(\hat{\alpha}_1, \hat{\alpha}_2)} \frac{\partial p_i(\hat{\alpha}_1, \hat{\alpha}_2)}{\partial \hat{\alpha}_1} - \sum_{i=1}^{k} \frac{n_i - x_i}{1 - p_i(\hat{\alpha}_1, \hat{\alpha}_2)} \frac{\partial p_i(\hat{\alpha}_1, \hat{\alpha}_2)}{\partial \hat{\alpha}_1} = 0 \quad (5.2.8)$$

and

$$\sum_{i=1}^{k} \frac{x_i}{p_i(\hat{\alpha}_1, \hat{\alpha}_2)} \frac{\partial p_i(\hat{\alpha}_1, \hat{\alpha}_2)}{\partial \hat{\alpha}_2} - \sum_{i=1}^{k} \frac{n_i - x_i}{1 - p_i(\hat{\alpha}_1, \hat{\alpha}_2)} \frac{\partial p_i(\hat{\alpha}_1, \hat{\alpha}_2)}{\partial \hat{\alpha}_2} = 0 \quad (5.2.9)$$

that maximizes $L(\alpha_1, \alpha_2)$, where

$$\frac{\partial p_i(\hat{\alpha}_1, \hat{\alpha}_2)}{\partial \hat{\alpha}_j} \equiv \frac{\partial p_i(\alpha_1, \alpha_2)}{\partial \alpha_j} \bigg|_{\substack{\alpha_1 = \hat{\alpha}_1 \\ \alpha_2 = \hat{\alpha}_2}}, \quad j = 1, 2.$$

Estimation for Model 1: $p_i = p_\infty - \alpha G(i)$. Suppose $p_i(\alpha_1, \alpha_2)$ is defined by (5.2.1), $i = 1, 2, \ldots, k$. In this case $\alpha_1 \equiv p_\infty$ and $\alpha_2 \equiv \alpha$. Assuming that the success ratios for the k stages are, respectively, $x_1/n_1, x_2/n_2, \ldots, x_k/n_k$, the least squares equation is

$$\psi(p_\infty, \alpha) = \sum_{i=1}^{k} \left[\frac{x_i}{n_i} - p_\infty + \alpha G(i) \right]^2. \quad (5.2.10)$$

Differentiating $\psi(p_\infty, \alpha)$ with respect to p_∞ and α, the solution of these equations is the vector (p_∞^*, α^*), which is obtained by solving

$$\sum_{i=1}^{k} \left[\frac{x_i}{n_i} - p_\infty^* + \alpha^* G(i) \right] = 0 \quad (5.2.11)$$

and

$$\sum_{i=1}^{k}\left[\left(\frac{x_i}{n_i}\right)G(i) - p_\infty^* G(i) + \alpha^* G^2(i)\right] = 0, \tag{5.2.12}$$

Equations 5.2.11 and 5.2.12 can also be found using (5.2.5) and (5.2.6), respectively, noting that

$$\frac{\partial p_i(p_\infty^*, \alpha^*)}{\partial p_\infty^*} = -1 \tag{5.2.13}$$

and

$$\frac{\partial p_i(p_\infty^*, \alpha^*)}{\partial \alpha^*} = G(i). \tag{5.2.14}$$

The least squares estimators p_∞^* and α^* are thus found by solving (5.2.11) and (5.2.12). Hence

$$p_\infty^* = \frac{\left(\sum_{i=1}^{k} G^2(i)\right)\left(\sum_{i=1}^{k} x_i/n_i\right) - \left(\sum_{i=1}^{k} G(i)\right)\left(\sum_{i=1}^{k} G(i)x_i/n_i\right)}{k\sum_{i=1}^{k} G^2(i) - \left(\sum_{i=1}^{k} G(i)\right)^2} \tag{5.2.15}$$

and

$$\alpha^* = \frac{\left(\sum_{i=1}^{k} G(i)\right)\left(\sum_{i=1}^{k} x_i/n_i\right) - k\left(\sum_{i=1}^{k} G(i)x_i/n_i\right)}{k\sum_{i=1}^{k} G^2(i) - \left(\sum_{i=1}^{k} G(i)\right)^2}. \tag{5.2.16}$$

The natural logarithm of the likelihood function is given by

$$\log_e L = C + \sum_{i=1}^{k} x_i \log_e(p_\infty - \alpha G(i)) + \sum_{i=1}^{k} (n_i - x_i)\log_e(1 - p_\infty + \alpha G(i)), \tag{5.2.17}$$

where C is a constant. The maximum likelihood estimators $\hat{p}_\infty$ and $\hat{\alpha}$ are then obtained as the unique solution to

$$\sum_{i=1}^{k} \frac{x_i}{(\hat{p}_\infty - \hat{\alpha} G(i))} - \sum_{i=1}^{k} \frac{n_i - x_i}{(1 - \hat{p}_\infty + \hat{\alpha} G(i))} = 0 \tag{5.2.18}$$

and

$$-\sum_{i=1}^{k} \frac{x_i G(i)}{(\hat{p}_\infty - \hat{\alpha}G(i))} + \sum_{i=1}^{k} \frac{(n_i - x_i)G(i)}{(1 - \hat{p}_\infty + \hat{\alpha}G(i))} = 0. \qquad (5.2.19)$$

The uniqueness of $\hat{p}_\infty$ and $\hat{\alpha}$ is established by examining the matrix of the second partial derivatives

$$M = \begin{bmatrix} f_{11} & f_{12} \\ f_{12} & f_{22} \end{bmatrix}, \qquad (5.2.20)$$

where we define $f_{11} \equiv \partial^2 \log_e L / \partial p_\infty^2$, $f_{12} \equiv \partial^2 \log_e L / \partial p_\infty \partial\alpha$, and $f_{22} \equiv \partial^2 \log_e L / \partial\alpha^2$. It is easy to see $f_{11} < 0$, $f_{22} < 0$, but $f_{11}f_{22} - f_{12}^2 > 0$, ensuring the unique solution $(\hat{p}_\infty, \hat{\alpha})$ of the vector parameter (p_∞, α) produces a maximum. (See Taylor [1955], p. 232.)

Unfortunately $\hat{p}_\infty$ and $\hat{\alpha}$ cannot be obtained in closed form, like the least squares estimators p_∞^* and α^*. For this reason p_∞^* and α^* are often preferred to $\hat{p}_\infty$ and $\hat{\alpha}$, especially if p_∞^*, and α^* provide a good fit to the data. On the other hand $\hat{p}_\infty$ and $\hat{\alpha}$ have desirable large sample properties, and ordinarily they should be obtained.

Determination of $\hat{p}_\infty$ and $\hat{\alpha}$ can be made using a two-dimensional Newton-Raphson method that in practice usually converges very rapidly, if the least squares estimators p_∞^* and α^* are taken as the initial estimates in a given problem. To this end let us define

$$\hat{g}_{1,n-1} = \sum_{i=1}^{k} \frac{x_i}{(\hat{p}_{\infty,n-1} - \hat{\alpha}_{n-1}G(i))} - \sum_{i=1}^{k} \frac{n_i - x_i}{(1 - \hat{p}_{\infty,n-1} + \hat{\alpha}_{n-1}G(i))},$$

$$(5.2.21)$$

$$\hat{g}_{2,n-1} = -\sum_{i=1}^{k} \frac{x_i G(i)}{(\hat{p}_\infty - \hat{\alpha}_{n-1}G(i))} + \sum_{i=1}^{k} \frac{(n_i - x_i)G(i)}{(1 - \hat{p}_{\infty,n-1} + \hat{\alpha}_{n-1}G(i))}, \qquad (5.2.22)$$

$$\hat{f}_{11,n-1} = -\left[\sum_{i=1}^{k} \frac{x_i}{(\hat{p}_{\infty,n-1} - \hat{\alpha}_{n-1}G(i))^2} + \sum_{i=1}^{k} \frac{n_i - x_i}{(1 - \hat{p}_{\infty,n-1} + \hat{\alpha}_{n-1}G(i))^2} \right],$$

$$(5.2.23)$$

$$\hat{f}_{12,n-1} = \left[\sum_{i=1}^{k} \frac{x_i G(i)}{\left(\hat{p}_{\infty,n-1} - \hat{\alpha}_{n-1} G(i)\right)^2} + \sum_{i=1}^{k} \frac{(n_i - x_i) G(i)}{\left(1 - \hat{p}_{\infty,n-1} + \hat{\alpha}_{n-1} G(i)\right)^2} \right],$$

(5.2.24)

and

$$\hat{f}_{22,n-1} = -\left[\sum_{i=1}^{k} \frac{x_i G^2(i)}{\left(\hat{p}_{\infty,n-1} - \hat{\alpha}_{n-1} G(i)\right)^2} + \sum_{i=1}^{k} \frac{(n_i - x_i) G^2(i)}{\left(1 - \hat{p}_{\infty,n-1} + \hat{\alpha}_{n-1} G(i)\right)^2} \right],$$

(5.2.25)

where $\hat{p}_{\infty,n-1}$ and $\hat{\alpha}_{n-1}$ are the values of $\hat{p}_{\infty}$ and $\hat{\alpha}$ after $n-1$ iterations, and $\hat{p}_{\infty,0} \equiv p_{\infty}^*$ and $\hat{\alpha}_0 \equiv \alpha^*$ by definition. It then follows that $\hat{p}_{\infty,n}$ and $\hat{\alpha}_n$ are found by solving the vector equation

$$\begin{pmatrix} \hat{p}_{\infty,n} \\ \hat{\alpha}_n \end{pmatrix} = \begin{pmatrix} \hat{p}_{\infty,n-1} \\ \hat{\alpha}_{n-1} \end{pmatrix} - \begin{bmatrix} \hat{f}_{11,n-1} & \hat{f}_{12,n-1} \\ \hat{f}_{12,n-1} & \hat{f}_{22,n-1} \end{bmatrix}^{-1} \begin{bmatrix} \hat{g}_{1,n-1} \\ \hat{g}_{2,n-1} \end{bmatrix}.$$

(5.2.26)

Individual values $\hat{p}_{\infty,n}$ and $\hat{\alpha}_n$ can then be obtained as

$$\hat{p}_{\infty,n} = \hat{p}_{\infty,n-1} - \frac{\hat{g}_{1,n-1}\hat{f}_{22,n-1} - \hat{g}_{2,n-1}\hat{f}_{12,n-1}}{\Delta_{n-1}}$$

(5.2.27)

and

$$\hat{\alpha}_n = \hat{\alpha}_{n-1} - \frac{\hat{g}_{2,n-1}\hat{f}_{11,n-1} - \hat{g}_{1,n-1}\hat{f}_{12,n-1}}{\Delta_{n-1}},$$

(5.2.28)

where $\Delta_{n-1} \equiv \hat{f}_{11,n-1}\hat{f}_{22,n-1} - \hat{f}_{12,n-1}^2$. Further explanation of the use of Newton-Raphson method is given in Chapter 6.

Example 5.2.1. A clinical trial is conducted on patients with acute leukemia. The trial consists of five stages with 10 new patients selected at each stage. The proportions of patients achieving remission at each stage are 3/10, 6/10, 7/10, 7/10, and 8/10, for stages 1 through 5. We hypothesize that the probability of a patient remission at the ith stage is given by $p_i = p_{\infty} - \alpha/i$, $i = 1,2,3,4,5$. That is, we hypothesize model 1 (5.2.1) with $G(i) = 1/i$. Our aim is to obtain the maximum likelihood estimates $\hat{p}_{\infty}$ and $\hat{\alpha}$. First we obtain the least squares estimates p_{∞}^* and α^* using (5.2.15) and (5.2.16), respectively. To facilitate computations, we use Table 5.1.

Table 5.1 Least squares computation layout, model 1

Stage i	$1/i$	$1/i^2$	x_i/n_i	$(x_i/n_i)(1/i)$
1	1	1	3/10	3/10
2	1/2	1/4	6/10	3/10
3	1/3	1/9	7/10	7/30
4	1/4	1/16	7/10	7/40
5	1/5	1/25	8/10	8/50
Totals	2.2833	1.4636	3.1000	1.1683

From Table 5.1 and (5.2.15) and (5.2.16) with $G(i)=1/i$, we have

$$p_\infty^* = \frac{(1.4636)(3.1000)-(2.2833)(1.1683)}{5(1.4636)-(2.2833)^2} = 0.888$$

and

$$\alpha^* = \frac{(2.2833)(3.1000)-5(1.1683)}{5(1.4636)-(2.2833)^2} = 0.588.$$

We take p_∞^* and α^* as $\hat{p}_{\infty,0}$ and $\hat{\alpha}_0$ for the Newton-Raphson iteration procedure to obtain $\hat{p}_\infty$ and $\hat{\alpha}$. To facilitate these computations, Table 5.2 is given. Here $\hat{g}_{10i}$, $\hat{g}_{20i}$, $-\hat{f}_{110i}$, $\hat{f}_{120i}$, and $-\hat{f}_{220i}$ are, respectively, $[x_i/\hat{p}_{i0}-(n_i-x_i)/(1-\hat{p}_{i0})]$, $-[x_i/i\hat{p}_{i0}-(n_i-x_i)/i(1-\hat{p}_{i0})]$, $[x_i/\hat{p}_{i0}^2+(n_i-x_i)/(1-\hat{p}_{i0})^2]$, $[x_i/i\hat{p}_{i0}^2+(n_i-x_i)/i(1-\hat{p}_{i0})^2]$, and $[x_i/i^2\hat{p}_{i0}^2+(n_i-x_i)/i^2(1-\hat{p}_{i0})^2]$, where $\hat{p}_{i0}=\hat{p}_{\infty,0}-\hat{\alpha}_0/i$.

Table 5.2 Maximum likelihood computation layout (first iteration), model 1

Stage i	$\hat{g}_{10i}$	$\hat{g}_{20i}$	$-\hat{f}_{110i}$	$\hat{f}_{120i}$	$-\hat{f}_{220i}$
1	0.0000	0.0000	47.6190	47.6190	47.6190
2	0.2488	-0.1244	41.2716	20.6358	10.3179
3	0.3753	-0.1251	46.2421	15.4140	5.1380
4	-2.1363	0.5341	57.4706	14.3677	3.5919
5	1.6734	-0.3347	51.4180	10.2836	2.0567
Totals	0.1612	-0.0501	244.0215	108.3202	68.7236

By (5.2.27) and (5.2.28) we then see that

$$\hat{p}_{\infty,1} = 0.8880 + \frac{(0.1612)(68.7236)-(0.0501)(108.3202)}{(244.0215)(68.7236)-(108.3202)^2} = 0.8891$$

$$\hat{\alpha}_1 = 0.5880 - \frac{(0.0501)(244.0215)-(0.1612)(108.3202)}{(244.0215)(68.7236)-(108.3202)^2} = 0.5891.$$

Examining these results indicates that another iteration is necessary. Table 5.3 is exactly the same as Table 5.2, but $\hat{p}_{\infty,1}$ and $\hat{\alpha}_1$ replace p^*_∞ and α^*, respectively.

Table 5.3 Maximum likelihood computation layout (second iteration), model 1

Stage	$\hat{g}_{11i}$	$\hat{g}_{21i}$	$-\hat{f}_{111i}$	$\hat{f}_{121i}$	$-\hat{f}_{221i}$
1	−0.0039	0.0039	47.6041	47.6041	47.6041
2	0.2239	−0.1120	41.3093	20.6547	10.3273
3	0.3394	−0.1131	46.3693	15.4564	5.1521
4	−2.1860	0.5465	57.7402	14.4351	3.6088
5	1.6263	−0.3253	51.6900	10.3380	2.0676
Totals	−0.0002	0.0001	244.7129	108.4882	68.7599

An application of (5.2.27) and (5.2.28) again shows

$$\hat{p}_{\infty2} = 0.8891 - \frac{(0.0002)(68.7599)-(0.0001)(108.4882)}{(244.7129)(68.7599)-(108.4882)^2} = 0.8891$$

and

$$\hat{\alpha}_2 = 0.5891 + \frac{(0.0001)(244.7129)-(0.0002)(108.4882)}{(244.7129)(68.7599)-(108.4882)^2} = 0.5892.$$

This second iteration gives the requisite maximum likelihood estimates, since further iterations do not change the results. Thus $\hat{p}_\infty = 0.8891$ and $\hat{\alpha} = 0.5892$.

Having obtained both the least squares and maximum likelihood estimates for p_∞ and α, it is instructive to use Table 5.4 to compare the observed and expected remission probabilities in each stage. Note that both sets of estimates fit the data quite well and in fact are very close together. However, both the least squares estimates and the maximum likelihood estimates are slightly conservative compared with the observed probabilities.

Table 5.4 Expected and observed remission probabilities for acute leukemia patients at each stage when clinical trial is conducted in five stages

Stage i	Observed Probability	Expected MLE[a] Probability	Expected LSE[a] Probability
1	0.3000	0.2999	0.3000
2	0.6000	0.5945	0.5940
3	0.7000	0.6927	0.6920
4	0.7000	0.7418	0.7410
5	0.8000	0.7713	0.7704

[a]MLE = maximum likelihood estimate, LSE = least squares estimate.

Methods of obtaining joint confidence regions for the parameters α_1 and α_2 as well as confidence intervals for $p_{k+1}(\alpha_1, \alpha_2)$ are similar to the methods discussed in Chapters 3 and 4. (See Sections 3.3, 3.4, 4.3, and 4.4.)

In general then, following the development of Mood [1950], a joint approximate $(1 - \alpha)$ 100 percent confidence region for α_1 and α_2, based on their maximum likelihood estimators $\hat{\alpha}_1$ and $\hat{\alpha}_2$, is given by

$$\sigma^{11}(\hat{\alpha}_1 - \alpha_1)^2 + \sigma^{22}(\hat{\alpha}_2 - \alpha_2)^2 + 2\sigma^{12}(\hat{\alpha}_1 - \alpha_1)(\hat{\alpha}_2 - \alpha_2) = \chi^2(1 - \alpha; 2),$$

$$(5.2.29)$$

where

$$\sigma^{ij} = -E\left[\frac{\partial^2 \log_e L(\alpha_1, \alpha_2)}{\partial \alpha_i \partial \alpha_j}\right], \qquad i, j = 1, 2,$$

$L(\alpha_1, \alpha_2)$ is given by (5.2.7), and $\chi^2(1 - \alpha; 2)$ is the $(1 - \alpha)$th percentage point of the chi-square distribution with 2 degrees of freedom.

To obtain a $(1 - \alpha)$ 100 percent confidence interval for $p_{k+1}(\alpha_1, \alpha_2)$, the predicted probability of survival at the $(k + 1)$th stage of the clinical trial, we must first find Var $p_{k+1}(\hat{\alpha}_1, \hat{\alpha}_2)$. Since we may find Var $p_i(\hat{\alpha}_1, \hat{\alpha}_2)$—or at least an approximate value thereof—at *any* stage i of the clinical trial, we do not restrict our procedure to finding only Var $p_{k+1}(\hat{\alpha}_1, \hat{\alpha}_2)$. As a first step we expand $p_i(\hat{\alpha}_1, \hat{\alpha}_2)$ about $p_i(\alpha_1, \alpha_2)$ in a Taylor series, ignoring terms of higher than the first order. Thus we have

$$p_i(\hat{\alpha}_1, \hat{\alpha}_2) \doteq (\hat{\alpha}_1 - \alpha_1) \frac{\partial p_i(\alpha_1, \alpha_2)}{\partial \alpha_1} + (\hat{\alpha}_2 - \alpha_2) \frac{\partial p_i(\alpha_1, \alpha_2)}{\partial \alpha_2}. \tag{5.2.30}$$

Then the approximate variance of $p_i(\hat{\alpha}_1, \alpha_2)$ is

$$\text{Var}\, p_i(\hat{\alpha}_1, \hat{\alpha}_2) \doteq \sigma_{\hat{\alpha}_1}^2 \theta_{1i}^2 + \sigma_{\hat{\alpha}_2}^2 \theta_{2i}^2 + 2\sigma_{\hat{\alpha}_1 \hat{\alpha}_2} \theta_{1i} \theta_{2i}, \tag{5.2.31}$$

where $\theta_{mi} = \partial p_i(\alpha_1, \alpha_2)/\partial \alpha_m$, $m = 1, 2$, and the values $\sigma_{\hat{\alpha}_1}^2$, $\sigma_{\hat{\alpha}_2}^2$, and $\sigma_{\hat{\alpha}_1 \hat{\alpha}_2}$ are the elements of the matrix

$$\begin{bmatrix} \sigma_{\hat{\alpha}_1}^2 & \sigma_{\hat{\alpha}_1 \hat{\alpha}_2} \\ \sigma_{\hat{\alpha}_1 \hat{\alpha}_2} & \sigma_{\hat{\alpha}_2}^2 \end{bmatrix} = \begin{bmatrix} \sigma^{11} & \sigma^{12} \\ \sigma^{12} & \sigma^{22} \end{bmatrix}^{-1} = \begin{bmatrix} \dfrac{\sigma^{22}}{\Delta} & \dfrac{-\sigma^{12}}{\Delta} \\ \dfrac{-\sigma^{12}}{\Delta} & \dfrac{\sigma^{11}}{\Delta} \end{bmatrix}, \tag{5.2.32}$$

$\Delta = [\sigma^{22}\sigma^{11} - (\sigma^{12})^2]$, where σ^{ij}, $i, j = 1, 2$, is defined in (5.2.29). Noting now that $p_i(\hat{\alpha}_1, \hat{\alpha}_2)$ is approximately normally distributed with mean $p_i(\alpha_1, \alpha_2)$ and approximate variance Var $p_i(\hat{\alpha}_1, \hat{\alpha}_2)$ given by (5.2.31), an approximate $(1 - \alpha)$ 100 percent lower confidence limit for $p_i(\alpha_1, \alpha_2)$, which we define as $p_{Li}(\hat{\alpha}_1, \hat{\alpha}_2)$, is

$$p_{Li}(\hat{\alpha}_1, \hat{\alpha}_2) \doteq p_i(\hat{\alpha}_1, \hat{\alpha}_2) - Z(1 - \alpha)\sqrt{\hat{\text{Var}}\, p_i(\hat{\alpha}_1, \hat{\alpha}_2)}, \tag{5.2.33}$$

where $Z(1 - \alpha)$ is the upper $(1 - \alpha)$ 100th percentage point of the standard normal distribution and $\hat{\text{Var}}\, p_i(\hat{\alpha}_1, \hat{\alpha}_2)$ is the same as Var $p_i(\hat{\alpha}_1, \hat{\alpha}_2)$ in (5.2.31), except that the unknown parameters are replaced by their maximum likelihood estimators. At this point we note that if $p_{Li}(\hat{\alpha}_1, \hat{\alpha}_2)$ is negative, the lower limit is taken to be zero.

Confidence Regions and Intervals for Model 1: $p_i = p_\infty - \alpha G(i)$. To obtain a joint elliptical confidence region for p_∞ and α, we must find σ^{11}, σ^{22}, and σ^{12} as given in (5.2.29). Here we observe that

$$\frac{\partial^2 \log_e L}{\partial p_\infty^2} = -\left[\sum_{i=1}^k \frac{x_i}{(p_\infty - \alpha G(i))^2} + \sum_{i=1}^k \frac{n_i - x_i}{(1 - p_\infty + \alpha G(i))^2} \right]. \tag{5.2.34}$$

Furthermore, we see that at the ith stage x_i is a binomial variable whose mean is $n_i(p_\infty - \alpha G(i))$. Thus it follows that

$$\sigma^{11} \equiv -E\left(\frac{\partial^2 \log_e L}{\partial p_\infty^2}\right) = \sum_{i=1}^{k} \frac{n_i}{[p_\infty - \alpha G(i)][1 - p_\infty + \alpha G(i)]}. \quad (5.2.35)$$

The other values σ^{12} and σ^{22} are similarly obtained:

$$\sigma^{12} = -\sum_{i=1}^{k} \frac{n_i G(i)}{[p_\infty - \alpha G(i)][1 - p_\infty + \alpha G(i)]} \quad (5.2.36)$$

and

$$\sigma^{22} = \sum_{i=1}^{k} \frac{n_i G^2(i)}{[p_\infty - \alpha G(i)][1 - p_\infty + \alpha G(i)]}. \quad (5.2.37)$$

An approximate $(1 - \gamma)$ 100 percent elliptical confidence region for the vector parameter $\begin{pmatrix} p_\infty \\ \alpha \end{pmatrix}$ is given by

$$\sigma^{11}(\hat{p}_\infty - p_\infty)^2 + \sigma^{22}(\hat{\alpha} - \alpha)^2 + 2\sigma^{12}(\hat{p}_\infty - p_\infty)(\hat{\alpha} - \alpha) \doteq \chi^2(1 - \gamma; 2),$$

$$(5.2.38)$$

where $\chi^2(1 - \gamma; 2)$ is the $(1 - \gamma)$ 100 percentage point for the chi-square distribution with 2 degrees of freedom. Noting that σ^{11}, σ^{22}, and σ^{12} contain values of the unknown parameters, in practice we use $\hat{\sigma}^{11}$, $\hat{\sigma}^{22}$, and $\hat{\sigma}^{12}$, where $\hat{p}_\infty$ and $\hat{\alpha}$ replace p_∞ and α, respectively, in (5.2.35) through (5.2.37).

As the first step in obtaining $p_{Li}(\hat{\alpha}_1, \hat{\alpha}_2)$ for Model 1 we compute its approximate variance. First, θ_{1i} and θ_{2i} given by (5.2.31) are easily calculated and are left as an exercise for the reader. These values are $\theta_{1i} = 1$ and $\theta_{2i} = -G(i)$. Thus we write

$$\text{Var } p_i(\hat{p}_\infty, \hat{\alpha}) \doteq \sigma_{\hat{p}_\infty}^2 + G^2(i)\sigma_{\hat{\alpha}}^2 - 2G(i)\sigma_{\hat{p}_\infty \hat{\alpha}}, \quad (5.2.39)$$

where the values $\sigma_{\hat{p}_\infty}^2$, $\sigma_{\hat{\alpha}}^2$, and $\sigma_{\hat{p}_\infty \hat{\alpha}}$ are given by the matrix (5.2.32), and values σ^{11}, σ^{22}, and σ^{12} are given by (5.2.35) through (5.2.37). To obtain the approximate value of the $(1 - \gamma)$ 100 percent lower confidence limit for

$p_i(\hat{p}_\infty, \hat{\alpha})$ we apply (5.2.33) directly:

$$p_{Li}(\hat{p}_\infty, \hat{\alpha}) \doteq (\hat{p}_\infty - \hat{\alpha} G(i)) - Z(1-\gamma)\sqrt{\hat{\sigma}_{\hat{p}_\infty}^2 + G^2(i)\hat{\sigma}_{\hat{\alpha}}^2 - 2G(i)\hat{\sigma}_{\hat{p}_\infty\hat{\alpha}}} \; ,$$

$$(5.2.40)$$

where $\hat{\sigma}_{\hat{p}_\infty}^2$, $\hat{\sigma}_{\hat{\alpha}}^2$, and $\hat{\sigma}_{\hat{p}_\infty\hat{\alpha}}$ are the same as $\sigma_{\hat{p}_\infty}^2$, $\sigma_{\hat{\alpha}}^2$, and $\sigma_{\hat{p}_\infty\hat{\alpha}}$, respectively, replacing the unknown values p_∞ and α by their maximum likelihood estimators $\hat{p}_\infty$ and $\hat{\alpha}$.

Example 5.2.2. The maximum likelihood estimates for the parameters p_∞ and α from the leukemia remission data in Example 5.2.1 are $\hat{p}_\infty = 0.8891$ and $\hat{\alpha} = 0.5892$. Using these for p_∞ and α, the estimated values $\hat{\sigma}^{11}$, $\hat{\sigma}^{12}$, and $\hat{\sigma}^{22}$ obtained from (5.2.35) through (5.2.37) are found to be $\hat{\sigma}^{11} = 244.9822$, $\hat{\sigma}^{12} = -108.4053$, and $\hat{\sigma}^{22} = 68.7327$. To obtain an approximate 95 percent joint elliptical confidence region for $\begin{pmatrix} p_\infty \\ \alpha \end{pmatrix}$ we apply (5.2.38), recalling that $\chi^2(0.95; 2) = 5.99$. Thus the 95 percent joint elliptical confidence region for $\begin{pmatrix} p_\infty \\ \alpha \end{pmatrix}$ is given by

$$244.9822(p_\infty - 0.8891)^2 + 68.7327(\alpha - 0.5892)^2$$

$$- 216.8106(p_\infty - 0.8891)(\alpha - 0.5892) \doteq 5.99. \quad (5.2.41)$$

Simplifying, we then have

$$2.2599(p_\infty - 0.8891)^2 + 0.6340(\alpha - 0.5892)^2 - 2(p_\infty - 0.8891)(\alpha - 0.5892)$$

$$\doteq 0.0553. \quad (5.2.41')$$

Furthermore, it is not hard to show the separate 95 percent confidence intervals for p_∞ and α are $0.6614 < p_\infty < 1$ and $0.1562 < \alpha < 0.6614$. We leave this as an exercise to the reader, noting that the upper limit of the confidence interval for p_∞ cannot exceed 1, and also the upper limit of the confidence interval for α cannot exceed the lower limit of the confidence interval of p_∞. Figure 5.1 shows the elliptical confidence region for $\begin{pmatrix} p_\infty \\ \alpha \end{pmatrix}$, with the angle of rotation $\phi = 34°10'$.

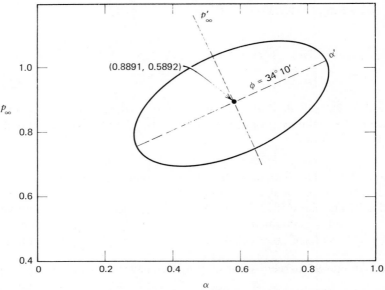

Figure 5.1 An approximate 95 percent joint confidence region for the leukemia data in Example 5.2.1: $2.2599(p_\infty - 0.8891)^2 + 0.6340(\alpha - 0.5892)^2 - 2(p_\infty - 0.8891)(\alpha - 0.5892) \doteq 0.0553$.

The values $\hat{\sigma}^2_{\hat{p}_\infty}$, $\hat{\sigma}^2_{\hat{\alpha}}$, and $\hat{\sigma}_{\hat{p}_\infty\hat{\alpha}}$ are obtained by writing the matrix of elements $\hat{\sigma}^{11}$, $\hat{\sigma}^{22}$, and $\hat{\sigma}^{12}$:

$$\begin{bmatrix} \hat{\sigma}^2_{\hat{p}\infty} & \hat{\sigma}_{\hat{p}\infty\hat{\alpha}} \\ \hat{\sigma}_{\hat{p}\infty\hat{\alpha}} & \hat{\sigma}^2_{\hat{\alpha}} \end{bmatrix} = \begin{bmatrix} 244.9822 & -108.4053 \\ -108.4053 & 68.7327 \end{bmatrix}^{-1} = \begin{bmatrix} 0.0135 & 0.0213 \\ 0.0213 & 0.0482 \end{bmatrix}.$$

(5.2.42)

By (5.2.2) with $G(i) = 1/i$, we have

$$\text{Vâr}\, p_i(\hat{p}_\infty, \hat{\alpha}) \doteq 0.0135 + \frac{0.0482}{i^2} - \frac{0.0426}{i},$$

(5.2.43)

at each stage i of the clinical trial. The approximate $(1 - \gamma)$ 100 percent lower limit $p_{Li}(\hat{p}_\infty, \hat{\alpha})$, which is given for Model 1 in general by (5.2.40), is in this case

$$P_{Li}(\hat{p}_\infty, \hat{\alpha}) \doteq \left(0.8891 - \frac{0.5892}{i}\right) - Z(1-\gamma)\sqrt{0.0135 + \frac{0.0482}{i^2} - \frac{0.0426}{i}},$$

(5.2.44)

where $Z(1-\gamma)$ is again the upper $(1-\gamma)$ 100 percentage point of the standard normal distribution. Table 5.5 gives the values of $p_{Li}(\hat{p}_\infty, \hat{\alpha})$ at each of the five stages of the clinical trial plus the predicted value $p_{L6}(\hat{p}_\infty, \hat{\alpha})$ with 95 percent confidence.

Table 5.5 95 percent lower confidence limits for probability of remission in acute leukemia at each of five stages of clinical trial, plus predicted sixth stage lower limit

Stage i	$p_{Li}(\hat{p}_\infty, \hat{\alpha})$
1	0.0727
2	0.3800
3	0.5815
4	0.6158
5	0.6346
6	0.6462

EXERCISES

1. Verify that p_∞^* and α^*, the least squares estimators of p_∞ and α, respectively, are given by (5.2.15) and (5.2.16).

2. Let us define $f_{11} \equiv \partial^2 \log_e L / \partial p_\infty^2$, $f_{12} \equiv \partial^2 \log_e L / \partial p_\infty \partial \alpha$, and $f_{22} \equiv \partial^2 \log_e L / \partial \alpha^2$, where $L \equiv L(p_\infty, \alpha)$ is the likelihood function for model 1. Prove that $f_{11} < 0$, $f_{22} < 0$, but that $f_{11}f_{22} - f_{12}^2 > 0$, thus ensuring that the unique solution $(\hat{p}_\infty, \hat{\alpha})$ of the vector parameter (p_∞, α) produces a maximum.

3. A clinical trial is conducted on patients with acute leukemia. The trial consists of five stages with 10 new patients selected at each stage. The proportions of patients achieving remission at each stage are 2/10, 4/10, 5/10, 6/10, and 7/10, for stages 1 through 5. Using model 1 (i.e., $p_i = p_\infty - \alpha/i$), obtain the least squares and maximum likelihood estimates of p_∞ and α.

4. Obtain a 95 percent joint elliptical confidence region for $\begin{pmatrix} p_\infty \\ \alpha \end{pmatrix}$ based on the maximum likelihood estimates $\hat{p}_\infty$ and $\hat{\alpha}$ in Exercise 3. Also obtain $P_{Li}(\hat{p}_\infty, \hat{\alpha})$, the approximate 95 percent lower confidence limit at each of the five stages of the clinical trial described in Exercise 3. Also obtain the estimated sixth stage 95 percent lower confidence limit.

5. Show that the solution of vector equation (5.2.26) yields (5.2.27) and (5.2.28).

6. Show that the separate 95 percent confidence intervals for p_∞ and α in Example 5.2.2 are $0.6614 < p_\infty < 1$ and $0.1562 < \alpha < 0.6614$.

5.3 EXPONENTIAL AND LOGISTIC GROWTH MODELS

In this section we discuss two more specific growth models that are important in applications—the exponential growth model and the logistic growth model, which are called Models 2 and 3, respectively.

Model 2, the exponential growth model, is used often as an alternative to Model 1, the hyperbolic growth model, when $G(i) = 1/i$. We now must determine the circumstances under which an experimenter would prefer Model 2 to Model 1. The answer is found by examining the types of survival growth represented by each model. When $G(i) = 1/i$, Model 1 is characterized by rapid growth in survival in the early stages, but tapering off occurs rather quickly, usually beyond the third stage. On the other hand, Model 2 depicts a slower growth in the early stages which sustains itself in the later stages. Figure 5.2 exhibits the difference in Models 1 and 2 for a specific choice of the model parameters. Thus when faced with the problem of which model to use, the experimenter should plot his survival probabilities at each stage of the clinical trial. He is then likely able to decide between the two models or perhaps to develop another model or even not to use a growth model to analyze his survival data.

Model 3, the logistic model, was pioneered by Berkson [1953]. Model 3 is discussed at some length later in this section. It should be noted now, however, that there are no parameter restrictions associated with using Model 3 as there are for Models 1 and 2. As was pointed out in the preceding paragraph, a plot of the survival probabilities is invaluable in assisting the experimenter to decide on the appropriate model.

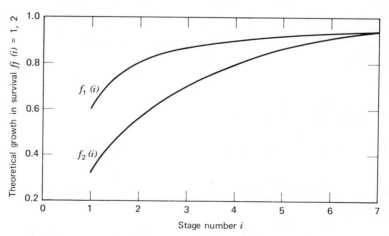

Figure 5.2 Exponential survivability versus hyperbolic survivability. Hyperbolic model: $f_1(i) = 1 - 1/2.5i$. Exponential model: $f_2(i) = 1 - \exp(-i/2.5)$.

Estimation for Model 2: $p_i = 1 - \alpha_1 \exp(-\alpha_2 i)$

Let $p_i(\alpha_1, \alpha_2)$ be defined by (5.2.2). That is,

$$p_i(\alpha_1, \alpha_2) = 1 - \alpha_1 \exp(-\alpha_2 i), \qquad i = 1, 2, \ldots, k. \tag{5.3.1}$$

Since for Model 2

$$1 - p_i(\alpha_1, \alpha_2) = \alpha_1 \exp(-\alpha_2 i), \qquad i = 1, 2, \ldots, k, \tag{5.3.2}$$

the least squares estimators α_1^* and α_2^* are obtained in the following way. First

$$E\left(\frac{n_i - x_i}{n_i}\right) = \alpha_1 \exp(-\alpha_2 i), \qquad i = 1, 2, \ldots, k,$$

where $x_i \leqslant n_i - 1$. $\tag{5.3.3}$

As a result of (5.3.3) a least squares fit on the logarithms is used to obtain α_1^* and α_2^*. The least squares equation* is then

$$\psi(\alpha_1, \alpha_2) = \sum_{i=1}^{k} \left[\log_e \frac{n_i - x_i}{n_i + 1} - \log_e \alpha_1 + \alpha_2 i \right]^2. \tag{5.3.4}$$

Let us set $z_i \equiv \log_e \{(n_i - x_i)/(n_i + 1)\}$. Differentiating $\psi(\alpha_1, \alpha_2)$ with respect to α_1 and α_2 and setting the resulting equations equal to zero, we obtain

$$\sum_{i=1}^{k} z_i - k \log_e \alpha_1^* + \frac{k(k+1)}{2} \alpha_2^* = 0 \tag{5.3.5}$$

and

$$\sum_{i=1}^{k} i z_i - \frac{k(k+1)}{2} \log_e \alpha_1^* + \frac{k(k+1)(2k+1)}{6} \alpha_2^* = 0. \tag{5.3.6}$$

Solving (5.3.5) and (5.3.6) in terms of $\log_e \alpha_1^*$ and α_2^*, we find

$$\log_e \alpha_1^* = 2[k(k-1)]^{-1} \left\{ (2k+1) \sum_{i=1}^{k} z_i - 3 \sum_{i=1}^{k} i z_i \right\} \tag{5.3.7}$$

---

*We use in (5.3.4) $n_i + 1$ instead of n_i in the denominator to avoid the possibility of a zero value.

and

$$\alpha_2^* = 6[k(k-1)]^{-1}\left\{ \sum_{i=1}^{k} z_i - 2\frac{\sum_{i=1}^{k} i z_i}{k+1} \right\}. \qquad (5.3.8)$$

Note that the least squares estimators for the exponential model are not obtained by using (5.2.5) and (5.2.6) because these equations do not yield explicit closed form values as do (5.3.7) and (5.3.8).

The maximum likelihood estimators for model 2 are obtained as follows. Let $L(\alpha_1, \alpha_2)$ be the likelihood function for all k stages of the clinical trial. Then we have

$$L(\alpha_1, \alpha_2) = \prod_{i=1}^{k} \binom{n_i}{x_i} (1 - \alpha_1\exp(-\alpha_2 i))^{x_i}(\alpha_1\exp(-\alpha_2 i))^{n_i - x_i}. \qquad (5.3.9)$$

Taking logarithms and then differentiating with respect to α_1 and α_2, the maximum likelihood estimators $\hat{\alpha}_1$ and $\hat{\alpha}_2$ are obtained as the unique solution to

$$-\sum_{i=1}^{k} \frac{x_i}{[\exp(\hat{\alpha}_2 i) - \hat{\alpha}_1]} + \frac{\sum_{i=1}^{k}(n_i - x_i)}{\hat{\alpha}_1} = 0 \qquad (5.3.10)$$

and

$$\hat{\alpha}_1 \sum_{i=1}^{k} \frac{i x_i}{[\exp(\hat{\alpha}_2 i) - \hat{\alpha}_1]} - \sum_{i=1}^{k} i(n_i - x_i) = 0. \qquad (5.3.11)$$

Since $\hat{\alpha}_1$ and $\hat{\alpha}_2$ cannot be obtained in closed form, the Newton-Raphson iterative technique in two dimensions is used to obtain these estimates. The details, which are completely analogous to those used in obtaining $\hat{p}_\infty$ and $\hat{\alpha}$ in Model 1, are left as an exercise to the reader. Here the initial estimates $\hat{\alpha}_{10}$ and $\hat{\alpha}_{20}$ are α_1^* and α_2^*, which are found from (5.3.7) and (5.3.8), respectively.

To apply the Newton-Raphson procedure to find $\hat{\alpha}_1$ and $\hat{\alpha}_2$ in Example

5.3.1, we include the second partial derivatives of $\log_e L(\alpha_1, \alpha_2)$.

$$\frac{\partial^2 \log_e L}{\partial \alpha_1^2} = -\left[\sum_{i=1}^{k} \frac{x_i}{[\exp(\alpha_2 i) - \alpha_1]^2} + \sum_{i=1}^{k} \frac{n_i - x_i}{\alpha_1^2}\right], \quad (5.3.12)$$

$$\frac{\partial^2 \log_e L}{\partial \alpha_1 \partial \alpha_2} = \sum_{i=1}^{k} \frac{i x_i \exp(\alpha_2 i)}{[\exp(\alpha_2 i) - \alpha_1]^2}, \quad (5.3.13)$$

and

$$\frac{\partial^2 \log_e L}{\partial \alpha_2^2} = -\alpha_1 \sum_{i=1}^{k} \frac{i^2 x_i \exp(\alpha_2 i)}{[\exp(\alpha_2 i) - \alpha_1]^2}. \quad (5.3.14)$$

Example 5.3.1. A clinical trial is conducted in five stages with the aim of improving the probability that patients suffering from arthritis will receive relief when they undergo therapy. There are eight patients in each of the five stages; the proportions of patients receiving relief at each of the stages are 2/8, 3/8, 3/8, 5/8, and 6/8. We assume an exponential growth model to describe the stage-to-stage improvement that is visible in the data.

We compute $\hat{\alpha}_1$ and $\hat{\alpha}_2$, the maximum likelihood estimates of α_1 and α_2, respectively, for this model. To this end, let us obtain the initial estimates $\hat{\alpha}_{10}$ and $\hat{\alpha}_{20}$ by the least squares equations (5.3.7) and (5.3.8). These estimates $\hat{\alpha}_{10}$ and $\hat{\alpha}_{20}$ are obtained from the layout in Table 5.6.

From Table 5.6 and (5.3.7) and (5.3.8), it follows that $\hat{\alpha}_{10} = 0.9760$ and $\hat{\alpha}_{20} = 0.2708$. To facilitate the computations for the Newton-Raphson procedure used to obtain $\hat{\alpha}_1$ and $\hat{\alpha}_2$, Table 5.7 is given. For

Table 5.6 Least squares computation layout, model 2

Stage i	$(n_i - x_i)/(n_i + 1)$	$z_i = \log_e(n_i - x_i)/(n_i + 1)$	$i z_i$
1	6/9	−0.4055	−0.4055
2	5/9	−0.5878	−1.1756
3	5/9	−0.5878	−1.7634
4	3/9	−1.0986	−4.3944
5	2/9	−1.5041	−7.5204
Totals		−4.1838	−15.2593

Model 2, $\hat{g}_{10i}$, $\hat{g}_{20i}$, $-\hat{f}_{110i}$, $\hat{f}_{120i}$, and $-\hat{f}_{220i}$ are, respectively,

$$-\frac{x_i}{\exp(\hat{\alpha}_{20}i) - \hat{\alpha}_{10}} + \frac{n_i - x_i}{\hat{\alpha}_{10}},$$

$$\frac{\hat{\alpha}_{10}ix_i}{\exp(\hat{\alpha}_{20}i - \hat{\alpha}_{10})} - i(n_i - x_i),$$

$$-\left(\frac{x_i}{[\exp(\hat{\alpha}_{20}i) - \hat{\alpha}_{10}]^2} + \frac{n_i - x_i}{\hat{\alpha}_{10}^2}\right),$$

$$\frac{ix_i\exp(\hat{\alpha}_{20}i)}{[\exp(\hat{\alpha}_{20}i) - \hat{\alpha}_{10}]^2},$$

and

$$-\hat{\alpha}_{10}\frac{i^2x_i\exp(\hat{\alpha}_{20}i)}{[\exp(\hat{\alpha}_{20}i) - \hat{\alpha}_{10}]^2}.$$

Table 5.7 Maximum likelihood computation layout (first iteration), model 2

Stage i	$\hat{g}_{10i}$	$\hat{g}_{20i}$	$-\hat{f}_{110i}$	$\hat{f}_{120i}$	$-\hat{f}_{220i}$
1	0.1774	−0.1731	24.1239	23.3690	22.8081
2	1.0842	−2.1163	10.6855	18.6894	36.4817
3	2.7742	−8.1230	7.0876	12.4301	36.3953
4	0.5461	−2.1319	4.4271	15.0993	58.9477
5	−0.0220	0.1074	2.8145	13.8450	67.5636
Totals	4.5599	−12.4369	49.1386	83.4328	222.1964

The values $\hat{\alpha}_{11}$ and $\hat{\alpha}_{21}$ are then found by solving the matrix equation

$$\begin{pmatrix} \hat{\alpha}_{11} \\ \hat{\alpha}_{21} \end{pmatrix} = \begin{pmatrix} 0.9760 \\ 0.2708 \end{pmatrix} - \begin{bmatrix} -49.1386 & 83.4328 \\ 83.4328 & -222.1964 \end{bmatrix}^{-1} \begin{bmatrix} 4.5599 \\ -12.4369 \end{bmatrix}.$$

$$(5.3.15)$$

We thus find $\hat{\alpha}_{11} = 0.9716$ and $\hat{\alpha}_{21} = 0.2136$. It took a total of five iterations before the Newton-Raphson procedure converged to the maximum likelihood estimates. Table 5.8 shows the results at each step. Hence the maximum likelihood estimates of α_1 and α_2 are taken as $\hat{\alpha}_1 = 0.9719$ and $\hat{\alpha}_2 = 0.2232$.

Table 5.8 Values of $\hat{\alpha}_{1j}$ and $\hat{\alpha}_{2j}$ at each iteration of the Newton-Raphson procedure for Example 5.3.1

Iteration j	$\hat{\alpha}_{1j}$	$\hat{\alpha}_{2j}$
0	0.9760	0.2708
1	0.9716	0.2136
2	1.0994	0.2701
3	0.9815	0.2246
4	0.9720	0.2232
5	0.9719	0.2232

As with Example 5.2.1, it is instructive to compare the observed and expected relief probabilities at each stage using both the least squares and maximum likeihood estimates. These comparisons appear in Table 5.9.

Table 5.9 Expected and observed relief probabilities for arthritis patients at each stage when clinical trial is conducted in five stages

Stage i	Observed Probability	Expected MLE Probability	Expected LSE Probability
1	0.2500	0.2225	0.2555
2	0.3750	0.3781	0.4321
3	0.3750	0.5025	0.5669
4	0.6250	0.6020	0.6696
5	0.7500	0.6816	0.7480

As Table 5.9 indicates, the least squares estimates (LSE) tend to overestimate, and the maximum likelihood estimates (MLE) tend to underestimate the observed probabilities of relief at each stage. Thus the least squares estimates are somewhat optimistic, whereas the maximum likelihood estimates are somewhat conservative in this application.

Confidence Regions and Intervals for Model 2: $p_i = 1 - \alpha_1 \exp(-\alpha_2 i)$

A joint elliptical confidence region for the parameters α_1 and α_2 follows as for Model 1. We leave the details to the reader, merely writing down the equation. Thus an approximate $(1-\alpha)100$ percent elliptical confidence region for the parameter vector $\begin{pmatrix} \alpha_1 \\ \alpha_2 \end{pmatrix}$ is

$$\sigma^{11}(\hat{\alpha}_1 - \alpha_1)^2 + \sigma^{22}(\hat{\alpha}_2 - \alpha_2)^2 + 2\sigma^{12}(\hat{\alpha}_1 - \alpha_1)(\hat{\alpha}_2 - \alpha_2) \doteq \chi^2(1-\alpha; 2),$$

(5.3.16)

where $\chi^2(1-\alpha; 2)$ is the $(1-\alpha)100$ percentage point of the chi-square distribution with 2 degrees of freedom, and

$$\sigma^{ij} = -E\left[\frac{\partial^2 \log_e L}{\partial \alpha_i \partial \alpha_j}\right], \qquad i,j = 1,2.$$

(5.3.17)

For Model 2 it is not difficult to show that

$$\sigma^{11} = \alpha_1^{-1} \sum_{i=1}^{k} \frac{n_i}{\exp(\alpha_2 i) - \alpha_1},$$

(5.3.18)

$$\sigma^{12} = -\sum_{i=1}^{k} \frac{i n_i}{\exp(\alpha_2 i) - \alpha_1},$$

(5.3.19)

and

$$\sigma^{22} = \alpha_1 \sum_{i=1}^{k} \frac{i^2 n_i}{\exp(\alpha_2 i) - \alpha_1}.$$

(5.3.20)

Again we should be aware that σ^{11}, σ^{12}, and σ^{22} contain the unknown parameters α_1 and α_2; therefore in practice we use $\hat{\sigma}^{11}$, $\hat{\sigma}^{12}$, and $\hat{\sigma}^{22}$, where $\hat{\alpha}_1$ and $\hat{\alpha}_2$ replace α_1 and α_2, respectively, in (5.3.18) through (5.3.20).

As with Model 1 we may obtain an approximate $(1-\alpha)100$ percent lower confidence limit $p_{Li}(\hat{\alpha}_1, \hat{\alpha}_2)$ for Model 2 by means of (5.3.21). We leave the details of the derivation as an exercise to the reader. We can thus write

$$p_{Li}(\hat{\alpha}_1, \hat{\alpha}_2) \doteq (1 - \hat{\alpha}_1 \exp(-\hat{\alpha}_2 i)) - Z(1-\alpha)\sqrt{\text{Vâr}(1 - \hat{\alpha}_1 \exp(-\hat{\alpha}_2 i))} \ ,$$

$$(5.3.21)$$

where

(i) $\text{Var}(1 - \hat{\alpha}_1 \exp(-\hat{\alpha}_2 i))$ is given as

$$\text{Var}(1 - \hat{\alpha}_1 \exp(-\hat{\alpha}_2 i)) \doteq \exp(-2\alpha_2 i)\left[\sigma_{\hat{\alpha}_1}^2 + i^2 \alpha_1^2 \sigma_{\hat{\alpha}_2}^2 - 2\alpha_1 i \sigma_{\hat{\alpha}_1 \hat{\alpha}_2}\right], \quad (5.3.22)$$

(ii)   the values $\sigma_{\hat{\alpha}_1}^2$, $\sigma_{\hat{\alpha}_2}^2$, and $\sigma_{\hat{\alpha}_1 \hat{\alpha}_2}$ are the elements of the matrix

$$\begin{bmatrix} \sigma_{\hat{\alpha}_1}^2 & \sigma_{\hat{\alpha}_1 \hat{\alpha}_2} \\ \sigma_{\hat{\alpha}_1 \hat{\alpha}_2} & \sigma_{\hat{\alpha}_2}^2 \end{bmatrix} = \begin{bmatrix} \sigma^{11} & \sigma^{12} \\ \sigma^{12} & \sigma^{22} \end{bmatrix}^{-1}, \qquad (5.2.23)$$

(iii) σ^{11}, σ^{12}, and σ^{22} are given by (5.3.18) through (5.3.20), and
(iv) $\text{Vâr}(1 - \hat{\alpha}_1 \exp(-\hat{\alpha}_2 i)$ replaces $\text{Var}(1 - \hat{\alpha}_1 \exp(\hat{\alpha}_2 i))$ in practice as we replace α_1 and α_2 by their respective maximum likelihood estimates $\hat{\alpha}_1$ and $\hat{\alpha}_2$, $i = 1, 2, \ldots, k+1$.

Example 5.3.2. From the arthritis relief data in Example 5.3.1, the maximum likelihood estimates for the parameters α_1 and α_2 are $\hat{\alpha}_1 = 0.9719$ and $\hat{\alpha}_2 = 0.2232$. Using these values and (5.3.18) through (5.3.20), we find $\hat{\sigma}^{11} = 60.3100$, $\hat{\sigma}^{12} = -120.1625$, and $\hat{\sigma}^{22} = 328.8402$. With the aid of (5.3.16) and recalling that $\chi^2(0.95; 2)$, the 95th percentage point of the chi-square distribution with 2 degrees of freedom is 5.99, an approximate 95 percent joint elliptical confidence region for the vector parameter $\begin{pmatrix} \alpha_1 \\ \alpha_2 \end{pmatrix}$ is given by

$$60.3100(\alpha_1 - 0.9719)^2 + 328.8402(\alpha_2 - 0.2232)^2 + 2(120.1625)(\alpha_1 - 0.9719)$$

$$\times (\alpha_2 - 0.2232) \doteq 5.99. \quad (5.3.24)$$

The separate 95 percent confidence intervals for α_1 and α_2 are found to be $0.4878 < \alpha_1 < 1.0159$ and $0.0158 < \alpha_2 < 0.4306$. The upper limit 1.0159 for α_1 follows because $\alpha_1 < \exp(\alpha_2)$ is a physical upper limit for α_1 and $1.0159 = \exp(0.0158)$. The details of (5.3.24) and the graph of the elliptical confidence region are left as an exercise to the reader.

The values $\hat{\sigma}_{\hat{\alpha}_1}^2$, $\hat{\sigma}_{\hat{\alpha}_2}^2$ and $\hat{\sigma}_{\hat{\alpha}_1\hat{\alpha}_2}$ are obtained by inverting the matrix of values $\hat{\sigma}^{11}$, $\hat{\sigma}^{12}$, and $\hat{\sigma}^{22}$. Thus we see that

$$
\begin{bmatrix} \hat{\sigma}_{\hat{\alpha}_1}^2 & \hat{\sigma}_{\hat{\alpha}_1\hat{\alpha}_2} \\ \hat{\sigma}_{\hat{\alpha}_1\hat{\alpha}_2} & \hat{\sigma}_{\hat{\alpha}_2}^2 \end{bmatrix} = \begin{bmatrix} 60.3100 & -120.1625 \\ -120.1625 & 328.8402 \end{bmatrix}^{-1} = \begin{bmatrix} 0.0610 & 0.0223 \\ 0.0223 & 0.0112 \end{bmatrix}.
$$

$$(5.3.25)$$

By (5.3.21) through (5.3.23) we find that

$$
p_{Li}(\hat{\alpha}_1,\hat{\alpha}_2) = (1 - 0.9719\exp(-0.2232i))
$$

$$
- (1.645)\sqrt{\exp(-0.4464i)\{0.0610 + 0.01056i^2 - 0.0433i\}}
$$

$$(5.3.26)$$

is a 95 percent lower confidence limit for $p_i(\hat{\alpha}_1,\hat{\alpha}_2)$, $i=1,2,\ldots,6$, noting that 1.645 is the upper 95th percentage point of the standard normal distribution. Table 5.10 gives the values of $p_{Li}(\hat{\alpha}_1,\hat{\alpha}_2)$ for Model 2 at each of the five stages of the clinical trial and, in addition, the predicted value $p_{L6}(\hat{\alpha}_1,\hat{\alpha}_2)$ with confidence 0.95.

Table 5.10 95 percent lower confidence limits for the probability of relief from arthritis at each of five stages of clinical trial, plus predicted sixth stage lower limit

Stage i	$P_{Li}(\hat{\alpha}_1,\hat{\alpha}_2)$
1	0.0014
2	0.2426
3	0.3665
4	0.4416
5	0.5041
6	0.5617

Special Parameter Considerations for Models 1 and 2

In some instances the maximum likelihood and/or the least squares methods of estimation yield values of p_∞ (in the case of Model 1) and α_1 (in the case of Model 2) that violate the physical restraints imposed on them. That is, we observe occasionally values $\hat{p}_\infty > 1$ for Model 1 and

$\hat{\alpha}_1 > \exp(\hat{\alpha}_2)$ for Model 2. The remedy in these situations is fairly straight-forward.

In the case of Model 1, after the experimenter has ascertained graphi-cally that the growth in survivability he observes has the form (5.3.27), the procedure is to set $p_\infty = 1$. The model then becomes

$$p_i = [1 - G(i)\alpha], \qquad (5.3.27)$$

where in the case of hyperbolic growth $G(i) = 1/i$, $i = 1, \ldots, k$. The like-lihood function for α based on x_i surviving patients among the n_i patients in the ith stage, $i = 1, 2, \ldots, k$, is

$$L(\alpha) = \prod_{i=1}^{k} \binom{n_i}{x_i} [1 - G(i)\alpha]^{x_i} [G(i)\alpha]^{n_i - x_i}. \qquad (5.3.28)$$

The maximum likelihood estimator $\hat{\alpha}$ is then found by solving

$$-\sum_{i=1}^{k} \frac{x_i G(i)}{[1 - \hat{\alpha}G(i)]} + \sum_{i=1}^{k} \frac{n_i - x_i}{\hat{\alpha}} = 0 \qquad (5.3.29)$$

in terms of $\hat{\alpha}$. As in the two-parameter case, $\hat{\alpha}$ must be solved iteratively—utilizing, for example, the Newton-Raphson procedure. The initial value $\hat{\alpha}_0$ can be taken as the least squares estimator that is obtainable in closed form. The details of the estimation procedure are left as an exercise to the reader.

For Model 2, it may happen in practice that $\hat{\alpha}_1 > \exp(\hat{\alpha}_2)$, which implies $\hat{p}_1 = 1 - \hat{\alpha}_1 \exp(-\hat{\alpha}_2) < 0$. If this apparently anomalous situation occurs, we set $\alpha_1 = 1$, and Model 2 becomes

$$p_i = 1 - \exp(-\alpha_2 i), \qquad i = 1, 2, \ldots, k, \qquad (5.3.30)$$

and the likelihood function to be maximized with respect to α_2 is

$$L(\alpha_2) = \prod_{i=1}^{k} \binom{n_i}{x_i} [1 - \exp(-\alpha_2 i)]^{x_i} [\exp(-\alpha_2 i)]^{n_i - x_i}. \qquad (5.3.31)$$

The maximum likelihood estimator $\hat{\alpha}_2$ is then found as the solution to

$$\sum_{i=1}^{k} \left[\frac{ix_i}{\exp(\hat{\alpha}_2 i) - 1} \right] - \sum_{i=1}^{k} i(n_i - x_i) = 0 \qquad (5.3.32)$$

in terms of $\hat{\alpha}_2$. The details of the estimation procedure are left as an exercise to the reader.

Estimation for Model 3: $p_i = \{1 + \exp[-(\alpha_1 + \alpha_2 i)]\}^{-1}$

Let $p_i(\alpha_1, \alpha_2)$ be defined by (5.2.3). That is,

$$p_i(\alpha_1, \alpha_2) = \{1 + \exp[-(\alpha_1 + \alpha_2 i)]\}^{-1}, \qquad i = 1, \ldots, k. \qquad (5.3.33)$$

Model 3, also termed the logistic model, is discussed by Berkson [1953] in considerable detail. It is no more computationally difficult than either Model 1 or Model 2, and $p_i(\alpha_1, \alpha_2)$ always lies in the interval $0 < p_i(\alpha_1, \alpha_2) < 1$, regardless of the values of α_1, α_2, and i. However, as was stated previously, before an experimenter chooses from among these three models the one that best fits his data, he should plot the survival probabilities stage by stage on graph paper and try to determine roughly by eye which is the appropriate model. Model 3 can be rewritten as

$$\log_e \frac{p_i}{1 - p_i} = \alpha_1 + \alpha_2 i, \qquad i = 1, 2, \ldots, k. \qquad (5.3.33')$$

Thus if the experimenter plots the stage-by-stage log odds—that is, $\log_e[x_i/(n_i - x_i)]$—and these plot roughly as a straight line as a function of i, he has empirical justification for utilizing Model 3.

The parameters α_1 and α_2 can be estimated by their maximum likelihood estimators as was done for the previous models. Development of the maximum likelihood estimation procedure is straightforward, differing little from that discussed for Models 1 and 2. However, to obtain closed form expressions for our estimators of α_1 and α_2 and to take advantage of the theory of linear regression, we consider a different method of estimation here. This procedure appears in Zellner and Lee [1965] and Cook and Gross [1968].

Let us define

$$r_i = p_i + u_i, \qquad (5.3.34)$$

where $r_i \equiv x_i/n_i$, $0 < r_i < 1$, is the ratio of survivors to patients in the ith stage of the trial, $p_i = \{1 + \exp[-(\alpha_1 + \alpha_2 i)]\}^{-1}$, and u_i is a variable having the binomial distribution with mean zero and variance $p_i(1 - p_i)/n_i$, such that $-p_i < u_i < 1 - p_i$, $i = 1, 2, \ldots, k$. A little algebra shows that we can write (5.3.34) as

$$\frac{r_i}{1 - r_i} = \left(\frac{p_i}{1 - p_i}\right)\left(\frac{1 + u_i/p_i}{1 - u_i/(1 - p_i)}\right), \qquad i = 1, 2, \ldots, k. \qquad (5.3.35)$$

Define $z_i \equiv \log_e\{r_i/(1-r_i)\}$, where we assume $0 < r_i < 1$. Then from the definition of p_i, it follows that

$$z_i = (\alpha_1 + \alpha_2 i) + \log_e\left(1 + \frac{u_i}{p_i}\right) - \log_e\left(1 - \frac{u_i}{1-p_i}\right), \qquad i = 1,2,\ldots,k. \quad (5.3.36)$$

Since $|u_i/p_i| < 1$ and $u_i/(1-p_i) < 1$, we can expand the last two terms of (5.3.36) using the logarithmic series expansion $\log_e(1+x) = x - x^2/2 + x^3/3$, for $|x| < 1$. Thus

$$z_i = \alpha_1 + \alpha_2 i + \frac{u_i}{p_i(1-p_i)} + R_i, \qquad (5.3.37)$$

where R_i is the remainder term whose omission will not seriously affect the model for values $n_i \geqslant 10$, say. Let us now set $\varepsilon_i = u_i/p_i(1-p_i)$ and write

$$z_i \doteq \alpha_1 + \alpha_2 i + \varepsilon_i, \qquad i = 1,2,\ldots,k. \qquad (5.3.38)$$

Estimating α_1 and α_2 is now carried out by the method of weighted least squares, since $\operatorname{Var} \varepsilon_i = [n_i p_i(1-p_i)]^{-1}$, hence $\varepsilon_1,\ldots,\varepsilon_k$, are not of equal variance (homoscedastic). Using Aitken's [1935] generalized least squares procedure, the least squares estimator vector $\begin{pmatrix} \alpha_1^* \\ \alpha_2^* \end{pmatrix}$ is given by the vector equation

$$\begin{pmatrix} \alpha_1^* \\ \alpha_2^* \end{pmatrix} = \left(X'\textstyle\sum^{-1}X\right)^{-1}X'\textstyle\sum^{-1}\mathbf{Z}, \qquad (5.3.39)$$

where

$$X' = \begin{pmatrix} 11 & \cdots & 1 \\ 12 & \cdots & k \end{pmatrix}.$$

Here X is the transpose of X', $\sum^{-1}$ is a diagonal matrix whose ith element $[n_i r_i(1-r_i)]$ and Z' is the row vector $(z_1, z_2,\ldots,z_k)$. Z being the corresponding column vector. Individually, then, the least squares estimators α_1^* and α_2^* are

$$\alpha_1^* = \frac{\left(\sum_{i=1}^{k} i^2 w_i^2\right)\left(\sum_{i=1}^{k} z_i w_i^2\right) - \left(\sum_{i=1}^{k} i w_i^2\right)\left(\sum_{i=1}^{k} i z_i w_i^2\right)}{\left(\sum_{i=1}^{k} w_i^2\right)\left(\sum_{i=1}^{k} i^2 w_i^2\right) - \left(\sum_{i=1}^{k} i w_i^2\right)^2} \qquad (5.3.40)$$

and

$$\alpha_2^* = \frac{\left(\sum_{i=1}^{k} w_i^2\right)\left(\sum_{i=1}^{k} iz_i w_i^2\right) - \left(\sum_{i=1}^{k} iw_i^2\right)\left(\sum_{i=1}^{k} z_i w_i^2\right)}{\left(\sum_{i=1}^{k} w_i^2\right)\left(\sum_{i=1}^{k} i^2 w_i^2\right) - \left(\sum_{i=1}^{k} iw_i^2\right)^2}, \qquad (5.3.41)$$

where we define $w_i = [n_i r_i (1 - r_i)]^{1/2}$ for $i = 1, 2, \ldots, k$.

The approximate variance–covariance matrix of the estimators α_1^* and α_2^* is

$$\mathrm{Var}\begin{pmatrix} \alpha_1^* \\ \alpha_2^* \end{pmatrix} \doteq \Delta^{-1} \begin{bmatrix} \sum_{i=1}^{k} i^2 w_i^2 & -\sum_{i=1}^{k} iw_i^2 \\[2mm] -\sum_{i=1}^{k} iw_i^2 & \sum_{i=1}^{k} w_i^2 \end{bmatrix}, \qquad (5.3.42)$$

where

$$\Delta = \left(\sum_{i=1}^{k} i^2 w_i^2\right)\left(\sum_{i=1}^{k} w_i^2\right) - \left(\sum_{i=1}^{k} iw_i^2\right)^2.$$

Thus to test the hypothesis $H_0 : \alpha_2 = 0$ against the alternative $H_1 : \alpha_2 \neq 0$, we consider the statistic

$$z = \frac{\alpha_2^*}{\sqrt{\sum_{i=1}^{k} w_i^2 / \Delta}}. \qquad (5.3.43)$$

When H_0 is true, z has approximately a standard normal distribution. Thus $H_0 : \alpha_2 = 0$ is rejected at level α only if $|z| > Z(1 - \alpha/2)$, where $Z(1 - \alpha/2)$ is the $(1 - \alpha/2)$ 100 percentage point of the standard normal distribution. Rejection of H_0 implies change in survival from stage to stage. Furthermore, if $\alpha_2^* > 0$, we can infer growth in survivability on rejection of H_0.

Finally, an approximate $(1 - \alpha)$ 100 percent lower confidence limit $p_{Li}(\alpha_1^*, \alpha_2^*)$ is given for each stage of the clinical trial. First of all, we note that an approximate $(1 - \alpha)$ 100 percent lower limit for $\alpha_1 + \alpha_2 i$ is $(\alpha_1^* + \alpha_2^* i) - Z(1 - \alpha)[\mathrm{Var}(\alpha_1^* + \alpha_2^* i)]^{\frac{1}{2}}$, where

$$\mathrm{Var}(\alpha_1^* + \alpha_2^* i) \doteq a_2^2 + a_0^2 i^2 - 2a_1^2 i, \qquad (5.3.44)$$

$a_m^2 = (\Sigma_{i=1}^k i^m w_i^2)/\Delta$, $m = 0, 1, 2$ for $i = 1, 2, \ldots, k$, and $Z(1 - \alpha)$ is the $(1 - \alpha)$ percentile of the standard normal distribution.
Hence

$$p_{Li}(\alpha_1^*, \alpha_2^*) = \left\{ 1 + \exp\left[-\left\{ (\alpha_1^* + \alpha_2^* i) - Z(1 - \alpha)[a_2^2 + a_0^2 i^2 - 2a_1^2 i]^{\frac{1}{2}} \right\} \right] \right\}^{-1},$$

$$1 = 1, 2, \ldots, k. \tag{5.3.45}$$

Example 5.3.3. A clinical trial on a new anticancer agent produced the following remission ratios for 20 patients on trial at each of the five stages of the trial: $9/20$, $12/20$, $14/20$, $15/20$, and $16/20$. Figure 5.3 is a graph of $\log_e r_i/(1 - r_i)$ as a function of the stage i, $i = 1, 2, 3, 4, 5$. From Figure 5.3 there appears to be empirical justification for utilizing Model 3 to describe the growth in the probability a patient achieves a remission as a function of the stage.

  To facilitate calculations involving α_1^*, α_2^*, their variances and covariances, as well as the calculations of the predicted remission probabilities and their lower bounds, we construct Table 5.11, from

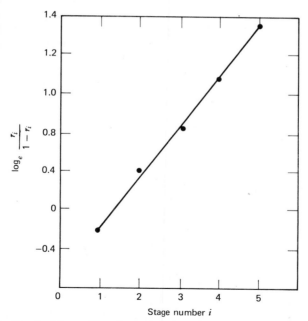

Figure 5.3 Graph of $\log_e r_i/(1 - r_i)$ as a function of stage i for cancer remission data (data from Example 5.3.3).

Table 5.11 Computational Layout of Model 3

Stage i	r_i	z_i	w_i^2	iw_i^2	$i^2w_i^2$	$z_iw_i^2$	$iz_iw_i^2$
1	0.4500	−0.2007	4.9500	4.9500	4.9500	−0.9935	−0.9935
2	0.6000	0.4055	4.8000	9.6000	19.2000	1.9464	3.8928
3	0.7000	0.8473	4.2000	12.6000	37.8000	3.5587	10.6761
4	0.7500	1.0986	3.7500	15.0000	60.0000	4.1198	16.4792
5	0.8000	1.3863	3.2000	16.0000	80.0000	4.4362	22.1810
Totals	—	—	20.9000	58.1500	201.9500	13.0676	52.2356

Table 5.12 Expected and observed remission probabilities for cancer patients at each stage when the clinical trial is conducted in five stages, including sixth stage predicted value and lower limits for all stages

Stage i	Observed Probability	Expected Probability	Lower 95 Percent Limit
1	0.4500	0.4802	0.3395
2	0.6000	0.5783	0.4758
3	0.7000	0.6714	0.5868
4	0.7500	0.7515	0.6521
5	0.8000	0.8179	0.6950
6	—	0.8696	0.7288

which it follows that

$$\alpha_1^* = \frac{(201.9500)(13.0676) - (58.1500)(52.2356)}{(201.9500)(20.9000) - (58.1500)^2} = -0.4748 \quad (5.3.46)$$

and

$$\alpha_2^* = \frac{(20.9000)(52.2356) - (58.1500)(13.0676)}{(201.9500)(20.9000) - (58.1500)^2} = 0.3954. \quad (5.3.47)$$

Furthermore, it follows from Table 5.11 that $a_0^2 = 0.0249$, $a_1^2 = 0.0693$, and $a_2^2 = 0.2406$. A test of the adequacy of the model shows $Z = 0.3954/\sqrt{0.0249} = 2.505$. This value of Z, significant at almost level

0.01, establishes a significant stage-to-stage growth in the proportion of patients achieving remission according to Model 3.

Table 5.12 shows the observed remission proportions and the expected remission proportions according to Model 3, including the predicted proportion of remissions at stage 6. Also included are the lower limits for the expected proportions with confidence 0.95 as obtained from (5.3.45).

EXERCISES

1. A clinical trial is conducted in five stages with the aim of improving the probability that patients suffering from arthritis receive relief when they undergo therapy. The proportions of patients who receive relief at each of five stages are $3/10$, $4/10$, $4/10$, $6/10$, and $8/10$, respectively. Assuming that Model 2 holds, obtain the least squares and maximum likelihood estimates of α_1 and α_2.

2. Obtain a 95 percent joint elliptical confidence region for $\begin{pmatrix} \alpha_1 \\ \alpha_2 \end{pmatrix}$ based on the maximum likelihood estimates in Exercise 1. Sketch the region. Finally, obtain the $p_{Li}(\hat{\alpha}_1, \hat{\alpha}_2)$, the approximate 95 percent lower confidence limit at each of the five stages of the clinical trial described in Exercise 1. Also obtain the estimated sixth stage 95 percent lower confidence limit.

3. Provide the details of the Newton-Raphson method for obtaining the maximum likelihood estimates $\hat{\alpha}_1$ and $\hat{\alpha}_2$ in Model 2.

4. Show, in general, how the elliptical confidence region for Model 2, equation (5.3.16), is obtained. Furthermore, obtain in general $p_{Li}(\hat{\alpha}_1, \hat{\alpha}_2)$ the lower $(1 - \alpha)100$ percent confidence limit for $p_i(\alpha_1, \alpha_2)$ at each stage i, $i = 1, \ldots, k$, as well as the predicted lower confidence limit $p_{L,k+1}(\hat{\alpha}_1, \hat{\alpha}_2)$ for Model 2.

5. Obtain the elliptical confidence region for the data in Example 5.3.2. That is, verify (5.3.24).

6. Outline the details for obtaining $\hat{\alpha}$ in the special case of Model 1, given by (5.3.27), by means of the Newton-Raphson procedure. What initial value would you choose, and how would you obtain it?

7. Outline the details for obtaining $\hat{\alpha}_2$ in the special case of Model 2, given by (5.3.30), by means of the Newton-Raphson procedure. What initial value would you choose, and how would you obtain it?

8. A clinical trial on a new anticancer agent produced the following remission ratios for 25 patients on trial at each of the five stages of the trial: $11/25$, $14/25$, $17/25$, $18/25$, and $20/25$. Assuming the validity of Model 3, obtain α_1^* and α_2^*. Test the hypothesis $H_0 : \alpha_2 = 0$ against $H_1 : \alpha_2 \neq 0$ at level 0.05. Finally, obtain a table similar to Table 5.12 showing the observed, expected, and 95 percent lower limit probabilities at each of the five stages, as well as the expected and 95 percent lower limit probability at stage 6.

9. Develop the maximum likelihood estimation procedure for Model 3 including the Newton-Raphson technique required to obtain the maximum likelihood estimates $\hat{\alpha}_1$ and $\hat{\alpha}_2$. Suggest initial estimates to be used for the Newton-Raphson procedure.

10. The regression technique for obtaining α_1^* and α_2^* in model 3 does not work if $r_i = 1$ or 0 at any stage i of the clinical trial. If faced with this situation in practice, how would you estimate α_1 and α_2 for Model 3?

5.4 AN ASSESSMENT MODEL FOR THE OVERALL PROBABILITY OF SUCCESS IN A CLINICAL TRIAL

Often in clinical trials patient response to treatment is classified as "success" or "failure." Models developed in this chapter thus far have been concerned in estimating the probability of success (failure) at each stage for clinical trials conducted in stages. In this section we consider the problem of assessing the overall probability of success (failure) in a clinical trial that is conducted in stages.

Lloyd and Lipow [1962] mention exponential smoothing in conjunction with reliability assessment; however, they do not develop the smoothing model that is put forth. In this section we consider a smoothing model developed by Gross [1971b] that is quite similar to the Lloyd and Lipow model.

Suppose a clinical trial is conducted in k stages, and we are to assess the overall probability of success at the end of each stage. One simple way to assess the probability of success at the end of each stage is to look at the ratio of the total number of patients who respond to the total number of patients for all stages up to that stage.

More generally, we consider the following approach. At the first stage of the trial $\tilde{R}_1$, the estimate of patient response is given by

$$\tilde{R}_1 = r_1 = \frac{x_1}{n_1}, \tag{5.4.1}$$

where n_1 patients enter stage 1 and x_1 are those who respond. At each succeeding stage i, $i = 2, \ldots, k$,

$$\tilde{R}_i = \alpha r_i + (1 - \alpha)\tilde{R}_{i-1}, \tag{5.4.2}$$

where r_i is the proportion of patients who respond at the ith stage, α is a constant to be chosen in the interval $0 \leqslant \alpha \leqslant 1$, and $\tilde{R}_{i-1}$ is the assessment of patient response at the end of the $(i-1)$th stage. It is not difficult to show that

$$\tilde{R}_i = \alpha \sum_{j=0}^{i-2} (1-\alpha)^j r_{i-j} + (1-\alpha)^{i-1} r_1, \qquad i = 2, \ldots, k. \qquad (5.4.3)$$

It is evident that the overall assessment of the probability of success is weighted more heavily toward the current stage of the trial than to the previous stages for $\alpha \cong 1$. In this case the early stages of the trial are discounted exponentially fast. If α is near zero, the first stage of testing dominates the overall estimate of assessment of the success of the clinical trial. Thus an appropriate choice of the smoothing constant, α, is of extreme importance for this model.

The experimenter ordinarily chooses α to be near unity if he believes the current stage of the clinical trial to be the most important. That is, if a major breakthrough has been made just prior to the current stage, it should be reflected in the overall assessment of the clinical trial. On the other hand, if the early stages of the clinical trial are not deemed any less important than the later stages, α should be chosen considerably less than 1, preventing the importance of early work from being diminished too quickly. More precise methods of estimating α may be found in Parzen [1967], who uses time series procedures to obtain estimators of α. These methods are generally beyond the scope of this book and are not discussed here. However, it is of interest to illustrate how changes in α influence the change in the overall assessment of a clinical trial.

Example 5.4.1. A clinical trial of an anticancer agent was conducted in five stages. The proportions of remissions in each of the stages were $r_1 = 0.50$, $r_2 = 0.60$, $r_3 = 0.70$, $r_4 = 0.80$, and $r_5 = 0.90$. To assess the overall proportion of remissions, the quantity $\tilde{R}_5$ is used, where $\tilde{R}_5$ is given by (5.4.3) for $i = 5$. Table 5.13 shows the values of $\tilde{R}_5$ for values of α between 0.10 and 0.90 in increments of 0.10. Clearly the assessed remission probability depends quite strongly on the value of α. Hence the judicious choice of α is very important.

We now introduce a variation of the Lloyd-Lipow model that is more flexible because the smoothing constant is allowed to change at each stage of the trial. We define

$$\hat{R}_1 = r_1 \qquad (5.4.4)$$

and

$$\hat{R}_i = \alpha_i r_i + (1 - \alpha_i)\hat{R}_{i-1}, \qquad i = 2, \ldots, k, \qquad (5.4.5)$$

Table 5.13 Values of $\tilde{R}_5$, the assessed overall remission probability of an anticancer agent, as a function of α, the smoothing constant.

α	$\tilde{R}_5$
0.10	0.590
0.20	0.664
0.30	0.723
0.40	0.769
0.50	0.806
0.60	0.835
0.70	0.857
0.80	0.875
0.90	0.889

where r_i, $i = 1, 2, \ldots, k$, is once more the proportion of responders in the ith stage of the trial; α_i is a constant chosen to lie in the interval $0 \leqslant \alpha_i \leqslant 1$, $i = 2, \ldots, k$. Thus the model represented by (5.4.5) has more flexibility than the Lloyd-Lipow model (5.4.2) in that the smoothing constant may change from stage to stage.

Gross [1971b] discusses two methods for determining the constants $\alpha_2, \ldots, \alpha_k$—a Bayesian approach as well as an empirical method. We summarize only the results of the empirical approach, since at each stage of the clinical trial the Bayesian technique requires subjective assessments, which in practice may be difficult to obtain.

We make the following assumptions:

1. $\hat{R}_i$ is an unbiased estimator of R_i, where R_i is the theoretical assessment of the overall probability of success after the ith trial, $i = 1, 2, \ldots, k$.
2. $p_i = E(r_i)$ (i.e., p_i is the probability that a patient on study during the ith stage responds).
3. All patients within a stage have the same independent probability of response. Furthermore, the stages are independent.

If the criterion for the choice of each smoothing constant α_i is to minimize the variance of its associated value $\hat{R}_i$, by assumptions 1 through 3, we have

$$\text{Var } \hat{R}_i = \alpha_i^2 \sigma_i^2 + (1 - \alpha_i)^2 \text{Var } \hat{R}_{i-1}, \qquad i = 2, \ldots, k, \qquad (5.4.6)$$

where

$$\sigma_i^2 = \frac{p_i(1 - p_i)}{n_i} . \qquad (5.4.7)$$

Let us minimize (5.4.6) with respect to α_i. Thus differentiating (5.4.6) we have

$$\frac{\partial \operatorname{Var}\hat{R}_i}{\partial \alpha_i} = 2\alpha_i\sigma_i^2 - 2(1-\alpha_i)\operatorname{Var}\hat{R}_{i-1}, \tag{5.4.8}$$

under the assumption that $\alpha_1,\ldots,\alpha_k$ are mathematically independent. The value α_i, α_i^0 (say), which minimizes (5.4.8), is obtained by setting (5.4.8) equal to zero and solving for α_i^0. Thus we have

$$\alpha_i^0 = \frac{\operatorname{Var}\hat{R}_{i-1}}{\sigma_i^2 + \operatorname{Var}\hat{R}_{i-1}}. \tag{5.4.9}$$

Using mathematical induction it is not difficult to verify that

$$\alpha_i^0 = \frac{\sigma_i^{-2}}{\sum\limits_{j=1}^{i}\sigma_j^{-2}}, \qquad i=2,\ldots,k. \tag{5.4.10}$$

In practice $\sigma_1^2,\ldots,\sigma_k^2$ are unknown. Thus we use the unbiased estimators $\hat{\sigma}_1^2,\ldots,\hat{\sigma}_k^2$ in any practical application, where

$$\hat{\sigma}_i^2 = \frac{r_i(1-r_i)}{n_i-1}, \qquad i=1,2,\ldots,k. \tag{5.4.11}$$

Thus the smoothing constants employed in practice are given by

$$\hat{\sigma}_i^0 = \frac{\hat{\sigma}_i^{-2}}{\sum\limits_{j=1}^{i}\hat{\sigma}_j^{-2}}, \qquad i=1,2,\ldots,k, \tag{5.4.12}$$

and the model is

$$\hat{R}_i = \hat{\alpha}_i^0 r_i + (1-\hat{\alpha}_i^0)\hat{R}_{i-1}, \qquad i=1,2,\ldots,k. \tag{5.4.13}$$

Example 5.4.2. Suppose a clinical trial is conducted in five stages and provisions are made to record the number of patients and number of responders. The survival ratios at each stage are $6/10$, $7/10$, $7/10$, $8/10$, and $9/10$. We wish to assess the overall probability of survival at each stage of the clinical trial. Thus we need to obtain the values $\hat{\sigma}_i^{-2}$, $\hat{\alpha}_i^0$, and $\hat{R}_i$ at each stage of the clinical trial. For example, at stages 1 and 2, $\hat{\sigma}_1^{-2} = (10-1)/(0.6)(0.4) = 37.50$, and $\hat{\sigma}_2^{-2} = (10-1)/(0.7)(0.3) = 42.86$.

Thus

$$\hat{\alpha}_2^0 = \frac{42.86}{37.50 + 42.86} = 0.533. \tag{5.4.14}$$

Hence

$$\hat{R}_2 = (0.533)(0.7) + (0.467)(0.6) = 0.653. \tag{5.4.15}$$

The other values $\hat{\alpha}_3^0$, $\hat{\alpha}_4^0$, $\hat{\alpha}_5^0$, $\hat{R}_3$, $\hat{R}_4$, and $\hat{R}_5$ are obtained in a similar manner. Table 5.14 contains the overall assessed probability of patient response after each stage, as well as the inverse estimated variance and smoothing constant for each stage.

Table 5.14 Assessed overall survival probabilities at each stage of a clinical trial

Stage i	$\hat{\sigma}_i^{-2}$	$\hat{\alpha}_i^0$	$\hat{R}_i$
1	37.50	—	0.600
2	42.86	0.533	0.653
3	42.86	0.348	0.669
4	56.25	0.313	0.710
5	100.00	0.358	0.778

Chapter 5 considers three different models for assessing and predicting the growth in the probability of survival in a stage-by-stage clinical trial. The experimenter who at various stages of a clinical trial has survival probabilities that appear to be increasing from stage to stage, should first plot these probabilities on arithmetic graph paper. If the growth is rapid in the initial stages but tapers off in the later stages, a hyperbolic growth model may provide a good fit to the data. If the growth is slow initially but sustained in the later stages, an exponential growth model is a better choice. If neither the hyperbolic nor the exponential model provides a good fit to the data, the experimenter can try fitting a logistic growth model to the data. If no model among the three fits the data, nonparametric procedures are available for assessing growth of survival probabilities. Barlow and Scheuer [1966] describe such a nonparametric approach. Unfortunately prediction of later stage survival probabilities is not possible with their method.

In Section 5.4 we present a method for assessing the overall probability of survival in a k-stage clinical trial. The approach is based on an exponential smoothing model proposed by Lloyd and Lipow [1962].

EXERCISES

1. Using the recursive relation defined by (5.4.2), derive (5.4.3).

2. Derive a formula for $\hat{R}_i$ in terms of $r_{i-j}, j = 0, 1, \ldots, i-1;\ i = 2, \ldots, k$, similar to (5.4.2) using the recursive relation (5.4.5).

3. Prove that (5.4.10) follows from (5.4.9) using mathematical induction.

4. Prove that $\hat{\sigma}_i^2$ is an unbiased estimator of σ_i^2 where σ_i^2 is given by (5.4.7) and $\hat{\sigma}_i^2$ by (5.4.11).

5. Verify the numerical values in Table 5.14.

6. What are the difficulties associated with obtaining a minimum variance solution for $\tilde{R}_i$ given by (5.4.3)?

CHAPTER 6

Numerical Maximization of Likelihood Functions

6.1 INTRODUCTORY REMARKS

A common method for estimating parameters of a survival distribution is the method of maximum likelihood. In Chapters 3 and 4 we used maximum likelihood techniques to estimate parameters in a variety of situations —noncensored distributions (Sections 3.1, 4.1, 4.2, and 4.8), censored distributions (Sections 3.2, 3.4, 4.3, and 4.8), truncated distributions (Section 4.5), and the exponential survival distribution with concomitant information (Section 3.5).

Maximum likelihood estimators for the exponential distribution, both censored and uncensored, are easily obtained directly (Sections 3.1 and 3.2). In most other cases, however, it is not possible to obtain explicit maximum likelihood estimators for the parameters of survival distributions; hence numerical techniques must be used. Numerical techniques are available for two cases: (1) the case of no constraints on values the parameters assume, and (2) the case for which the parameters are subject to constraint.

Examples of the no-constraints case are found in Chapter 3, whereas the growth model parameter estimation problem discussed in Chapter 5 contains examples of constraint parameter cases. In maximizing the logarithm of likelihood functions, the constraint, if there is one, is typically as follows: the value of a parameter must lie interior to a particular region (e.g., $0 < p < 1$) and must not lie on the boundary of that region (e.g., $p \neq 0, 1$). Numerical techniques that do not allow for constraints can be used as long as successive approximations to the estimators do not approach or cross the boundary, (i.e., as long as the successive maximum likelihood estimator lies in the interior of the region).

We describe maximizing techniques that do not consider constraints explicitly, and also we describe briefly methods that do account for constraints. However, the case of the maximum occurring on the boundary is not discussed. (See Saaty [1959].)

The numerical techniques described in this chapter can be put into two classes—direct and indirect. In the direct technique class, a starting value is determined at what is thought to be a good approximation to the desired value. One then proceeds in a stepwise fashion upward toward the peak or maximum. An example of this class is the steepest ascent or gradient method of Cauchy. In the indirect method class, one obtains the first derivatives of the logarithm of the likelihood function with respect to each parameter. The ensuing equations are set equal to zero, and an attempt is made to find the values of the parameters (in terms of the observations) that simultaneously satisfy these equations. Two examples of this class are the Newton-Raphson procedure and the method of scoring.

In this chapter we investigate the two indirect procedures, the Newton-Raphson technique in Section 6.2 and the method of scoring in Section 6.3. The latter approach is also discussed by Rao [1952] and Kale [1961, 1962]. We also describe the steepest ascent method in Section 6.4.

Although more sophisticated and modern methods are available for solving nonlinear equations of the type with which we are confronted, the methods under consideration, particularly the Newton-Raphson procedure and the method of scoring, are very practical to implement and conceptually are not difficult. Furthermore, these procedures are well documented.

We conclude the chapter in Section 6.5 with a description of the use of two readily available computer programs that enable the reader to avoid writing his own program. A brief discussion of boundary value considerations is also given.

6.2 THE NEWTON–RAPHSON METHOD

We illustrate the Newton-Raphson technique first solving for a single θ and then presenting the case for θ_i, $i=1,2,\ldots,k$. In most cases the maximum likelihood estimator $\hat{\theta}$ of the parameter θ with corresponding death density function $f(t;\theta)$ is obtained by differentiating $\log L(\theta)$ with respect to θ, setting the derivative equal to zero, and solving the ensuing equation in terms of $\hat{\theta}$, where $L(\theta)=\prod_{i=1}^{n} f(t_i;\theta)$. For the cases in which there is censoring, the reader is referred to Sections 3.2 and 3.3. Let

$$g(\theta)=\frac{\partial \log_e L(\theta)}{\partial \theta}. \tag{6.2.1}$$

The problem is then to find that value of θ, $\hat{\theta}$ (say), such that

$$g(\hat{\theta}) = 0. \tag{6.2.2}$$

Thus $\hat{\theta}$ is the requisite maximum likelihood estimator of θ.* If $\hat{\theta}$ cannot be obtained explicitly from solving (6.2.2) we may attempt a solution by means of the Newton-Raphson procedure for which the function $g(\hat{\theta})$ is approximated by the first two terms in a Taylor series expansion. (We ignore the contribution of the higher order terms, since their deviations from the solution value $\hat{\theta}$ are raised to higher powers, hence diminish rapidly.) Let $\hat{\theta}_0$ be an initial estimate of $\hat{\theta}$, then $\hat{\theta}_\nu$, the νth iteration or approximation of $\hat{\theta}$, is given by

$$\hat{\theta}_\nu = \hat{\theta}_{\nu-1} - \frac{g(\hat{\theta}_{\nu-1})}{g'(\hat{\theta}_{\nu-1})}, \tag{6.2.3}$$

which follows by rearranging the first two terms in the Taylor series, and

$$g'(\hat{\theta}_{\nu-1}) = \frac{dg(\theta)}{d\theta}\bigg|_{\theta=\hat{\theta}_{\nu-1}}. \tag{6.2.4}$$

Figure 6.1 is a graphical representation of the Newton-Raphson method which converges.

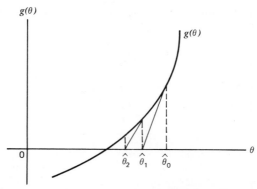

Figure 6.1   Graphical representation of the Newton-Raphson technique, which converges.

*If there is more than one value $\hat{\theta}$ such that $g(\hat{\theta}) = 0$, judicious choice of an initial value is very important. In most cases, however, the initial value obtained is in the neighborhood of the maximum likelihood estimate. If in doubt, try several different initial values.

Note that $g'(\theta) = (\partial^2 \log_e L(\theta)) / \partial \theta^2$ is the second derivative of the log likelihood function.

Figure 6.1 shows a function of θ for which the successive approximations obtained by the Newton-Raphson method converge to the root $\hat{\theta}$. Note that since $g'(\hat{\theta}_0) \neq 0$, the tangent line is not parallel to the θ axis. Where this line crosses the θ axis we find our next approximation $\hat{\theta}_1$, and so on.

Ordinarily, we stop the iteration procedure when $\log_e L(\theta)$ stops increasing appreciably. The value of $\log_e L(\theta)$ should be calculated at each step (with a high-speed computer, this is no problem), which permits us to monitor the stopping procedure. In this way the effects of different initial estimates can be assessed. If a high-speed computer is used, double precision is recommended.

Example 6.2.1. Suppose in a clinical trial 25 patients succumb (out of an unknown number on the study) within the first 100 days of the trial; thus the death density of the ith patient, $i = 1, 2, \dots, 25$ (only these 25 among the unknown total number), is

$$f(t_i; \lambda) = \frac{\lambda \exp(-\lambda t_i)}{1 - \exp(-100\lambda)}, \qquad 0 < t_i < 100. \tag{6.2.5}$$

Suppose further that the mean survival time for these 25 patients is 16.8 days. The problem is to obtain numerically the maximum likelihood estimate $\hat{\lambda}$ by the Newton-Raphson procedure.

The first derivative of the log likelihood function as a function of λ is

$$g(\lambda) = \frac{25}{\lambda} - \frac{2500 \exp(-100\lambda)}{1 - \exp(-100\lambda)} - 420. \tag{6.2.6}$$

We can simplify (6.2.6) by dividing the right hand side of the equation by 25 and multiplying both the numerator and denominator by $\exp(100\lambda)$. Thus denoting $g(\lambda)/25$ by $g^*(\lambda)$, we write

$$g^*(\lambda) = \frac{1}{\lambda} - \frac{100}{\exp(100\lambda) - 1} - 16.8. \tag{6.2.7}$$

As an initial guess $\hat{\lambda}_0$, we use $1/16.8 = 0.0595$:

$$g^*(\hat{\lambda}_0) = \frac{1}{\hat{\lambda}_0} - \frac{100}{\exp(100\hat{\lambda}_0) - 1} - 16.8 = -0.261. \tag{6.2.8}$$

Furthermore,

$$g^{*\prime}\left(\hat{\lambda}_0\right) = \frac{-1}{\hat{\lambda}_0^2} + \left[\frac{100\exp\left(50\hat{\lambda}_0\right)}{\exp\left(100\hat{\lambda}_0\right) - 1}\right]^2$$

$$= -282.24 + 26.19 = -256.05. \tag{6.2.9}$$

Now using the Newton-Raphson procedure, we have

$$\hat{\lambda}_1 = \hat{\lambda}_0 - \frac{g^*\left(\hat{\lambda}_0\right)}{g^{*\prime}\left(\hat{\lambda}_0\right)},$$

$$= 0.0595 - \frac{0.261}{256.05} = 0.0585. \tag{6.2.10}$$

Evaluating $g^*(\hat{\lambda}_1)$, we find

$$g^*\left(\hat{\lambda}_1\right) = \frac{1}{\hat{\lambda}_1} - \frac{100}{\exp\left(100\hat{\lambda}_1\right) - 1} - 16.8 = 0.005 \tag{6.2.11}$$

and

$$g^{*\prime}\left(\hat{\lambda}_1\right) = \frac{-1}{\hat{\lambda}_1^2} + \left[\frac{100\exp\left(50\hat{\lambda}_1\right)}{\exp\left(100\hat{\lambda}_1\right) - 1}\right]^2$$

$$= -292.21 + 28.97 = 263.24. \tag{6.2.12}$$

Iterating again, we find that $\hat{\lambda}_2 = \hat{\lambda}_1$ to four decimal places. We thus conclude that $\hat{\lambda} = 0.0585$ is a root of $g^*(\hat{\lambda}) = 0$. Finally, we note that since $g^*(\lambda)$ is monotonic in λ, the Newton-Raphson method has obtained the maximum likelihood estimate $\hat{\lambda} = 0.0585$.

Suppose $f(t; \theta_1, \ldots, \theta_k)$ is a death density function containing k parameters, $\theta_1, \theta_2, \ldots, \theta_k$, $k \geqslant 2$. Furthermore, suppose the maximum likelihood estimators $\hat{\theta}_1, \hat{\theta}_2, \ldots, \hat{\theta}_k$ of $\theta_1, \theta_2, \ldots, \theta_k$ (respectively) are found by differentiating the logarithm of the likelihood function, with respect to $\theta_1, \theta_2, \ldots, \theta_k$, setting these derivatives equal to zero and solving the resulting equations in terms of $\hat{\theta}_1, \hat{\theta}_2, \ldots, \hat{\theta}_k$. This often leads to a system of k equations in k unknowns which cannot be solved directly. We can, however, employ a direct extension of the Newton-Raphson procedure to k dimensions.

Let $L(\theta_1, \theta_2, \ldots, \theta_k)$ be the likelihood function and let us suppose that the maximum likelihood estimators $\hat{\theta}_1, \hat{\theta}_2, \ldots, \hat{\theta}_k$ are found by solving simultaneously the vector equation

$$g'(\hat{\theta}_1, \ldots, \hat{\theta}_k) = 0' \qquad (6.2.13)$$

in terms of $\hat{\theta}_1, \hat{\theta}_2, \ldots, \hat{\theta}_k$, where $g'(\hat{\theta}_1, \ldots, \hat{\theta}_k) = (g_1(\hat{\theta}_1, \ldots, \hat{\theta}_k), \ldots, g_k(\hat{\theta}_1, \ldots, \hat{\theta}_k))$, $0' = (0, 0, \ldots, 0)$, and

$$g_i(\hat{\theta}_1, \hat{\theta}_2, \ldots, \hat{\theta}_k) = \left. \frac{\partial \log_e L(\theta_1, \ldots, \theta_k)}{\partial \theta_i} \right|_{\theta_j = \hat{\theta}_j}, \qquad i, j = 1, 2, \ldots, k. \quad (6.2.14)$$

The prime notation here indicates the transpose of a matrix.

Suppose $\hat{\theta}_{10}, \hat{\theta}_{20}, \ldots, \hat{\theta}_{k0}$ is the set of initial estimates of $\hat{\theta}_1, \hat{\theta}_2, \ldots, \hat{\theta}_k$, respectively. Then the νth iteration $\hat{\theta}_{1\nu}, \hat{\theta}_{2\nu}, \ldots, \hat{\theta}_{k\nu}$ to the solution $\hat{\theta}_1, \hat{\theta}_2, \ldots, \hat{\theta}_k$ is

$$\hat{\theta}'_\nu = \hat{\theta}'_{\nu-1} - g'^{(\nu-1)} \|v_{ij}\|^{-1}_{(\nu-1)}, \qquad (6.2.15)$$

where $\hat{\theta}'_\nu = (\hat{\theta}_{1\nu}, \hat{\theta}_{2\nu}, \ldots, \hat{\theta}_{k\nu})$, $\hat{\theta}'_{\nu-1} = (\hat{\theta}_{1,\nu-1}, \hat{\theta}_{2,\nu-1}, \ldots, \hat{\theta}_{k,\nu-1})$,

$$g'^{(\nu-1)} = (g_1^{(\nu-1)}, g_2^{(\nu-1)}, \ldots, g_k^{(\nu-1)}), \quad g_i^{(\nu-1)} \equiv g_i(\hat{\theta}_{1,\nu-1}, \hat{\theta}_{2,\nu-1}, \ldots, \hat{\theta}_{k,\nu-1}),$$

$$i = 1, 2, \ldots, k,$$

and $\|v_{ij}\|^{-1}_{(\nu-1)}$ is the $k \times k$ matrix whose ijth element is

$$v_{\nu-1}^{ij} = \left. \frac{\partial g_i(\theta_1, \ldots, \theta_k)}{\partial \theta_j} \right|_{(\theta_1, \ldots, \theta_k) = (\hat{\theta}_{1,\nu-1}, \ldots, \hat{\theta}_{k,\nu-1})}, \qquad i, j = 1, 2, \ldots, k \quad (6.2.16)$$

or the second derivative of the log likelihood.

The choice of the initial estimates $\hat{\theta}_{10}, \hat{\theta}_{20}, \ldots, \hat{\theta}_{k0}$ is very important, because it is possible that the Newton-Raphson procedure will converge to a value that is not the maximum of $L(\theta_1, \ldots, \theta_k)$. Hence we suggest that several sets of initial values be looked at, and each time the convergent values $\hat{\theta}_{1\nu}, \hat{\theta}_{2\nu}, \ldots, \hat{\theta}_{k\nu}$ are obtained, these values be substituted into $L(\theta_1, \ldots, \theta_k)$. The overall maximum likelihood estimates $\hat{\theta}_1, \ldots, \hat{\theta}_k$ are taken as that set of convergent values which maximizes $L(\theta_1, \ldots, \theta_k)$. Although this method does not always guarantee a maximum, it is a safeguard in that more than one set of initial values is considered, which means that any peculiarities in convergence can be uncovered.

Often there are restrictions on the parameters, with the result that for some subset of the parameters $\theta_{s_1}, \theta_{s_2}, \ldots, \theta_{sw}$,

$$\theta_{s_j}^{(1)} \leqslant \theta_{s_j} \leqslant \theta_{s_j}^{(2)}, \qquad j = 1, 2, \ldots, w. \tag{6.2.17}$$

For example, Model 1, given by (5.2.1), requires that $0 < p_\infty \leqslant 1$.

It may happen that the maximum likelihood estimates of some (or all) of these w parameters do not satisfy the restrictions required by (6.2.17). Should this occur, estimates falling outside their restricted boundary are set equal to their nearest boundary point and the remaining nonrestricted maximum likelihood estimates are computed in the usual way. (See the discussion concerning Model 2, Section 5.3.)

There is no guarantee that a set of initial values or subsequent values obtained by the iteration procedure yields a positive definite matrix $\|v_{ij}\|^{-1}$, $i, j = 1, 2, \ldots, k$. If a nonpositive definite matrix occurs, one should first attempt a set of initial values different from the set yielding the nonpositive definite. If the change in initial values does not remedy the situation, another procedure such as gradient method (discussed in Section 6.4) should be used.

Consider the clinical trial described in Example 5.2.1. There we showed that the least squares estimates of p_∞ and α were $p_\infty^* = 0.888$ and $\alpha^* = 0.588$. Using these as the initial estimates for determining $\hat{p}_\infty$ and $\hat{\alpha}$, the requisite maximum likelihood estimates, we showed that after two iterations of the two-dimensional Newton-Raphson technique $\hat{p}_\infty = 0.8891$ and $\hat{\alpha} = 0.5892$. For the details, review Example 5.2.1. Other examples of this procedure are found in Section 3.5, Examples 3.5.1 and 3.5.2.

EXERCISES

1. Suppose a clinical trial is conducted in six stages with proportions of survivors $2/9$, $3/7$, $5/10$, $7/11$, $8/10$, and $7/8$ in stages 1 through 6. Find the maximum likelihood estimates $\hat{p}_\infty$ and $\hat{\alpha}$ of p_∞ and α to four-decimal-place accuracy, assuming that $p_i = p_\infty - \alpha/i$. Use the Newton-Raphson procedure, with the least squares estimates serving as starting values.

2. In another clinical trial, conducted in six stages, the following proportions of survivors were observed for each stage: $2/7$, $3/10$, $4/11$, $3/7$, $5/10$, $6/9$. Assuming that $p_i = 1 - \alpha_1 \exp(-\alpha_2 i)$ is the growth model for this trial (model 2, Chapter 5) use the Newton-Raphson procedure to obtain the maximum likelihood estimates $\hat{\alpha}_1$ and $\hat{\alpha}_2$ of α_1 and α_2, respectively, to four decimal places. *Hint*: Reread Example 5.3.1.

3. Suppose we observe a sample of size 10 of times to death of patients with leukemia in an unknown total sample size. Suppose that these times are ordered as 2.7, 3.4, 3.7, 3.9, 4.2, 4.3, 4.3, 4.6, 4.9, and 5.1 years. If the death

density of the ith individual in the sample is given by the truncated exponential death density with $T = 5.5$, use the Newton-Raphson procedure to obtain the maximum likelihood estimate $\hat{\lambda}$ of the parameter λ.

6.3 THE SCORING METHOD

The method of scoring (Rao [1952]) is similar to the Newton-Raphson technique for obtaining maximum likelihood estimates of parameters of distributions when the requisite system of equations for solution is non-linear. The only difference between these procedures is that the matrix of second derivatives used in the Newton-Raphson technique is replaced by the matrix of the expected values of second derivatives in the method of scoring.

More precisely, suppose that $L(\theta_1, \theta_2, \ldots, \theta_k)$ is the likelihood function whose maximum likelihood estimators $\hat{\theta}_1, \hat{\theta}_2, \ldots, \hat{\theta}_k$ are found by solving (6.2.13). Suppose $\hat{\theta}_{10}, \hat{\theta}_{20}, \ldots, \hat{\theta}_{k0}$ are the set of initial estimates of $\hat{\theta}_1, \hat{\theta}_2, \ldots, \hat{\theta}_k$, respectively. Then the νth iteration $\hat{\theta}_{1\nu}, \hat{\theta}_{2\nu}, \ldots, \hat{\theta}_{k\nu}$ to the solution $\hat{\theta}_1, \hat{\theta}_2, \ldots, \hat{\theta}_k$ is

$$\hat{\boldsymbol{\theta}}'_\nu = \hat{\boldsymbol{\theta}}'_{\nu-1} - \mathbf{g}'^{(\nu-1)} \| E(v_{ij}) \|^{-1}, \tag{6.3.1}$$

where $\hat{\boldsymbol{\theta}}'_\nu$, $\hat{\boldsymbol{\theta}}'_{\nu-1}$, and $\mathbf{g}'^{(\nu-1)}$ are defined in (6.2.15), and $\| E(v_{ij}) \|^{-1}$ is the matrix whose ijth element is $E(v^{ij})$ where v^{ij} is defined as $\partial^2 \log_e L / \partial\theta_i \partial\theta_j$. That is, the expected value of each element in the matrix is with respect to the sample values $t_1, \ldots, t_n$. For example, in the continuous case

$$E(v^{ij}) = E\left(\frac{\partial^2 \log_e L}{\partial\theta_i \, \partial\theta_j} \right). \tag{6.3.2}$$

However, $L = \prod_{\nu=1}^n f(t_\nu; \boldsymbol{\theta}')$. Hence $\log_e L = \sum_{\nu=1}^n \log_e f(t_\nu; \boldsymbol{\theta}')$. It then follows, since $t_1, \ldots, t_n$ are identically distributed, that

$$E\left(\frac{\partial^2 \log_e L}{\partial\theta_i \partial\theta_j} \right) = nE\left(\frac{\partial^2 \log_e f(t; \boldsymbol{\theta}')}{\partial\theta_i \partial\theta_j} \right) \qquad \text{for all } i, j = 1, 2, \ldots, k.$$

That is,

$$E(v^{ij}) = n \int_0^\infty \frac{\partial^2 \log_e f(t; \boldsymbol{\theta}')}{\partial\theta_i \, \partial\theta_j} f(t; \boldsymbol{\theta}') \, dt. \tag{6.3.3}$$

A similar argument with the sum replacing the integral in (6.3.3) holds in the discrete case. For some distributions, determining (6.3.3) is quite simple and the method of scoring is useful. For others, the integration is difficult (involving series expansions, numerical integration, etc.), and in those cases the Newton-Raphson technique is usually selected.

As with the Newton-Raphson method, a judicious choice of the starting values $\hat{\theta}_{10}, \ldots, \hat{\theta}_{k0}$ is quite important, and for the same reason—namely, (6.3.1) may converge to a value that is not the maximum of $L(\theta_1, \ldots, \theta_k)$ or $\|E(v_{ij})\|^{-1}$ may not be positive definite for some starting values. Thus several sets of starting values should be used in any numerical application of this technique.

Usually the scoring method is a bit slower in its convergence rate than the Newton-Raphson method. Further discussion concerning the variance–covariance matrix of maximum likelihood estimators can be found in Mood and Graybill [1963]. A discussion of the Newton-Raphson method and the method of scoring is given by Kale [1961, 1962].

Finally, if there are restrictions on the parameters, we follow the same modification procedures outlined for the Newton-Raphson method. (See Section 6.2.)

Example 6.3.1.   Consider a clinical trial conducted in five stages with the following survival ratios: 5/10, 7/10, 8/10, 8/10, and 9/10. The ratios indicate rapid growth in survivability in the early stages, with a tapering off in the later stages; thus the hyperbolic growth model

$$p_i = 1 - \frac{\alpha}{i}, \qquad i = 1, 2, 3, 4, 5, \tag{6.3.4}$$

was indicated as the appropriate growth model, and $p_\infty = 1$. To estimate α, we wish to obtain its maximum likelihood estimate, which is the solution in terms of $\hat{\alpha}$ of

$$- \sum_{i=1}^{5} \frac{x_i/i}{1 - \hat{\alpha}/i} + \sum_{i=1}^{5} \frac{n_i - x_i}{\hat{\alpha}} = 0 \tag{6.3.5}$$

This follows from (5.3.29), with $G(i) = 1/i$.

To find $\hat{\alpha}$ we use the Method of Scoring. That is, if $\hat{\alpha}_n$ is the nth iteration of the scoring procedure, we have

$$\hat{\alpha}_n = \hat{\alpha}_{n-1} - \frac{f(\hat{\alpha}_{n-1})}{E\{f'(\hat{\alpha}_{n-1})\}}, \tag{6.3.6}$$

where

$$f(\hat{\alpha}_{n-1}) = -\sum_{i=1}^{5} \frac{x_i/i}{1-\hat{\alpha}_{n-1}/i} + \sum_{i=1}^{5} \frac{n_i - x_i}{\hat{\alpha}_{n-1}}, \qquad (6.3.7)$$

$$f'(\hat{\alpha}_{n-1}) = -\left[\sum_{i=1}^{5} \frac{x_i/i^2}{(1-\hat{\alpha}_{n-1}/i)^2} + \sum_{i=1}^{5} \frac{n_i - x_i}{\hat{\alpha}_{n-1}^2}\right], \qquad (6.3.8)$$

and $E(f'(\hat{\alpha}_{n-1}))$ is the expected value of $f'(\hat{\alpha}_{n-1})$ with respect to x_i, $i=1,2,3,4,5$. Thus substituting $E(x_i)=n_ip_i=n_i(1-\alpha/i)$ and $E(n_i - x_i) = n_i\alpha/i$ into (6.3.8), it is easy to show that

$$E(f'(\hat{\alpha}_{n-1})) = -\sum_{i=1}^{5} \frac{n_i}{\hat{\alpha}_{n-1}(i-\hat{\alpha}_{n-1})}. \qquad (6.3.9)$$

The actual verification is left as an exercise to the reader.

The initial estimate $\hat{\alpha}_0$ is taken as the least squares estimate α^*, which is obtained by minimizing

$$\psi(\alpha) = \sum_{i=1}^{5} \left(\frac{x_i}{n_i} - 1 + \frac{\alpha}{i}\right)^2 \qquad (6.3.10)$$

with respect to α. We thus find

$$\hat{\alpha}_0 \equiv \alpha^* = \frac{\displaystyle\sum_{i=1}^{5} \frac{1}{i} - \sum_{i=1}^{5} \frac{x_i}{n_i i}}{\displaystyle\sum_{i=1}^{5} \frac{1}{i^2}} = 0.5374. \qquad (6.3.11)$$

Using (6.3.6) we can obtain $\hat{\alpha}$ by successive iterations. The results of the iterations are given in Table 6.1. Thus after three iterations the scoring procedure converges, providing us with the maximum likelihood estimate $\hat{\alpha} = 0.5512$, to four decimal places. Table 6.2 is a comparison of the observed survival probabilities and the corresponding expected survival probabilities using both the least squares and maximum likelihood estimates.

Table 6.1 Values of $\hat{\alpha}_j$ at each iteration of the scoring procedure for example 6.3.1

Iteration j	$\hat{\alpha}_j$
1	0.5520
2	0.5513
3	0.5512

Table 6.2 Observed and expected survival probabilities at each of five stages of clinical trial of example 6.3.1.

Stage i	Observed Probability	Expected LSE Probability	Expected MLE Probability
1	0.5000	0.4626	0.4488
2	0.7000	0.7313	0.7244
3	0.8000	0.8209	0.8163
4	0.8000	0.8656	0.8622
5	0.9000	0.8925	0.8898

Example 6.3.2. Consider a group of patients with intractable, bilateral renal calculi. It is assumed that patients with this kidney disease will lose the use of one kidney; ultimately both kidneys will cease to function, and death will occur. When both kidneys are functioning it is assumed each has a failure rate λ_0. After the failure of one kidney, the failure rate of the remaining functioning organ is λ_1. Both failure rates are assumed to be constant. The death density function of patients with this disease, derived by Gross et al. [1971], is

$$f(t;\lambda_0,\lambda_1) = \frac{2\lambda_0\lambda_1}{\lambda_1 - 2\lambda_0}[\exp(-2\lambda_0 t) - \exp(-\lambda_1 t)], \qquad t \geqslant 0, \quad (6.3.12)$$

where without loss of generality, $\lambda_1 > 2\lambda_0$ is assumed. On the average, patients with this disease lose the use of one kidney 15 years after the onset of symptoms. They have a mean time to death due to loss of the remaining kidney 3 years later.

The likelihood function for a sample of n times to death of patients inflicted with intractable, bilateral renal calculi is

$$L(\lambda_0, \lambda_1) = \left[\frac{2\lambda_0\lambda_1}{(\lambda_1 - 2\lambda_0)} \right]^n \prod_{i=1}^{n} \{\exp(-2\lambda_0 t_i) - \exp(-\lambda_1 t_i)\}. \quad (6.3.13)$$

Since λ_0 and λ_1 are known for this disease, a Monte Carlo study was conducted to compare the Newton-Raphson procedure and the method of scoring in obtaining the maximum likelihood estimates $\hat{\lambda}_0$ and $\hat{\lambda}_1$. The Monte Carlo study was as follows:

1. For sample sizes of 200, two sets of random numbers were drawn between zero and one.
2. The random numbers were converted to data having the survival distribution corresponding to the death density function (6.3.13) by computing the two survival times separately for the two kidneys:

$$t_0 = \frac{\log_e(1 - F_0(t))}{-2\lambda_0} \quad (6.3.14)$$

and

$$t_1 = t_0 + \frac{\log_e(1 - F_1(t))}{-\lambda_1}, \quad (6.3.15)$$

where a pair of random numbers was inserted in $F_0(t)$ and $F_1(t)$.
3. Noting that the mean and variance for the death density (6.3.12) are, respectively,

$$\mu = \lambda_1^{-1} + (2\lambda_0)^{-1} \quad (6.3.16)$$

and

$$\sigma^2 = \lambda_1^{-2} + (2\lambda_0)^{-2}, \quad (6.3.17)$$

initial values $\hat{\lambda}_0^0$ and $\hat{\lambda}_1^0$ were obtained by the method of moments.
4. These initial values were used in both the Newton-Raphson and scoring iteration techniques to obtain the maximum likelihood estimates $\hat{\lambda}_0$ and $\hat{\lambda}_1$. In addition, estimates of the variances, covariances, and correlations of $\hat{\lambda}_0$ and $\hat{\lambda}_1$ were obtained for both the Newton-Raphson and scoring methods.
5. The process was repeated 100 times. Averages and actual variances over the 100 samples were obtained for the estimated $\hat{\lambda}_0$ and $\hat{\lambda}_1$.

The results of the Monte-Carlo study (Table 6.3) reveal that the maximum likelihood estimates are virtually the same (as they should be) for both the Newton-Raphson and the scoring procedures. On the average, however, the variances are smaller for the scoring method, albeit the Newton-Raphson method converges more quickly to the appropriate values.

Table 6.3 Monte Carlo simulation comparison of method of moments and maximum likelihood estimates of λ_0 and λ_1 for 100 random samples of size $n = 200$: true values are $\lambda_0 = 0.0667$ and $\lambda_1 = 0.3333$.

	Average Sample Statistics		
	Method of Moments	Maximum Likelihood	
		Scoring	Newton-Raphson
$\hat{\lambda}_0$	0.06807	0.06874	0.06874
$\hat{\lambda}_1$	0.36506	0.34838	0.34842
Average number of iterations	—	4.0	3.2
$\text{Var}\,\hat{\lambda}_0$	—	0.000253	0.000344
$\text{Var}\,\hat{\lambda}_1$	—	0.013372	0.014989
$\text{Cov}(\hat{\lambda}_0, \hat{\lambda}_1)$	—	-0.001246	-0.001516
$\text{Corr}(\hat{\lambda}_0, \hat{\lambda}_1)$	—	-0.82	-0.82

EXERCISES

1. Redo Exercise 1, Section 6.2, using the method of scoring and compare your results to the Newton-Raphson procedure with respect to number of iterations and accuracy of the estimates. Use four-decimal-place accuracy for the comparisons.

2. Redo Exercise 2, Section 6.2, using the method of scoring and compare your results to the Newton-Raphson procedure with respect to number of iterations and accuracy of the estimates. Use four-decimal-place accuracy for the comparisons.

3. Verify that the mean and the variance for the death density function (6.3.12) are given by (6.3.16) and (6.3.17), respectively. Thus setting $\mu = \bar{t}$ and $\sigma^2 = s_t^2$, where $\bar{t}$ and s_t^2 are the sample mean and variance, respectively, for a sample of size n of times to death from (6.3.12), obtain the method of moment estimators $\tilde{\lambda}_0$ and $\tilde{\lambda}_1$ of λ_0 and λ_1.

4. Consider the likelihood function (6.3.13). If this model is reparameterized $\alpha = 2\lambda_0$ and $\beta = \lambda_1$, show that the two-dimensional Newton-Raphson procedure to obtain $\hat{\alpha}$ and $\hat{\beta}$ can be reduced to a one-dimensional Newton-Raphson procedure by examining the equations $\partial \log_e L / \partial \alpha$ and $\partial \log_e L / \partial \beta$.

5. Define

$$v = \sum_{i=1}^{k} \frac{x_i / i^2}{[1 - \alpha/i]^2} + \sum_{i=1}^{k} \frac{n_i - x_i}{\alpha^2} .$$

Prove that

$$E(v) = \sum_{i=1}^{k} \frac{n_i}{\alpha(i - \alpha)} ,$$

where $E(x_i) = n_i(1 - \alpha/i)$.

6. Verify the numerical computations for Example 6.3.1. Explicitly verify $\hat{\alpha}_0 = 0.5374$, $\hat{\alpha} = 0.5512$, and the corresponding expected values in Table 6.2.

6.4. DIRECT NUMERICAL SOLUTION METHODS

When the Newton-Raphson and scoring methods do not adequately find the appropriate maximum likelihood estimates $(\hat{\lambda}_1, \ldots, \hat{\lambda}_k)$ of the requisite death density parameters $(\lambda_1, \ldots, \lambda_k)$, other numerical solution methods must be applied. The gradient method or method of steepest ascent is another commonly used technique. Further discussion can be found in Saaty [1959] or Kowalik and Osborne [1968].

Suppose $\log_e L(\lambda_1, \lambda_2, \ldots, \lambda_k)$ is the surface whose maximum with respect to $(\lambda_1, \lambda_2, \ldots, \lambda_k)$ is to be found. By using the gradient of the function $\log_e L(\lambda_1, \lambda_2, \ldots, \lambda_k)$ it is often possible to proceed toward the maximum of the function in an iterative manner.

The gradient of the function $g(\lambda_1, \lambda_2, \ldots, \lambda_k) \equiv \log_e L(\lambda_1, \lambda_2, \ldots, \lambda_k)$ is the vector

$$\nabla g = \left(\frac{\partial g}{\partial \lambda_1}, \frac{\partial g}{\partial \lambda_2}, \ldots, \frac{\partial g}{\partial \lambda_k} \right). \qquad (6.4.1)$$

The tangent plane to the surface represented by $g(\lambda_1, \lambda_2, \ldots, \lambda_k)$ then passes through the point $(\lambda_1, \lambda_2, \ldots, \lambda_k)$ in the direction $(\partial g / \partial \lambda_1, \partial g / \partial \lambda_2, \ldots, \partial g / \partial \lambda_k)$. Figure 6.2 shows the tangent plane to the surface $g(\lambda_1, \lambda_2)$ at the point $(\lambda_1, \lambda_2, g(\lambda_1, \lambda_2))$.

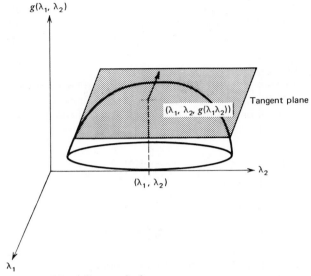

Figure 6.2 Depiction of the gradient method.

It is well known (e.g., see Saaty and Bram [1964]) that the gradient ∇g points locally in the direction of maximum increase. Thus by taking very small steps along the gradient, one can reach a maximum, albeit a local maximum.

The iterative vector equation for finding the maximum of $\log_e L(\lambda_1, \lambda_2, \ldots, \lambda_k)$ can then be expressed as

$$\hat{\boldsymbol{\lambda}}'^{i+1} = \hat{\boldsymbol{\lambda}}'^{i} + \theta \boldsymbol{\zeta}'^{i} \tag{6.4.2}$$

where

$$\hat{\boldsymbol{\lambda}}'^{i+1} = \left(\hat{\lambda}_1^{i+1}, \hat{\lambda}_2^{i+1}, \ldots, \hat{\lambda}_k^{i+1}\right), \ \hat{\boldsymbol{\lambda}}'^{i} = \left(\hat{\lambda}_1^{i}, \hat{\lambda}_2^{i}, \ldots, \hat{\lambda}_k^{i}\right),$$

$$\boldsymbol{\zeta}'^{i} = \left(\zeta_1^{i}, \quad \zeta_2^{i}, \ldots, \zeta_k^{i}\right), \zeta_j^{i} \equiv \frac{\partial \log_e L}{\partial \hat{\lambda}_j^{i}},$$

and

$$\frac{\partial \log_e L}{\partial \hat{\lambda}_j^{i}} \equiv \partial \log_e L / \partial \lambda_j \big|_{\boldsymbol{\lambda}' = \hat{\boldsymbol{\lambda}}'^{i}} , \qquad j = 1, 2, \ldots, k,$$

and i is the ith iteration of the procedure. A scalar quantity θ is the step size, chosen in advance. In fact θ can be regulated to permit reducing the step size if at first it is too large, causing an overshoot of the maximum. On the other hand, if the step size is too small, it can be increased making the reaching of the maximum a lengthy procedure.

The gradient method has the advantage of being very good at avoiding saddle points. However, if the maximum is located along a long, narrow ridge, convergence may be very slow. In these cases it is useful to attempt linear transformations when such transformations are possible, to obtain uncorrelated parameters. For two parameters, λ_1 and λ_2, Figure 6.3a would lead to slow convergence of the gradient method; Figure 6.3b would be quick. Further discussion of such transformations is given in Wilde and Beightler [1967].

Usually at least one of the three methods discussed in this chapter is applicable for finding requisite maximum likelihood estimates arising from the simultaneous solution of nonlinear equations. Excellent discussions of the newer techniques are given in the books by Kowalik and Osborne [1968] and Wilde and Beightler [1967].

Regardless of the method used to derive the maximum likelihood estimates, their values should be used to obtain the requisite variance–covariance matrix estimate associated with the maximum likelihood estimates.

Although the variance–covariance estimate is based on the expected values of the second derivatives of the likelihood function, in practice these expected values are not always obtainable. Should this occur, the raw second derivatives can be used as a substitute. As illustrated in Table 6.3, the numerical results can be quite satisfactory.

Methods that do not require first and second derivatives of the log likelihood function are not very useful in general, because these derivatives are necessary in determining the variance–covariance matrix.

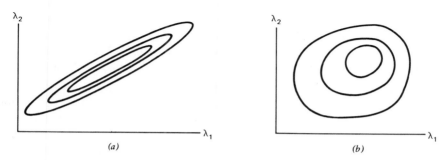

(a)   *(b)*

Figure 6.3 Contours of untransformed and transformed data.

EXERCISES

1. Obtain (6.4.2) if the step size θ is subject to change at each iteration.
2. Redo Exercise 1, Section 6.2, using the gradient method; compare your results with the scoring and Newton-Raphson methods.
3. Redo Exercise 2, Section 6.2, using the gradient method, compare your results with the scoring and Newton-Raphson methods.

6.5 COMPUTER PROGRAMS

Two readily available computer programs can be used to maximize the likelihood function. The IBM Scientific Subroutine Package [1970] includes a program that employs the Fletcher and Powell [1963] procedure, which can be thought of as an extension to the Newton-Raphson method. In the Newton-Raphson method, we approximate a function by using the first two terms of a Taylor series expansion. In the Fletcher and Powell method, an approximation to the third term is also included. The program is designed to find the minimum rather than the maximum of a function. Hence the user must supply the negative of the function to be maximized, as well as the number of parameters, first derivatives, initial values, maximum number of iterations, and error terms. This program tends to have fast convergence and has been widely used successfully.

The BMDX85 program available in the BMD series (Dixon [1973]) can also be readily used to maximize a log likelihood. This program was written to estimate the parameters of a nonlinear regression by minimizing the error variance using stepwise Gauss-Newton iterations. The technique is described by Jennrich and Sampson [1968]. Additional discussion is available in Hartley [1961]. That is, BMDX85 will minimize a function of the form

$$s^2 = \frac{1}{n-p} \sum_{i=1}^{n} (y_i - g(\mathbf{X}'; \boldsymbol{\theta}'))^2, \qquad (6.5.1)$$

where $\boldsymbol{\theta}' = (\theta_1, \theta_2, \ldots, \theta_p)$ and $\mathbf{X}' = (X_{i1}, X_{i2}, \ldots, X_{ip})$ for $i = 1, \ldots, n$.

This program can be used because maximizing $\sum_{i=1}^{n} \log_e f(t'; \boldsymbol{\theta}')$ is equivalent to minimizing $\sum_{i=1}^{n}(-\log_e f(t'; \boldsymbol{\theta}') + c)$ or minimizing $\sum_{i=1}^{n}(\sqrt{c - \log_e f(t'; \boldsymbol{\theta}')}\,)^2$. A constant $c \geq 0$ is added to ensure that the expression under the square root sign is positive for all values of t_i. This measure would not be necessary for most distributions. In addition, y_i is set equal to zero.

To run BMDX85, the user must furnish a subroutine with the terms in (6.5.1) and the first derivatives defined. To demonstrate the use of this program, we provide an example of the necessary subroutine statements to determine estimates of α and β for the Gompertz hazard rate.

The Gompertz hazard rate has been shown to fit the age-specific death rates of older adults as well as many species of organisms (see Whitman [1969] and Garg et al. [1970]). The hazard rate is given by

$$\lambda(t) = \alpha e^{\beta t} \tag{6.5.2}$$

and the cumulative hazard rate is

$$\Lambda(t) = \frac{\alpha e^{\beta t}}{\beta}. \tag{6.5.3}$$

Hence whether the distribution fits given data can be determined by plotting the cumulative on semilogarithm paper using the methods given in Section 4.7. The death density is skewed positively for $\alpha/\beta > 0.16$ with a single mode at $\beta^{-1}\log_e(\beta/\alpha)$ and a mean given by

$$E(t) = -\beta^{-1}e^{\alpha/\beta}\left[\gamma + \log_e\frac{\alpha}{\beta} + \sum_{i=1}^{\infty}(-1)^{i+1}\left(\frac{\alpha}{\beta}\right)^i / i\cdot i!\right], \tag{6.5.4}$$

where $\gamma = 0.577216$ is Euler's constant. The median is given by $\beta^{-1}\log_e(1-\beta/\alpha\log_e 0.5)$. The logarithm of the death density needed for the use of BMDX85 is given by

$$\log_e f(t;\alpha,\beta) = \log_e\alpha + \frac{\alpha}{\beta} + \beta t - \frac{\alpha}{\beta e^{\beta t}}, \tag{6.5.5}$$

$$\frac{\partial \log_e f}{\partial \alpha} = \frac{1}{\alpha} + \frac{1}{\beta} - \frac{1}{\beta e^{\beta t}}, \tag{6.5.6}$$

and

$$\frac{\partial \log_e f}{\partial \beta} = t - \frac{\alpha}{\beta^2}(1 + e^{\beta t}(\beta t - 1)). \tag{6.5.7}$$

Forty values of t have been generated with a distribution of (6.5.3) and population parameters $\alpha = 0.0005$ and $\beta = 0.0667$. In the following subroutine, $\alpha = P(1)$, $\beta = P(2)$, $t_i = X(1)$, $G = \log_e f(t;\theta)$, $F = \sqrt{C-G}$, $D(1) =$

$\partial F/\partial\alpha$, and $D(2)=\partial F/\partial\beta$, to comply with the notation needed by the program. Dimension statements are supplied by the program and do not need to be included by the user.

```
      C = 0.0
      G = ALOG(P(1)) + P(1)/P(2) + P(2)*X(1) − (P(1)/P(2))*EXP(P(2) *X(1))
      X(5) = G
    9 F = C − G
      IF(F.GT.O.) GO TO 10
      C = C + 1.
      GO TO 9
   10 F = SQRT(F)
      D(1) = ( − 1./(2.*F))*(1./P(1) + 1./P(2) − (1./P(2))*EXP(P(2)*X(1)))
      D(2) = ( − 1./(2.*F))*(X(1) − (P(1)/P(2)**2)
     C*(1. + EXP(P(2)*X(1))*(P(2)*X(1) − 1.)))
      X(3) = D(1)
      X(4) = D(2)
      X(6) = C
      RETURN
      END
```

The dependent variable is given in the program card as the second variable, $X(2) = Y = 0$. We read the value of Y from a blank column of data using the usual format card. It need not be defined in this subroutine. The remaining X's,

$$X(3) = D(1)$$

$$X(4) = D(2)$$

$$X(5) = G$$

$$X(6) = C$$

were simply made up to permit checking the magnitude of these expressions on the printout or feeding them into subsequent programs. These statements may be removed if no printout is desired: they are dummy variables and are not used in the computation.

Initial values, the maximum number of iterations, and convergence criterion are needed. The user can also furnish upper and lower limits for the parameters. In this example, we used 0.0 as a lower limit (because we did not desire negative parameters) and 1.0 as an upper limit.

Our sample estimates of α and β were as follows from BMDX85 and a previously written Newton-Raphson program.

Initial Values	True Parameter Values	Number of Iterations	Sample Estimates	
			Newton-Raphson (double precision)	BMDX85
$\hat{\alpha}_0 = 0.0001$	$\alpha = 0.0005$	6 Newton-Raphson	0.000544	0.000544
$\hat{\beta}_0 = 0.05$	$\beta = 0.0667$	8 BMDX85	0.062694	0.062697

Note that we would not expect $\hat{\alpha}$ and $\hat{\beta}$ to be precisely equal to α and β for $n = 40$.

With both the BMD and IBM programs, the covariance matrix would have to be evaluated as a separate step using the sample estimates of the parameters.

If the maximum is known to lie on the boundary, Lagrange multipliers can be used. Saaty [1959] devotes considerable discussion to the problem of determining maxima on boundaries. Kowalik and Osborne [1968] give techniques for the case where the parameters must be greater than zero. If a step would send a point over the boundary, the BMDX85 program halves step size until the point is inside the boundary and then iterates on other parameters. This technique has worked successfully in a wide range of examples (see Jennrich and Sampson [1968]). If the maximum is clearly in the interior, boundary values will have no effect on the solution.

Another application of BMDX85 can be found in Afifi and Azen [1972], who use BMDX85 to obtain least squares rather than maximum likelihood estimates. As we indicated, BMDX85 is designed to minimize (6.5.1), which is precisely the way Afifi and Azen use the program. Daniel and Wood [1971] also describe in detail a computer program for obtaining least squares estimates of parameters of a nonlinear regression.

We recommend caution in trusting the results when a high proportion of the cases have been censored. In preliminary simulation work using one distribution, we obtained estimates of the parameters whose distribution differed considerably from normality when the distribution was highly censored. Caution is also recommended if the log likelihood function has several maxima. If it is not possible to demonstrate that the log likelihood function is convex, several sets of initial values should be used to check to see whether the program iterates to the same answer. With only two parameters it is frequently useful to evaluate the log likelihood at a grid of points and draw contours. Transformations are helpful if a long ridge is encountered.

Chapter 6 presents an overview of nonlinear estimation techniques that are useful in obtaining maximum likelihood estimates that cannot be obtained in closed form. The three techniques discussed in some detail are the Newton-Raphson technique, the method of scoring, and the gradient method. Examples of both the Newton-Raphson method and the method of scoring are presented. The gradient method is also described but in less detail. The principal reasons for not describing the gradient method more thoroughly are as follows: (1) it is not used widely to obtain maximum likelihood estimates; and (2) since variance–covariance matrices usually accompany maximum likelihood estimates, methods that do not require the first and second derivatives of the log likelihood function are generally less useful than those that do (i.e., the Newton-Raphson technique and the method of scoring). Finally, we examine two computer programs that are useful in obtaining maximum likelihood estimates. The BMDX85 was originally ' designed for nonlinear least squares estimation but can be modified to obtain requisite maximum likelihood estimates.

EXERCISES

1. Given a hazard rate $\lambda(t) = \lambda_0 + 2\lambda_1 t$, write the FORTRAN subroutine that is needed to use BMDX85.

2. Given a hazard rate $\lambda(t) = \lambda t^{\alpha-1}$, write the FORTRAN subroutine that is needed to use BMDX85.

3. Verify (6.5.4).

CHAPTER 7

Comparison of Survival Distributions

7.1 INTRODUCTORY REMARKS

In many clinical trial situations, patients with an acute disease are randomly placed on two or more different treatment regimens to determine which (if any) of the treatments significantly lengthens patient survival or remission time. These clinical trials are common in cancer chemotherapy research having produced effective anticancer drugs such as cytoxan, 6-MP, vincristine, and methylgag.

In this chapter we focus on a comparison of survival distributions. For example, suppose $n_1 + n_2$ patients are allocated to two different treatments for an acute illness, such as carcinoma of the lung. Of these patients, n_1 receive standard treatment such as radiation therapy and/or surgery. The remaining n_2 patients, in addition to radiation and/or surgery, receive an experimental drug that is supposed to help prolong their lives. To determine whether the experimental drug indeed prolongs the lives of lung cancer patients, we compare the survival distributions of the patients in the two treatment groups: those receiving the experimental drug and those not receiving the drug.

Suppose we observe the $n_1 + n_2$ patients on the two treatment regimens until they all have succumbed. Assume that the times to death of patients are $t_1, t_2, \ldots, t_{n_1}$; and $t'_1, t'_2, \ldots, t'_{n_2}$ for the standard and experimental treatment regimens, respectively. The death distribution function for patients on the standard treatment is $F_1(t)$, and the death distribution function for patients on the experimental treatment is $F_2(t)$. We then wish to develop methodology for comparing either these two distributions directly or meaningful measures in terms of the parameters of the distributions, such as the mean survival times of patients. We use both parametric and nonparametric procedures to compare survival distributions. The parametric techniques are an outgrowth of the methodology developed in the previous chapters, whereas the nonparametric procedures (mainly Gehan's generalization of the Wilcoxon test [1965a]) are different. In Section 7.2 we discuss general methods, both parametric and nonparametric, for comparing survival distributions when complete information is available on all patients under treatment. We apply this to testing the equality of the exponential, gamma, and Weibull distributions in two populations. Note that in the noncensored case, standard nonparametric comparisons such as Wilcoxon rank sums can be used for any of the continuous survival distributions. In addition, the noncensored Wilcoxon test is discussed. In Section 7.3 we consider the same problem when there is censoring in the data. That is, we consider the censored cumulative death distribution functions $F_1(t)$ and $F_2(t)$ which are censored in that not all patients on study are dead or still in remission.

7.2 NONPARAMETRIC AND PARAMETRIC PROCEDURES FOR COMPARING TWO SURVIVAL DISTRIBUTIONS: COMPLETE SAMPLES

The Sign Test

Suppose $n_1 + n_2$ patients with an acute illness are allocated randomly to two different treatment regimens: n_1 patients receive a standard treatment, and n_2 patients receive an experimental drug that is designed to increase survival for patients with this illness.

Suppose that patients matched in pairs are randomly assigned to each of the two treatments with the result that $n_1 = n_2 = n$ (say). Let t_{ij} be the survival time of the ith patient receiving treatment j, $i = 1, 2, \ldots, n$; $j = 1, 2$ (i.e., for the ith patient receiving treatment 1, the standard treatment is paired with the ith patient receiving treatment 2, the experimental treatment). Let x_i be a random variable such that*

$$x_i = \begin{cases} 1 & \text{if } t_{i2} > t_{i1} \\ 0 & \text{if } t_{i2} < t_{i1}, \quad \text{for } i = 1, 2, \ldots, n. \end{cases} \tag{7.2.1}$$

If it turns out in practice that $t_{i1} = t_{i2}$ for many pairs of patients, procedures developed in Section 7.3 may be more useful. If $t_{i1} = t_{i2}$ only occasionally, that pair is dropped. If we now set

$$x = \sum_{i=1}^{n} x_i, \tag{7.2.2}$$

*Since t_{i1} and t_{i2} are continuous measurements, $\text{pr}\{t_{i1} = t_{i2}\} = 0$.

x has a binomial distribution with parameter p, where

$$p = \mathrm{pr}\{t_{i2} > t_{i1}\}. \tag{7.2.3}$$

If there is no difference between the experimental and standard treatments with regard to prolonging survival, $p = \frac{1}{2}$. On the other hand, if the experimental treatment is superior (inferior) to the standard treatment, $p > (<)\frac{1}{2}$.

Determining whether the experimental treatment significantly increases survival with respect to the standard treatment is then equivalent to testing the hypothesis $H_0 : p \leqslant \frac{1}{2}$, against the alternative $H_a : p > \frac{1}{2}$. On the other hand, determining whether there is a significant difference in survival times between the two treatments is equivalent to testing the hypothesis $H_0 : p = \frac{1}{2}$, against the alternative $H_a : p \neq \frac{1}{2}$.

There are two cases to consider: n small (e.g., $n \leqslant 25$) and n large ($n > 25$). For small values of n an α-level test of the hypothesis $H_0 : p \leqslant \frac{1}{2}$ against the alternative $H_a : p > \frac{1}{2}$ is given by the rule: reject $H_0 : p \leqslant \frac{1}{2}$ only if $x \geqslant x_{1-\alpha} + 1$ where $x_{1-\alpha}$ is the smallest integer value such that

$$\sum_{x=0}^{x_{1-\alpha}} \binom{n}{x} \geqslant 2^n(1-\alpha), \tag{7.2.4}$$

where as usual

$$\binom{n}{x} = \frac{n!}{x!(n-x)!} \tag{7.2.5}$$

and

$$\sum_{x=0}^{n} \binom{n}{x} = 2^n. \tag{7.2.6}$$

Tables of sums of binomial coefficients are readily available. See, for example, Dixon and Massey [1969]. For large values of n an α-level test of H_0 is given by the rule: reject H_0 only if $Z > Z(1-\alpha)$, where $Z(1-\alpha)$ is the $1-\alpha$ percentage point of the standard normal distribution and

$$Z = \frac{x - n/2}{\sqrt{n/4}}. \tag{7.2.7}$$

This follows because for large n, Z has an approximate standard normal distribution.

Note that this test is nonparametric for patients receiving the standard and experimental treatments, respectively. Patients must be paired to allow investigators to compare the ith patient receiving the standard treatment with the ith patient receiving the experimental treatment. In Section 7.3 we discuss Gehan's generalization of the Wilcoxon procedure, which allows for unequal sample sizes as well as censoring while still retaining the basic nonparametric structure of the test. The noncensored Wilcoxon test is discussed at the end of this section. In conclusion, it should be noted that the test described here is a sign test and can be advantageously used in "before and after" situations for the same sample of n patients. That is, if the standard treatment is first given to a group of patients to measure a time response (e.g., length of remission) and if it is followed by an experimental treatment at a later time on the identical patients, the sign test can be used to test the difference in time responses (i.e., lengths of remissions) between the two treatments in the same sample of patients.

The Likelihood Ratio Test

Another method for testing the equality of two survival distributions $S_1(t)$ and $S_2(t)$ is the likelihood ratio test, which is parametric. Although the likelihood ratio test does not require equal sample sizes in the two groups, nor pairing of observations, it can be adapted and used in these situations.

The likelihood ratio test is a procedure for testing a hypothesis against a composite alternative utilizing maximum likelihood estimation. The general formulation of the likelihood ratio test is as follows. Let $x_1, x_2, \ldots, x_n$ be a random sample of size n from a population with frequency function $f(x; \theta_1, \theta_2, \ldots, \theta_k)$, $n > k$. On the basis of this sample it is required to test the hypothesis $H_0 : r$ of the k parameters $\theta_1, \theta_2, \ldots, \theta_k$ are specified, against the alternative H_a : all the parameters are unspecified. Let ω represent the $(k - r)$-dimensional subspace of the unspecified parameters under H_0, and let Ω represent the k-dimensional subspace of the unspecified parameters under H_a. Without loss of generality, assume under H_0 that the parameters $\theta_1, \ldots, \theta_r$ are specified and take the values $\theta_1^0, \ldots, \theta_r^0$. Replace the remaining unspecified parameters $\theta_{r+1}, \ldots, \theta_k$ by their maximum likelihood estimators $\hat{\theta}_{r+1}, \ldots, \hat{\theta}_k$ and form the likelihood function $L(\hat{\omega})$, where

$$L(\hat{\omega}) = \prod_{i=1}^{n} f\left(x_i; \theta_1^0, \ldots, \theta_r^0, \hat{\theta}_{r+1}, \ldots, \hat{\theta}_k\right). \qquad (7.2.8)$$

Under H_a all the parameters are replaced by their maximum likelihood estimators. Thus the likelihood function $L(\hat{\Omega})$ is given by

$$L(\hat{\Omega}) = \prod_{i=1}^{n} f\left(x_i; \hat{\theta}_1, \ldots, \hat{\theta}_k\right). \qquad (7.2.9)$$

The statistic Λ, called the likelihood ratio statistic, is defined by

$$\Lambda = \frac{L(\hat{\omega})}{L(\hat{\Omega})}, \tag{7.2.10}$$

which does not depend on any of the unknown parameters. It is not difficult to see that $0 < \Lambda < 1$. First, since $L(\hat{\omega})$ and $L(\hat{\Omega})$ are likelihood functions, their ratio is positive. Second, since $L(\hat{\omega})$ is the restricted maximum and $L(\hat{\Omega})$ is the unrestricted maximum, $L(\hat{\Omega}) > L(\hat{\omega})$, which proves $\Lambda < 1$.

The best critical region for Λ is the left tail of the frequency function of Λ. This follows because when the restricted and unrestricted maxima $L(\hat{\omega})$ and $L(\hat{\Omega})$, respectively, are close together, there is the implication that H_0 cannot be rejected, whereas when the maxima are not close together, the rejection of H_0 is implied. To prescribe the critical region for Λ, it is necessary to know its distribution when H_0 is true. This is usually very difficult, and its solution depends on the frequency function $f(x; \theta_1, \ldots, \theta_k)$. However, if n is sufficiently large ($n \geqslant 25$, say), a very satisfactory solution to this problem exists. This solution depends on $f(x; \theta_1, \ldots, \theta_k)$ and $L(\Omega)$ satisfying the regularity conditions embodied in the following theorem.

Theorem 7.2.1

Suppose $f(x; \theta_1, \ldots, \theta_k)$ is a frequency function that satisfies the following conditions.

1. It is permissible to interchange the operations of integration with respect to x and differentiation with respect to θ_i, $i = 1, 2, \ldots, k$.
2. The term $\partial^2 f / \partial \theta_i \partial \theta_j$ has a finite variance, where $f \equiv f(x; \theta_1, \ldots, \theta_k)$.
3. The likelihood function $L(\Omega)$ has bounded derivatives of the third order with respect to $\theta_1, \ldots, \theta_k$.
4. The derivative of $L(\Omega)$ with respect to $\theta_1, \ldots, \theta_k$ vanishes at its maximum.

Furthermore, given that the dimensionality of Ω, the unrestricted parameter space, is k and that of ω, the restricted parameter space is $k - r$, then $-2 \log_e \Lambda$ is approximately distributed as the chi-square distribution with $r < k$ degrees of freedom when H_0 is true and n is sufficiently large ($n \geqslant 25$, say). See Mood [1950] for a detailed discussion of Theorem 7.2.1.

The appropriate critical region for $-2 \log_e \Lambda$ is easily seen to be the right tail of the chi-square distribution with r degrees of freedom.

Comparison of Two Exponential Survival Distributions

We can compare two exponential survival distributions by using the following tests.

Nonparametric comparison matched samples: The sign test can be used directly to compare two exponentially distributed populations in which the observations can be paired as (t_{i1}, t_{i2}), where t_{i1} is the survival time of the ith patient receiving the standard treatment and t_{i2} is the survival time of the ith patient receiving the experimental treatment, these patients being matched so that under the null hypothesis of no treatment difference, $\mathrm{pr}\{t_{i1} > t_{i2}\} = \mathrm{pr}\{t_{i1} < t_{i2}\} = \frac{1}{2}$, $i = 1, 2, \ldots, n$. This is the sign test given in Section 7.2.

Nonparametric comparison independent samples: In the case of independent samples, with complete survival information, the Wilcoxon test, discussed at the end of this section is used.

Likelihood ratio test: Suppose n_1 patients receiving a standard treatment and n_2 patients receiving an experimental treatment (independent samples, where n_1 and n_2 may be different) have as their death density function

$$f(t_{ij}, \lambda_j) = \lambda_j \exp(-\lambda_j t_{ij}), \qquad i = 1, 2, \ldots, n_j; j = 1, 2. \qquad (7.2.11)$$

Under $H_0 : \lambda_1 = \lambda_2 = \lambda$ (say), the likelihood function is

$$L(\lambda) = \lambda^{n_1 + n_2} \exp\left(-\lambda \sum_{j=1}^{2} \sum_{i=1}^{n_j} t_{ij}\right). \qquad (7.2.12)$$

Then $\hat{\lambda}$, the maximum likelihood estimator of λ, is

$$\hat{\lambda} = \frac{n_1 + n_2}{n_1 \bar{t}_1 + n_2 \bar{t}_2}, \qquad (7.2.13)$$

where

$$\bar{t}_j = \sum_{i=1}^{n_j} t_{ij} / n_j, \qquad j = 1, 2.$$

Under $H_a : \lambda_1 \neq \lambda_2$, we have

$$L(\lambda_1, \lambda_2) = \lambda_1^{n_1} \lambda_2^{n_2} \exp\left[-\lambda_1 \sum_{i=1}^{n_1} t_{i1} - \lambda_2 \sum_{i=1}^{n_2} t_{i2}\right]. \qquad (7.2.14)$$

The maximum likelihood estimator $\hat{\lambda}_j$ of λ_j is

$$\hat{\lambda}_j = \bar{t}_j^{-1}, \qquad j = 1, 2. \tag{7.2.15}$$

The likelihood ratio statistic Λ is then

$$\Lambda = \frac{\bar{t}_1^{n_1} \bar{t}_2^{n_2}}{\left[\dfrac{n_1 \bar{t}_1 + n_2 \bar{t}_2}{n_1 + n_2} \right]^{n_1 + n_2}}. \tag{7.2.16}$$

Thus to test the hypothesis $H_0 : \lambda_1 = \lambda_2$ against the alternative $H_a : \lambda_1 \neq \lambda_2$, we compute Λ given by (7.2.16). The H_0 is rejected at level α if and only if $-2 \log_e \Lambda$ exceeds $\chi^2(1 - \alpha; 1)$, which is the $100(1 - \alpha)$ percentage point of the chi-square distribution with 1 degree of freedom.

In conclusion, we note that the likelihood ratio test can be applied to any parametric distribution for which the maximum likelihood estimators can be obtained. However, the test does have its shortcomings for comparing two survival distributions. First, the test itself is approximate, as we noted in the previous paragraph. This makes its use of dubious value when only small samples are available. Second, one cannot obtain confidence intervals on the requisite parameters using this method. Third, the likelihood ratio test is difficult to apply if we are concerned with one-sided tests of significance. Finally, the power of likelihood ratio tests—that is, the probability of rejecting H_0 when it is false—is not high. Hence by using the likelihood ratio procedure on a regular basis, one is at times likely to accept the equality of treatment effects when they are not equal.

An Exact Test of Significance for Two Exponential Distributions. In Section 3.3 we observed that $2n_j \bar{t}_j \lambda_j$ was distributed as a chi-square variable exactly with $2n_j$ degrees of freedom, $j = 1, 2$. Assuming that we are taking independent samples from each population, it follows easily that $\bar{t}_1 \lambda_1 / \bar{t}_2 \lambda_2$ has an F distribution with $2n_1$ degrees of freedom for the numerator and $2n_2$ degrees of freedom for the denominator. Thus a two-sided test of the hypothesis $H_0 : \lambda_1 = \lambda_2$ against the alternative $H_a : \lambda_1 \neq \lambda_2$ at level α is to reject H_0 only if $\bar{t}_1 / \bar{t}_2 > F(1 - \alpha/2; 2n_1, 2n_2)$ or $\bar{t}_1 / \bar{t}_2 < F(\alpha/2; 2n_1, 2n_2)$, where $F(\alpha/2; 2n_1, 2n_2)$ and $F(1 - \alpha/2; 2n_1, 2n_2)$ are the $100(\alpha/2)$ and $100(1 - \alpha/2)$ percentage points of the F distribution with $2n_1$ degrees of freedom for the numerator and $2n_2$ degrees of freedom for the denominator. This follows because under $H_0 : \lambda_1 = \lambda_2, \bar{t}_1 / \bar{t}_2$ has an F distribution with

$2n_1$ degrees of freedom for the numerator and $2n_2$ degrees of freedom for the denominator.

Furthermore, since $\bar{t}_1\lambda_1/\bar{t}_2\lambda_2$ has an F distribution generally, a $(1-\alpha)$ 100 percent confidence interval for λ_1/λ_2 is given by

$$\left(\frac{\bar{t}_2}{\bar{t}_1}\right)F\left(\frac{\alpha}{2};2n_1,2n_2\right)<\frac{\lambda_1}{\lambda_2}<\left(\frac{\bar{t}_2}{\bar{t}_1}\right)F\left(1-\frac{\alpha}{2};2n_1,2n_2\right). \quad (7.2.17)$$

Excellent F tables for computing this confidence interval are given in Dixon and Massey [1969].

Example 7.2.1. To test whether a new analgesic produces a quicker mean response time in relieving headaches than a standard treatment, 10 patients subject to frequent spells of headaches were given each remedy on two successive headache episodes. The order was randomized and at the time the analgesics were administered it was unknown to both subject and experimenter which analgesic the subject received (a double-blind study). The times to relief for each treatment, which are assumed to be exponentially distributed, are given in Table 7.1.

Table 7.1 Response times (in minutes) to relief of headaches for standard and new treatments given to 10 patients

Patient	Treatment	
	Standard	New
1	8.4	6.9
2	7.7	6.8
3	10.1	10.3
4	9.6	9.4
5	9.3	8.0
6	9.1	8.8
7	9.0	6.1
8	7.7	7.4
9	8.1	8.0
10	5.3	5.1

First of all, if we compare these matched samples nonparametrically, we find that 9 patients show a preference for the new analgesic and there is one who prefers the standard remedy. If there is no difference between the treatments, the probability of 9 or more (10) patients showing preference for the new analgesic is

$$\text{pr}\{x \geqslant 9\} = 2^{-10}\left[\binom{10}{9} + \binom{10}{10}\right] = 0.011. \qquad (7.2.18)$$

Thus at level $\alpha = 0.05$ for a one-sided test we would conclude the new analgesic is superior to the standard. If a two-sided test were made, we would also conclude that the new analgesic is significantly different from the control (2) $(0.011) < 0.05$.

Let us ignore the matching of the samples and treat them as independent samples. If we apply the likelihood ratio test to the hypothesis of equality of the two treatments, we find $-2\log_e\Lambda = 0.06$, which would imply that the hypothesis should not be rejected. However, this test does not account for the preference for the new analgesic by 9 out of 10 patients.

Finally, the F ratio calculated statistic $F_{cal} = \bar{t}_2/\bar{t}_1 = 1.10$. Here $\bar{t}_1 = 8.43$ minutes is the mean response time for patients in the group receiving the standard analgesic and $\bar{t}_2 = 7.70$ minutes is the mean response time for patients receiving the new analgesic. Since $F(0.90; 20, 20) = 1.79$, the value F_{cal} is not significant again.

Only the sign test shows a significant difference between the samples because the sign test looks merely at the direction of the differences between pairs of observations on the 10 patients. The two parametric tests do account for both the direction and magnitude of the differences. However, the parametric tests do not account for the fact that the samples are repeated observations on the same 10 patients.

Comparison of Two Weibull Survival Distributions

Nonparametric Comparison. Using either the sign test or the Wilcoxon two-sample test we can compare two Weibull distributions in exactly the same way employed for the two exponential distributions in Example 7.2.1. In addition, other nonparametric procedures are available for either independent or paired samples for the two-sample case and the k-sample case ($k \geqslant 3$). The power of these tests compared to the t and F tests used in comparing normal distributions is quite high. The reader may gain further insight to such nonparametric procedures by referring to Siegel [1956] or Gibbons [1971].

Likelihood Ratio Test. The likelihood ratio test for testing the equality of two Weibull survival distributions can be formulated in a number of ways. For example, if it is known that the two shape parameters are equal (i.e., $\gamma_1 = \gamma_2 = \gamma$, say), the likelihood ratio test is formulated to test the equality of the two scale parameters γ_1 and γ_2. If γ_1 and γ_2 are known, the transformation $y = t^{\gamma_i}$, $i = 1, 2$, reduces the problem to testing for equality of two exponential populations. Alternatively, it may be known that $\lambda_1 = \lambda_2$. In this case, a test for equality of two distributions is a test of equality for the shape parameters γ_1 and γ_2. Finally, the most general case would be a test for the equality of λ_1 and λ_2 simultaneously with the equality of γ_1 and γ_2. The general case, although not difficult conceptually, is tedious to derive.

Thoman-Bain Tests. We define the standard Weibull death density function as that for which $\lambda = \gamma = 1$. That is, the standard Weibull death density function is

$$f(t) = e^{-t}, \qquad t \geq 0. \qquad (7.2.19)$$

In their paper Thoman and Bain [1969] show that $(\hat{\gamma}_1 / \gamma_1)/(\hat{\gamma}_2 / \gamma_2)$ is distributed independently of the parameters $\gamma_1, \gamma_2, \lambda_1$, and λ_2, the shape and scale parameters of the two respective Weibull death density functions. In fact, $(\hat{\gamma}_1 / \gamma_1)/(\hat{\gamma}_2 / \gamma_2)$ has the same distribution as $\hat{\gamma}_1^* / \hat{\gamma}_2^*$ where $\hat{\gamma}_i^*$ is the maximum likelihood estimator of γ_i when the sample is taken from the population having the standard Weibull density, $i = 1, 2$. Thoman and Bain obtain the distribution of $\hat{\gamma}_1^* / \hat{\gamma}_2^*$ by Monte Carlo methods and percentage points a_α such that $\text{pr}\{\hat{\gamma}_1^* / \hat{\gamma}_2^* < a_\alpha\} = \alpha$. These are based on equal samples for both populations and appear as Table 4 of the Appendix. Thus to test the null hypothesis $H_0 : \gamma_1 = \gamma_2$ against the one-sided alternative $H_a : \gamma_1 > \gamma_2$ at significance level α, we compute $a_{\text{cal}} = \hat{\gamma}_1 / \hat{\gamma}_2$ and reject H_0 only if $a_{\text{cal}} > a_{1-\alpha}$ where $\text{pr}\{a_{\text{cal}} < a_{1-\alpha} | H_0\} = 1 - \alpha$. This follows because $\hat{\gamma}_1 / \hat{\gamma}_2$ has the same distribution as γ_1^* / γ_2^* when $H_0 : \gamma_1 = \gamma_2$ is true.

Note that it is always possible to use a one-sided test even when the objective is to determine whether γ_1 and γ_2 are significantly different. This follows, as for the F test for the equality of exponential parameters, by requiring the larger of $\hat{\gamma}_1$ or $\hat{\gamma}_2$ to be the numerator and the smaller to be the denominator.

If the hypothesis $H_0 : \gamma_1 = \gamma_2$ is accepted, a test can be made for the equality of the scale parameters λ_1 and λ_2. Thoman and Bain show that when $\gamma_1 = \gamma_2$, a test at significance level α of the hypothesis $H_0' : \lambda_1 = \lambda_2$ against the one-sided alternative $H_a' : \lambda_1 < \lambda_2$ is given by the rule: reject H_0' only if

$$b_{\text{cal}} = \log_e \hat{\lambda}_2 - \log_e \hat{\lambda}_1 > b_{1-\alpha}, \qquad (7.2.20)$$

where $\mathrm{pr}\{(\log_e\hat{\lambda}_2 - \log_e\hat{\lambda}_1) > b_{1-\alpha}|H_0\} = 1 - \alpha$. Percentage points b_α such that $G(b_\alpha) = \alpha$ are given in Table 5 of the Appendix. Noting that $b_\alpha = -b_{1-\alpha}$, two-sided tests or one-sided tests for which the alternative hypothesis $H_a'' : \lambda_1 > \lambda_2$ can also be obtained using Table 5.

In using Tables 4 and 5 in the Appendix, equal samples from both populations are required. Thus until more extensive tables are available, the two-stage Thoman-Bain procedure for testing the equality of two populations, both having Weibull survival distributions, requires equal sample sizes.

Example 7.2.2. Harter and Moore [1965] collected simulated life test data from two populations whose survival distributions are Weibull. Table 7.2 illustrates the numerical values for both samples. Since we are concerned with two-parameter Weibull distributions, the estimates $\hat{\lambda}_1, \hat{\lambda}_2, \hat{\gamma}_1$, and $\hat{\gamma}_2$ for the two samples are obtained assuming that the location parameters ξ_1 and ξ_2 are known, $\xi_1 = 10$, $\xi_2 = 20$. From these samples the maximum likelihood estimates for the parameters are $\hat{\lambda}_1 = 1.49 \times 10^{-4}$, $\hat{\gamma}_1 = 1.94$, $\hat{\lambda}_2 = 4.21 \times 10^{-6}$, and $\hat{\gamma}_2 = 2.72$. A test of $H_0 : \gamma_1 = \gamma_2$ yields the value $a_{\mathrm{cal}} = 1.40$, which is significant at the 0.05 level ($a_{0.95,40} = 1.342$ from Table 5) but not at the 0.02 level ($a_{0.98,40} = 1.453$). Assuming that H_0 is accepted, we test $H_0' : \lambda_1 = \lambda_2$ against the alternative $H_a' : \lambda_1 > \lambda_2$. Here we find $b_{\mathrm{cal}} = 3.57$, which is highly significant. We thus conclude that the two samples come from distinct Weibull populations. Since these were simulated data from two different Weibull populations, the significant result is not surprising.

Table 7.2 Two Weibull samples, $n_1 = n_2 = 40$, with known location parameters $\xi_1 = 10$, $\xi_2 = 20$

Weibull 1				Weibull 2			
15	65	92	124	40.9	81.4	108.7	129.7
20	68	95	126	52.2	85.4	109.3	130.8
27	68	100	127	53.2	86.0	111.6	134.1
42	71	102	134	59.4	86.3	113.1	137.5
43	74	102	149	60.0	87.4	114.2	139.2
43	75	112	152	66.8	88.5	117.7	140.3
44	75	113	153	77.3	89.9	121.6	143.0
46	76	116	161	78.0	92.4	121.9	143.8
64	77	117	168	79.7	93.0	127.6	183.3
65	78	124	205	81.1	93.2	128.0	185.1

Source: Harter and Moore [1965].

Asymptotic Procedures. Various tests on the equality of parameters of two Weibull distributions, or even testing whether one or both parameters in a single Weibull distribution has (have) known value(s), can be accomplished for large samples by utilizing the asymptotic normality of the maximum likelihood estimators. The relationship between tests and elliptical confidence regions in Section 4.4 is important; note, however, that $\hat{\lambda}$ and $\hat{\gamma}$ are not independent as are $\bar{x}$ and s^2 for the normal distributions.

Comparison of Two Gamma Survival Distributions with Known Shape Parameters

Suppose $t_{11}, t_{21}, \ldots, t_{n1}$ and $t_{12}, t_{22}, \ldots, t_{n2}$ are the survival times of $2n$ patients randomly assigned to two different treatment regimens, with n patients on each treatment given, say, an experimental treatment and a control. Suppose, further, that the death density function of $t_{ij} \geq 0$ is the gamma; that is,

$$f(t_{ij}) = \frac{\lambda_j^{\gamma_j} t_{ij}^{\gamma_j - 1} \exp(-\lambda_j t_{ij})}{\Gamma(\gamma_j)}, \qquad i = 1, 2, \ldots, n; j = 1, 2. \qquad (7.2.21)$$

It is further assumed that γ_1 and γ_2 are known. (In most applications for which the experimenter wishes to test for the equality of the two distributions, it is further assumed that $\gamma_1 = \gamma_2$.) It is not difficult to show that $\bar{t}_1$ and $\bar{t}_2$, the sample mean survival times for patients receiving each treatment, also have gamma death density functions. In fact,

$$f(\bar{t}_j) = \frac{(n\lambda_j)^{n\gamma_j} (\bar{t}_j)^{n\gamma_j - 1} \exp(-n\lambda_j \bar{t}_j)}{\Gamma(n\gamma_j)}, \qquad j = 1, 2. \qquad (7.2.22)$$

If we let $z = \bar{t}_1/\bar{t}_2$, it follows that the density of z is

$$g(z) = \frac{\lambda_1^{n\gamma_1} \lambda_2^{n\gamma_2} \Gamma[n(\gamma_1 + \gamma_2)]}{\Gamma(n\gamma_1)\Gamma(n\gamma_2)} \frac{z^{n\gamma_1 - 1}}{(n\lambda_1 + n\lambda_2 z)^{n(\gamma_1 + \gamma_2)}}, \qquad z \geq 0. \qquad (7.2.23)$$

If γ_1 and γ_2 are known (not necessarily but usually assumed equal), we wish to test the hypothesis $H_0: \lambda_1 = \lambda_2$ against the alternative $H_a: \lambda_1 \neq \lambda_2$. When H_0 is true the density function of z is

$$g(z) = \frac{\Gamma[n(\gamma_1 + \gamma_2)]}{\Gamma(n\gamma_1)\Gamma(n\gamma_2)} \frac{z^{n\gamma_1 - 1}}{(1 + z)^{n(\gamma_1 + \gamma_2)}} \quad \text{with } \gamma_1, \gamma_2, \lambda_1, \text{ and } \lambda_2 > 0. \qquad (7.2.24)$$

That is, z has the F-ratio distribution with $2n\gamma_1$ degrees of freedom in the numerator and $2n\gamma_2$ degrees of freedom in the denominator. (See Rao [1952].) Hence an α level test of $H_0 : \lambda_1 = \lambda_2$ is to reject H_0 when the calculated value $z = \bar{t}_1 / \bar{t}_2$ exceeds $F(1 - \alpha; 2n\gamma_1, 2n\gamma_2)$ the $100(1 - \alpha)$th percentage point of the F distribution with $2n\gamma_1$ degrees of freedom in the numerator and $2n\gamma_2$ degrees of freedom in the denominator. Unless γ_1 and γ_2 are both integers, it is necessary to interpolate in a given F table of critical values. Furthermore, if γ_1 and γ_2 are both nonintegers, bilinear interpolation is required to obtain the requisite critical values.

Example 7.2.3.   In Example 4.2.1 we selected a random sample of 20 survival times (in weeks) of 208 male mice exposed to 240 rads of gamma radiation; the animals were first studied by Furth, Upton, and Kimball [1959]. We select another random sample of survival times of size 20. The data for the two random samples appear in Table 7.3. These survival times originally came from the same overall sample of 208 survival times, and the gamma death density function was assumed; thus we obtain an overall γ value for the shape parameter. Using (4.2.10) and Table 1 of the Appendix, it is not hard to show that $\hat{\gamma}$, the overall maximum likelihood estimate of γ for the data in Table 7.3, is 9.60. Thus for the illustrative purposes of this example, we assume that the data in each sample have a gamma death density function with common shape parameter $\gamma = 9.6$. Now $\bar{t}_1 = 113.45$ is the computed mean survival time for sample 1 and $\bar{t}_2 = 123.55$ is the computed mean survival time for sample 2. Under the null hypothesis, it follows that $z = \bar{t}_2 / \bar{t}_1$ has the F distribution, with 384 degrees of freedom for both the numerator and the denominator. The calculated value of z is 1.089. The 95 percent critical value of z is 1.183. Thus there is no significant difference between λ_1 and λ_2 at the 5 percent level of significance.

Table 7.3 Survival times (in weeks) of male mice exposed to 240 rads of radiation

Sample I	Sample II
152, 152, 115, 109,	56, 174, 134, 157,
137, 88, 94, 77,	166, 131, 147, 127,
160, 165, 125, 40,	156, 137, 91, 86,
128, 123, 136, 101,	96, 70, 87, 177,
62, 153, 83, 69	83, 128, 123, 145

A Nonparametric Test for Noncensored Samples

In this section we indicated that the Wilcoxon test can be used to test for the equality of two survival distributions. In this case, suppose the complete sample of survival times of patients who receive the standard treatment is $t_{11}, t_{21}, \ldots, t_{n_1 1}$ and the complete sample of survival times of patients who receive the experimental treatment is $t_{12}, t_{22}, \ldots, t_{n_2 2}$. The null hypothesis, as with Gehan's censored Wilcoxon test, is $H_0 : F_1(t) = F_2(t)$ (the treatments are equally effective). The alternative may be either one-sided or two-sided. We consider here the two-sided alternative $H_a : F_1(t) \neq F_2(t)$.

This time we define

$$U_{ij} = \begin{cases} -1 & \text{if } t_{j2} < t_{i1} \\ 0 & \text{if } t_{j2} = t_{i1} \\ 1 & \text{if } t_{j2} > t_{i1} \end{cases} \tag{7.2.25}$$

and calculate the statistic $W = \sum_{i=1}^{n_1} \sum_{j=1}^{n_2} U_{ij}$.

When the null hypothesis is true (i.e., when the two treatments are equally effective), W is approximately normally distributed, with mean zero and variance $n_1 n_2 (n_1 + n_2 + 1)/3$ (see Exercise 3 at the end of Section 7.3) for values n_1 and n_2, each greater than or equal to 5. Thus the statistic

$$Z = \frac{W}{\sqrt{n_1 n_2 (n_1 + n_2 + 1)/3}} \tag{7.2.26}$$

has an approximate standard normal distribution when H_0 is true. We then would reject H_0 at significance level α only if $|Z| > Z(1 - \alpha/2)$, where $Z(1 - \alpha/2)$ is the upper $(1 - \alpha/2)100$ percentage point of the standard normal distribution. Although the test presented is for two-sided alternatives, a one-sided alternative is considered in Example 7.2.4.

Example 7.2.4. Suppose the response times in Table 7.1 are for two patient groups, one group receiving the standard analgesic and the other group receiving the new drug. We wish to compare these response times by means of the Wilcoxon test. To this end let t_{i1} be the response time of the ith patient on the standard treatment and t_{j2} be the response time of the jth patient on the experimental treatment, $i, j = 1, \ldots, 10$. It follows that $W = -38$, which leads to a calculated value $Z = -1.43$, which is significant (one-sided) at level 0.10: $Z(0.10) = -1.28$ but not at level 0.05: $Z(0.05) = -1.64$.

Comparison of Other Pairs of Survival Distributions

To conclude this section we note that the two nonparametric tests as well as the likelihood ratio test are generally applicable to comparing other pairs of survival distributions such as the gamma distribution and the Rayleigh distribution. Estimating the parameters of these distributions is covered in Chapter 4. Furthermore, the likelihood ratio test is applicable when there is censoring so that tests for equality of censored Weibull, censored gamma, and other censored parametric distributions can be made. Other nonparametric tests are available in Siegel [1956] or Gibbons [1971].

EXERCISES

1. Suppose in two exponential populations we wish to test the hypothesis $H_0 : \lambda_1 = \lambda_2$ against the alternative $H_a : \lambda_1 = c\lambda_2$, where $0 < c < 1$. Compute the power of the F test as a function of c. That is, compute the probability of rejecting H_0 when H_a is true. If $n_1 = n_2 = 20$, and $\alpha = 0.05$, obtain a graph of the power as a function of c for $c = 0.1, 0.2, 0.3, 0.4, 0.5$.

2. Develop a likelihood ratio test for testing the hypothesis $H_0 : \lambda_1 \leqslant \lambda_2$ against the one-sided alternative $H_a : \lambda_1 > \lambda_2$ based on independent survival data whose sample sizes are n_1 and n_2, respectively from two exponential distributions. How does this alter Example 7.2.1 if we now test the null hypothesis H_0 (the new treatment is no better than the standard treatment) against H_a (the new treatment is better than the standard treatment)?

3. Develop a likelihood ratio test for testing the equality of two Weibull survival distributions. Consider all three possibilities: (1) $H_{01} : \lambda_1 = \lambda_2$, given $\gamma_1 = \gamma_2$; (2) $H_{02} : \gamma_1 = \gamma_2$, given $\lambda_1 = \lambda_2$; and (3) $H_{03} : \lambda_1 = \lambda_2$ and $\gamma_1 = \gamma_2$ simultaneously. For the data in Example 7.2.2 test H_{01} and then test H_{03} using a 5 percent level of significance each time.

4. Prove the following theorem of Thoman, Bain, and Antle [1969]. Suppose $\hat{\gamma}$ is the maximum likelihood estimator of γ, the shape parameter, based on an independent sample of size n from a Weibull survival distribution. Then $\hat{\gamma}$ has the same distribution as $\hat{\gamma}^*$, where $\hat{\gamma}^*$ is the maximum likelihood estimator of γ, when in fact the sample is taken from the standard Weibull survival distribution.

5. Suppose in two Weibull populations we wish to test the hypothesis $H_0 : \gamma_1 = \gamma_2$ against the alternative $H_a : \gamma_1 = c\gamma_2$, $0 < c < 1$. Obtain an expression for computing the power of this test, using the percentage points from Appendix Table 4. If $n_1 = n_2 = 20$, and $\alpha = 0.05$, obtain a graph of the power as a function of c for $c = 0.1, 0.2, 0.3, 0.4,$ and 0.5. (Thoman and Bain [1969].)

6. Suppose the survival times (in weeks) of male mice exposed to 240 rads of radiation appeared in two samples as follows: Sample 1: 152, 152, 115, 109, 56, 174, 134, 157, 137, 88, 94, 77, 166, 131, 147, 127, 160, 165, 125, and 40; sample 2: 156, 137, 91, 86, 128, 123, 136, 101, 96, 70, 87, 177, 62, 153, 83, 69, 83, 128, 123, and 145. Assuming $\gamma = 9.60$ as in Example 7.2.3, test the hypothesis $H_0 : \lambda_1 = \lambda_2$ against the alternative $H_a : \lambda_1 \neq \lambda_2$ at significance level $\alpha = 0.05$.

7.3 COMPARISON OF TWO SURVIVAL DISTRIBUTIONS: CENSORED OBSERVATIONS

As in Section 7.2 we assume that $n_1 + n_2$ patients are randomly allocated to two treatment regimens—n_1 patients receiving a standard course of therapy and n_2 patients receiving an experimental treatment. Patients are not necessarily matched in pairs. Thus it is not essential that $n_1 = n_2$. However in most clinical trials there are approximately equal numbers of patients in the treatment groups. Survival, the parameter of treatment efficacy being measured, is expected to be increased for patients receiving the experimental treatment.

We assume for the remainder of this chapter that the survival times in both treatment groups are progressively censored, with each surviving patient having a known time on study.

In the parametric case the likelihood ratio test is generally applicable for testing the equality of two survival distributions when the data are censored, even though the details of the test procedure are generally complicated, despite the straightforwardness of the method. The percentage of censored observations should be small, less than 20 percent, to ensure that the asymptotic properties of the test hold.

Comparison of Two Exponential Survival Distributions with Progressive Censoring of the Data

Suppose

$$S_j(t; \lambda_j) = \exp(-\lambda_j t), \qquad j = 1, 2, \qquad (7.3.1)$$

for patients assigned to the standard and experimental therapies. These patients are not matched. Furthermore, suppose the ith patient receiving the jth treatment is on study for a total time T_{ij} ($j = 1$ for standard therapy, and $j = 2$ for experimental treatment). If the patient dies prior to T_{ij}, his time to death t_{ij} is recorded.

The likelihood function $L(\Omega)$ is then given by

$$L(\Omega) = \lambda_1^{d_1} \lambda_2^{d_2} \exp\left(- \sum_{j=1}^{2} \sum_{i=1}^{n_j} \lambda_j \left\{ \delta_{ij} t_{ij} + (1 - \delta_{ij}) T_{ij} \right\} \right), \qquad (7.3.2)$$

where

$$\delta_{ij} = \begin{cases} 1 & \text{if the } i\text{th patient on the } j\text{th treatment} \\ & \quad \text{dies in the interval } 0 < t_{ij} \leqslant T_{ij}, \\ 0 & \text{if he dies in the interval } t_{ij} > T_{ij} \end{cases} \qquad (7.3.3)$$

$i = 1, \ldots, n_j, j = 1, 2.$

Under the null hypothesis $H_0 : \lambda_1 = \lambda_2 = \lambda$(say),

$$L(\omega) = \lambda^{-(d_1 + d_2)} \exp\left(-\lambda \sum_{j=1}^{2} \sum_{i=1}^{n_j} \left\{ \delta_{ij} t_{ij} + (1 - \delta_{ij}) T_{ij} \right\} \right). \qquad (7.3.4)$$

The maximum likelihood estimators $\hat{\lambda}_j$ and $\hat{\lambda}$ of λ_j and λ are, respectively,

$$\hat{\lambda}_j = \frac{d_j}{\displaystyle\sum_{i=1}^{n_j} \left(\delta_{ij} t_{ij} + (1 - \delta_{ij}) T_{ij} \right)} \qquad (7.3.5)$$

and

$$\hat{\lambda} = \frac{d_1 + d_2}{\displaystyle\sum_{j=1}^{2} \sum_{i=1}^{n_j} \left(\delta_{ij} t_{ij} + (1 - \delta_{ij}) T_{ij} \right)}, \qquad j = 1, 2. \qquad (7.3.6)$$

These estimators are derived in Section 3.3.

To test the hypothesis $H_0 : \lambda_1 = \lambda_2$ against the alternative $H_a : \lambda_1 \ne \lambda_2$, we form the ratio $L(\hat{\omega}) / L(\hat{\Omega})$, where $L(\hat{\Omega})$ and $L(\hat{\omega})$ are given by (7.3.2) and (7.3.4), respectively, with $\hat{\lambda}_j$ replacing λ_j, $j = 1, 2$, and $\hat{\lambda}$ replacing λ. As for the complete sample case if Λ is given by (7.2.10), $-2 \log_e \Lambda$ has an approximate chi-square distribution with a single degree of freedom for samples of size 25 or greater when H_0 is true. Thus an α-level test of H_0 is: reject H_0 if and only if $-2 \log_e \Lambda$ exceeds $\chi^2(1 - \alpha; 1)$, where $\chi^2(1 - \alpha; 1)$ is the $100(1 - \alpha)$ percentage point for the chi-square distribution with a single degree of freedom. The same shortcomings concerning the likelihood ratio test are present in the censored sample case that were found in the complete sample case. In particular, the power of this test is not at all high.

An extension of this test to the k-treatment situation is left as an exercise at the end of this section.

The Cox F Test

Cox [1953] proposed a test of the hypothesis $H_0 : \lambda_1 = \lambda_2$ against either the one-sided or two-sided alternative for exponential survival distributions with independent samples.

Let $t_{ij}, i = 1, 2, \ldots, d_j$, and $T_{ij}, i = d_j + 1, \ldots, n_j; j = 1, 2$, be defined as before with respect to the two treatments. If as in progressive censoring the numbers of observed treatment failures (deaths) are fixed in advance, the ratio $\hat{t}_1 / \hat{t}_2$ has an exact F distribution with $2d_1$ degrees of freedom for the numerator and $2d_2$ degrees of freedom for the denominator when $H_0 : \lambda_1 = \lambda_2$ is true, where $\hat{t}_j = \hat{\lambda}_j^{-1}, j = 1, 2$. If the observation times are fixed, the number of treatment failures (deaths) in each group is random. In this case, $\bar{t}_1/\bar{t}_2$ follows an approximate F distribution with $2d_1$ and $2d_2$ degrees of freedom for numerator and denominator, respectively. Furthermore, since $\bar{t}_1\lambda_1/\bar{t}_2\lambda_2$ obeys an F distribution even when H_0 is not true, a 100 $(1-\alpha)$ percent confidence interval for λ_1/λ_2 is given by

$$\left(\frac{\bar{t}_2}{\bar{t}_1}\right)F\left(\frac{\alpha}{2};2d_1,2d_2\right) < \frac{\lambda_1}{\lambda_2} < \left(\frac{\bar{t}_2}{\bar{t}_1}\right)F\left(1-\frac{\alpha}{2};2d_1,2d_2\right), \qquad (7.3.7)$$

where $F(\alpha/2;2d_1,2d_2)$ and $F(1-\alpha/2;2d_1,2d_2)$ are the $100\alpha/2$ and $100(1-\alpha/2)$ percentage points of the F distribution with $2d_1$ and $2d_2$ degrees of freedom for numerator and denominator, respectively.

Example 7.3.1. In a clinical trial reported by Freireich et al. [1963], 6-mercaptopurine (6–MP) was compared with a placebo in the maintenance of remissions in acute leukemia. One year after the start of the study, the following remission times were recorded on 42 patients, 21 of whom received the 6–MP treatment while the other 21 received placebo (see Table 7.4). This example is found in Gehan [1965a]. A plus sign after an observation means the observation is censored. We note that prior to analyzing these data it appears that 6–MP maintenance therapy is superior to placebo maintenance therapy for these patients.

Let us assume that patients receiving 6–MP who actually go into remission have a guaranteed remission period of 5 weeks. Thus we wish to test the hypothesis $H_0:\lambda_1=\lambda_2$ against the alternative $H_a:\lambda_1\neq\lambda_2$, where λ_1 is the hazard rate for patients receiving placebo therapy and λ_2

**Table 7.4 Remission times (in weeks)
leukemia of patients with acute
receiving 6-MP or placebo as treatments**

Treatment	
Placebo	6 – MP
1, 1, 2	6, 6, 6
2, 3, 4	7, 10, 13
4, 5, 5	16, 22, 23
8, 8, 8	6+, 9+, 10+
8, 11, 11	11+, 17+, 19+
12, 12, 15	20+, 25+, 32+
17, 22, 23	32+, 34+, 35+

is the hazard rate for patients receiving $6-MP$, assuming a guaranteed remission of 5 weeks. The maximum likelihood estimates are $\hat{\lambda}_1 = 0.115$ and $\hat{\lambda}_2 = 0.035$, respectively. Employing the likelihood ratio test, we find with the aid of (7.3.2) and (7.3.4) that $-2 \log_e \Lambda = 9.25$. This well exceeds $\chi^2(0.99; 1) = 6.63$, implying a significant difference between the two hazard rates. If we apply the Cox F test, $\bar{t}_2 / \bar{t}_1 = 3.28$, which exceeds $F(0.99; 18, 42) = 2.85$.

We thus conclude that patients who receive $6-MP$ have a significantly longer remission period than patients receiving placebo, even after we account for a 5-week guarantee period for the $6-MP$ patients.

Gehan's Nonparametric Test for Comparing Survival Distribution with Censored Data

We assume that $n_1 + n_2$ patients are allocated randomly to the two treatment regimens as before. We observe

$$\left. \begin{array}{l} t_{11}, \ldots, t_{d_1 1} \text{ deaths} \\[2mm] T_{d_1 + 1, 1, \ldots,} T_{n_1 1} \text{ censored} \end{array} \right\} \text{standard treatment}$$

and

$$\left. \begin{array}{l} t_{12}, \ldots, t_{d_2 2} \text{ deaths} \\[2mm] T_{d_2 + 1, 2, \ldots,} T_{n_2 2} \text{ censored} \end{array} \right\} \text{experimental treatment.}$$

Here t_{ij}, $i = 1, 2, \ldots, d_j; j = 1, 2$, are times to death and $T_{ij}, i = d_j + 1, \ldots, n_j; j = 1, 2$, are times to censoring (all measured from time of entry into study).

Thus as before we assume that d_j patients die and $n_j - d_j$ survive during the course of the study, $j = 1, 2$. Furthermore, the ith patient receiving the jth treatment is observed for a time T_{ij}, $i = 1, \ldots, n_j, j = 1, 2$. He is assigned this time only if he survives his observation period; otherwise he is assigned t_{ij}, the time to death.

The times to death are from cumulative distribution functions $F_1(t)$ (standard treatment) and $F_2(t)$ (experimental treatment), which are not specified. That is, Gehan's test [1965a], proposed here is nonparametric. The null hypothesis is $H_0 : F_1(t) = F_2(t)(t \leqslant T)$ (the treatments are equally effective). The alternative hypothesis of primary importance is $H_1 : F_1(t) > F_2(t)$, or equivalently, $S_1(t) < S_2(t)$ (the experimental treatment significantly improves survival over the standard treatment). However, the other one-sided alternative $H_2 : F_1(t) < F_2(t)$ or the two-sided alternative $H_3 : F_1(t) \neq F_2(t)$ may also be considered.

We define

$$
U_{ij} = \begin{cases} -1 & \text{if} \quad t_{j2} < t_{i1} \quad \text{or} \quad t_{j2} \leqslant T_{i1}, \\ 0 & \text{if} \quad t_{j2} = t_{i1} \quad \text{or} \quad (T_{i1}, T_{j2}) \text{ or } T_{j2} < t_{i1} \quad \text{or} \quad T_{i1} < t_{j2}, \\ 1 & \text{if} \quad t_{j2} > t_{i1} \quad \text{or} \quad T_{j2} \geqslant t_{i1}, \end{cases}
$$

$$(7.3.8)$$

and calculate the statistic

$$
W = \sum_{i=1}^{n_1} \sum_{j=1}^{n_2} U_{ij}.^{*}
$$

Hence there will be a nonzero contribution to W for all comparisons of the two samples where both patients have died (except for ties) and in all comparisons where a patient censored from observation has survived longer than one who has succumbed.

In this arrangement we have $n_1 + n_2$ observations, which can be placed in the general pattern illustrated in Figure 7.1, where m_i is the number of uncensored observations at rank i in the rank ordering of uncensored observations with distinct values, and l_i is the number of right-censored observations with values greater than observations at rank i but less than observations at rank $i + 1$.

The points on the vertical line in Figure 7.1 correspond to a rank ordering of the distinct values of the observed deaths, and these occur at s distinct points. Any set of observations that includes deaths and censored

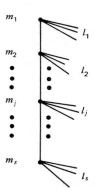

Figure 7.1 The arrangement of $n_1 + n_2$ observations in Gehan's non-parametric test.

---

*In line 2 of (7.3.8) (T_{i1}, T_{j2}) implies that the ith patient on the standard treatment and the jth patient on the experimental treatment are censored.

survivals can be represented according to Figure 7.1. If there are censored survivors prior to the first death, these could be included by counting them as l_1 with $m_1 = 0$. Ordinarily their inclusion or exclusion does not matter, since they provide no information on the difference between the two treatments. The calculation of the mean and variance is not affected because the calculation is conditional on the given pattern of observations.

Example 7.3.2.   As an example of a typical pattern of observations, let $a +$ represent a censored observation and suppose we have the following sample of ordered survival times (weeks): 7, 8, 9+, 10, 10, 10+, 10+, 11, 11, 12+, 14+. The pattern is given in Figure 7.2, where $m_1 = m_2 = 1$, since the first two observations (at ranks 1 and 2) are uncensored, and $l_1 = 0$, since there is no censored observation between 7 and 8. However $l_2 = 1$, since 9+ is censored. Similarly, $m_3 = 1$, $m_4 = 2$, and $l_3 = l_4 = 2$.

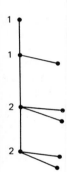

Figure 7.2   Arrangement of observations in Gehan's nonparametric test for Example 7.3.2.

Under the null hypothesis (H_0) that there is no difference between patients on the standard therapy and patients on the experimental therapy, and conditional with respect to P, the pattern of observations, the mean and variance of W are, respectively,

$$E(W|P, H_0) = 0 \qquad (7.3.9)$$

and

$$\mathrm{Var}(W|P, H_0) = \frac{n_1 n_2}{(n_1 + n_2)(n_1 + n_2 - 1)}$$

$$\times \left\{ \sum_{i=1}^{s} m_i M_{i-1}(M_{i-1} + 1) + \sum_{i=1}^{s} l_i M_i (M_i + 1) \right.$$

$$+ \sum_{i=1}^{s} m_i (n_1 + n_2 - M_i - L_{i-1})$$

$$\left. (n_1 + n_2 - 3 M_{i-1} - m_i - L_{i-1} - 1) \right\}, \qquad (7.3.10)$$

where

$$M_j = \sum_{i=1}^{j} m_i, \qquad M_0 = 0$$

$$L_j = \sum_{i=1}^{j} l_i, \qquad L_0 = 0.$$

Equation 7.3.9 follows easily from the symmetry between i and j in (7.3.8). Gehan [1965a] derives (7.3.10) in Appendix A of his article.

Gehan [1965] also demonstrates that when H_0 is true, the ratio

$$Z = \frac{W}{\sqrt{\mathrm{Var}(W|P, H_0)}} \qquad (7.3.11)$$

is approximately normally distributed with mean zero and variance unity. Gehan further demonstrates that this W test can be applied when sample sizes are as small as $n_1 = n_2 = 5$, as long as not more than 6 of the 10 observations are involved in ties or censoring and there are at least 5 distinct times to death. Thus to test H_0 against H_a, an alternative, we use (7.3.11) and reject when Z is found to lie in the appropriate critical region.

Calculation of W and $\mathrm{Var}(W|P, H_0)$ is lengthy if the sample sizes n_1 and n_2 are relatively large (say, n_1, n_2 both at least 25). The programming of W and $\mathrm{Var}(W|P, H_0)$ for a computer would not be a difficult task. Alternatively, we may choose to group the observations into s intervals similar to a life table as in Table 7.5, where

Table 7.5 Life table arrangement of grouped data for Gehan's test

Interval	Number of Deaths	Cumulative Number of Deaths	Number of Censored Observations
1	f_{1j}	F_{1j}	c_{1j}
2	f_{2j}	F_{2j}	c_{2j}
.	.	.	.
.	.	.	.
.	.	.	.
i	f_{ij}	F_{ij}	c_{ij}
.	.	.	.
.	.	.	.
.	.	.	.
s	f_{sj}	F_{sj}	c_{sj}

f_{ij} = number of deaths in interval i for patients receiving treatment $j, j = 1, 2$,

c_{ij} = number of censored observations in interval i for patients receiving treatment $j, j = 1, 2$,

$$F_{ij} = \sum_{v=1}^{i} f_{vj}.$$

The intervals, which need not be of equal length, are chosen similarly as those for ordinary frequency distributions. Information concerning the ordering of patients dying and censored within each interval, therefore, is lost.

The formula for W becomes

$$W = \sum_{i=1}^{s} \{[f_{i1} + c_{i1}]F_{i-1,2} - [f_{i2} + c_{i2}]F_{i-1,1}\} \qquad (7.3.12)$$

and the formula for $\text{Var}(W|P, H_0)$ is (7.3.10), with

$$m_i = f_{i1} + f_{i2}, \qquad l_i = c_{i+1,1} + c_{i+1,2}, \qquad i = 1, 2, \ldots, s; l_s = 0. \quad (7.3.13)$$

Gehan notes that $\text{Var}(W|P, H_0)$ and the expected value of W under one of the alternative hypotheses (in absolute value) tend to be smaller on the average for the grouped case than for the ungrouped case. This results from a loss of a proportion of the $n_1 n_2$ comparisons because of grouping. If the researcher is at all concerned about this loss, the data may be retested on an ungrouped basis.

Example 7.3.3. As a numerical example we apply Gehan's W-test to the data in Example 7.3.1. As was noted in Example 7.3.1, 6-MP therapy is superior to placebo maintenance therapy. The data are, however, analyzed to illustrate the test procedure.

Length of Remission[a] (weeks)

6-MP (21)   6, 6, 6, 7, 10, 13, 16, 22, 23
           6+, 9+, 10+, 11+, 17+, 19+, 20+, 25+, 32+, 32+, 34+, 35+

Placebo (21)   1, 1, 2, 2, 3, 4, 4, 5, 5, 8, 8, 8, 8, 11, 11, 12, 12, 15, 17, 22, 23

[a]Plus sign represents censored observation.

If W were to be calculated directly, it would require a total of 441 comparisons $(n_1 \times n_2 = 21 \times 21 = 441)$, using (7.3.8). Making the requisite computations we find

$$W = 271, \quad \text{and} \quad \sqrt{\text{Var}(W|P, H_0)} = 75.1.$$

Since n_1 and n_2 are moderately large, we calculate W and $\sqrt{\text{Var}(W|P, H_0)}$ by grouping. Using the format of Gehan we obtain Table 7.6, which indicates that $W = \sum_{i=1}^{6}(b_i - a_i) = 307 - 42 = 265$, where $a_i = [f_{i2} + c_{i2}]F_{i-1,1}$ and $b_i = [f_{i1} + c_{i1}]F_{i-1,2}$.

Table 7.6 Life table arrangement of leukemia remission times

Interval (weeks)	Placebo				6-MP			
	f_{i1}	F_{i1}	c_{i1}	a_i	f_{i2}	F_{i2}	c_{i2}	b_i
0–4	7	7	0	—	0	0	0	—
5–9	6	13	0	0	4	4	2	42
10–14	4	17	0	16	2	6	2	52
15–19	2	19	0	12	1	7	2	51
20–24	2	21	0	14	2	9	1	57
25–	0	21	0	0	0	9	5	105
Total	21	—	0	42	9	—	12	307

$\text{Var}(W|P, H_0)$ for grouped data is calculated by using (7.3.10) with m_i and l_i given by (7.3.13). In using Table 7.7 to carry out this calculation, the following values are obtained:

Table 7.7 Computational layout for leukemia remission data

i	m_i	M_i	d_i	$m_i \times d_{i-1}$	l_i	L_{i-1}	$l_i \times d_i$	e_i	g_i	$e_i \times g_i$	$m_i \times e_i \times g_i$
1	7	7	56	0	2	0	112	35	34	1190	8330
2	10	17	306	560	2	2	612	23	8	184	1840
3	6	23	552	1836	2	4	1104	15	−20	−300	−1800
4	3	26	702	1656	1	6	702	10	−37	−370	−1110
5	4	30	930	2808	5	7	4650	5	−48	−240	−960
Total	30	—	—	6860	12	—	7180	—	—	—	6300

$$d_i = M_i(M_i + 1), \quad d_0 = 0, \quad e_i = n_1 + n_2 - M_i - L_{i-1},$$
$$g_i = n_1 + n_2 - 3M_{i-1} - m_i - L_{i-1} - 1.$$

Making the required substitutions in (7.3.10), we find

$$\text{Var}(W|P,H_0) = \frac{n_1 n_2}{(n_1 + n_2)(n_1 + n_2 - 1)} \left\{ \sum_{i=1}^{5} m_i d_{i-1} + \sum_{i=1}^{5} l_i d_i + \sum_{i=1}^{5} m_i e_i g_i \right\}$$

$$= \frac{21 \times 21}{42 \times 41} \{6860 + 7180 + 6300\} = 5065.6. \qquad (7.3.14)$$

Thus

$$\sqrt{\text{Var}(W|P,H_0)} = 71.2.$$

We compute

$$Z = \frac{W}{\sqrt{\text{Var}(W|P,H_0)}} = 3.72, \qquad (7.3.15)$$

which is significant beyond the 0.005 level of significance, implying the existence of strong evidence that patients receiving 6-MP have longer remissions than those receiving placebo.

If the test is carried out with the ungrouped data, we find $Z = 271/75.1 = 3.61$, which is comparable to (7.3.15).

Mantel [1967] has presented an alternative method for computing the statistic W proposed by Gehan. He also has a discussion of the patterns of censoring needed to decide whether any two samples can be compared without bias. In most medical applications, the right tail of both distributions is censored and no problems of bias arises. Breslow [1970] has extended Gehan's results to the case of k samples with right censoring. Efron [1967] has proposed an alternative statistic to Gehan's statistic which estimates a measure of the difference between the two distributions when the null hypothesis is not truly independent of the censoring distribution. Breslow [1974] presents a technique for testing the equality of distributions with covariables present and with censoring. Peto and Peto [1972] present a logrank test that can be used with two or more groups with censoring.

To this point we have been concerned with fixed sample comparisons of survival distributions. That is, the number of patients allocated to two treatments in a clinical trial has been fixed in advance of the study.

The usual sequential clinical trial is one in which patients enter in pairs (in the case of two treatments). After the survival* information is recorded on each pair of patients one of the following three decisions is made:

---

*Survival time here is meant in a generic sense: remission times as well as other response times are also included.

1. Terminate the trial and choose in favor of one particular treatment (the standard treatment, say).

2. Terminate the trial and choose in favor of the other treatment (the experimental treatment).

3. Continue the trial by recording the survival times on the next pair of treatments.

Thus in the usual sequential clinical trial the number of patients in each treatment group is random.

The classical ·sequential procedure was developed by Abraham Wald and his associates during World War II. However, because of the economic value of sequential analysis, which enhances the value of statistical procedures in military applications, most of the results were not published until the war ended. In 1947 Wald published *Sequential Analysis*, which contains the requisite theoretical results pertaining to the sequential procedure. We now discuss briefly the sequential designs employed in clinical trials. Note that their use has mainly been restricted to problems where the results are obtained without delay. This is often not the case in clinical trials, where the outcome variable is survival time.

The study of sequential clinical trials is well documented by Armitage [1960]. His work, however, deals primarily with dichotomous responses to treatment, although for continuous responses, he considers the exponential, normal, and t distributions. In this section we outline the sequential designs that have been used in clinical trials. The interested reader will find the details in Armitage [1960].

Open Design

The basic open sequential design is the classical procedure developed by Wald [1947]. In this scheme as adapted to clinical trials, patients enter the trial in pairs; one person receives the standard treatment, and the other the new or experimental treatment. An open region can then be constructed, as in Figure 7.3. The parallel straight lines $a_i + bm$, $i = 1, 2$, are constructed based on the null hypothesis H_0, the alternative hypothesis H_1, and the sizes of the type I and type II errors (i.e., the probabilities of rejecting H_0 when H_0 is true and accepting H_0 when H_1 is true, respectively). As results from the clinical trial are recorded, each preference for the experimental treatment is plotted vertically beginning at the origin, whereas each preference for the standard treatment is plotted horizontally. (The distance of each plotted point from the previous point depends on whether we are dealing with survival times or are deciding between the two treatments on the basis of which survival time in the pair is larger.) The trial continues

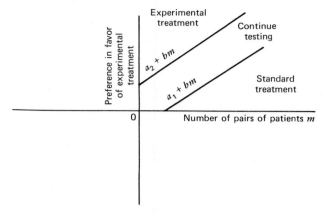

Figure 7.3   An open sequential region.

either until the upper boundary is crossed, in which case the experimental treatment is chosen, or until the lower boundary is crossed, in which case the standard treatment is chosen.

Modified Open Design

Armitage [1960] discusses the modified open design, depicted in Figure 7.4. The open design differs from the modified open design in that the latter allows the experimenter to end the trial concluding no preference between the standard and experimental treatment. (In actuality the experimenter may only wish to change to the new treatment, when preference is shown for it.)

As is the case with the open design (Figure 7.3), the constants a_i and b, $i = 1, 2$, are determined as functions of the null hypothesis, which this time is that there is no preference for either treatment, the two alternative hypotheses and the sizes of the type I and type II errors. (Note that the type I error for this design is based on a two-sided procedure.) As the results from the clinical trial are recorded, each preference for the experimental treatment is plotted at a 45° angle starting from 0, whereas each preference for the standard treatment is plotted at an angle −45°. (The distance of each point from the previous point again depends on the type of response—dichotomous or continuous.) If we join lines connecting consecutive pairs of points, we obtain a zigzag line that eventually crosses one of the three boundaries, corresponding to one of the three possible decisions as exhibited in Figure 7.4.

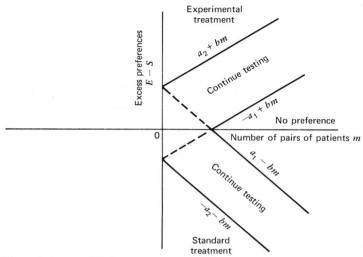

Figure 7.4 A modified open sequential region.

Closed and Restricted Designs

The number of patients is not known in advance of a sequential procedure; thus a sequential clinical trial using either the open design or the modified open design may continue for quite a long period and may require a large number of patients before a decision is reached. As a result the experimenter may decide prior to beginning the study to limit the maximum number of patients who may enter the trial. If he makes this decision, he has a closed sequential procedure. Armitage discusses several different closed sequential procedures, including the restricted sequential design. The interested reader will find a complete discussion of sequential designs in this reference.

In recent years the subject of sequential clinical trials has become quite popular. Thus many articles on the subject have described new approaches to the conduct of a sequential clinical trial. For example, Zelen [1969] and Sobel and Weiss [1970] describe play-the-winner rules for allocating patients to two different treatments in a clinical trial when a patient's response to treatment is dichotomous (i.e., success or failure). In their scheme the first patient to enter the trial is randomly allocated to one of the treatments. If his response to the treatment is a success, the next patient is placed on the same treatment. If the patient's response to treatment is a failure, the next patient is placed on the other treatment, and so on. We discuss Zelen's procedure in more detail in Chapter 8.

Procedures such as the play-the-winner rules have led to research into sequential procedures designed to reduce the number of patients receiving the inferior treatment or treatments in a clinical trial. In such sequential procedures, the aim is not only to make a decision between two treatments as quickly as possible, but also to minimize the number of patients who receive the inferior treatment. These procedures are discussed in the articles by Flehinger and Louis [1971], Flehinger et al. [1972], and Robbins and Siegmund [1972].

An experimenter must decide whether sequential procedures are appropriate for his particular clinical trial. Often this is difficult to determine in advance. The advantages of sequential trials include the following:

1. Sample sizes on the average are considerably less than for the corresponding fixed sample procedure.

2. Since one often realizes early in a clinical trial that a particular treatment is superior to the other treatment, use of sequential trials to reach an early decision in such situations is a sound ethical decision.

3. Although some of the underlying theory of sequential procedures is quite difficult, the practical design of a sequential procedure is generally straightforward and in most cases can be carried out with a minimum of statistical consultation.

4. Sequential clinical trials are easily graphed. Thus if the clinical trial is a large-scale cooperative study with many patients, it will be easier to present the clinical trial graphically.

Unfortunately there are also some disadvantages to sequential clinical trials:

1. It is true that sample sizes are less on the average for sequential procedures than for corresponding fixed sample size procedures. However, it is possible that a sequential clinical trial will require more samples.

2. The response in a clinical trial is often delayed, especially if the response is length of survival. In such clinical trials it is not feasible to wait for patient response before deciding whether to make further observations.

3. If an early decision is reached in favor of one treatment over the other, this decision could be based on a relatively small sample size. Thus estimation of treatment effects is based on a small sample, and it yields little information about these responses.

A final topic in this chapter is the basic similarity between testing for the equality of two or more treatment groups with respect to survival times (which is the topic of Chapter 7) and the study of survival times in the presence of concomitant information (which we considered in Section 3.5).

To this end let us consider a clinical trial of patients with an acute illness such as angina pectoris. Patients with this disease are frequently monitored to obtain readings on blood pressure, cardiogram, weight, age and other variables. Thus at a given point during the study we would have the following information for each patient: (1) his survival time, withdrawal time, or censored time (if he is still on trial); (2) his blood pressure reading; (3) his cardiogram results, (4) his current weight, and (5) his age. Within the clinical trial there may be two or more treatment groups. To test the null hypothesis that all treatments are equal in terms of survival times, it would be necessary to determine how the concomitant measurements (2) through (5) affect the survival times and to adjust these according to the null hypothesis.

Cox [1972] considers the general problem of analyzing censored failure times in the presence of one or more explanatory variables. The hazard function is considered as a function of the explanatory variables (which in turn may be functions of time) and the unknown regression coefficients multiplied by an arbitrary unknown function of time. Thus if in addition to a patient's survival time or time to censoring further measurements are available on the variables $z_1,\ldots,z_p$, the generalized hazard rate can be written

$$\lambda^*(t;\mathbf{z}') = h(\mathbf{z}',\boldsymbol{\beta}')\lambda(t), \qquad (7.3.16)$$

where $\boldsymbol{\beta}' = (\beta_1,\ldots,\beta_p)$ is a $p \times 1$ vector of the unknown parameters, $\mathbf{z}' = (z_1,\ldots,z_p)$, and $\lambda(t)$ is an unknown function. The $h(\mathbf{z}',\boldsymbol{\beta}')$ can be arbitrary, but we assume that $h(\mathbf{0}',\boldsymbol{\beta}') = 1$, giving $\lambda^*(t;\mathbf{0}') = \lambda(t)$. Cox specifies that $h(\mathbf{z}',\boldsymbol{\beta}') = \exp(\boldsymbol{\beta}'\mathbf{z})$, which covers a large class of situations. He discusses methods of estimating $\boldsymbol{\beta}'$ for the situations when t is a continuous as well as a discrete variable. Under the assumption that $\lambda(t)$ is arbitrary, he obtains the conditional likelihood function given that patient deaths (or remission failures) are singly censored. The conditional log likelihood is maximized to obtain $\hat{\boldsymbol{\beta}}$, the conditional maximum likelihood estimator of $\boldsymbol{\beta}$ given the particular failure times $t_{(1)} \leqslant t_{(2)} \leqslant \cdots \leqslant t_{(d)}$. In this way an estimator of the regression parameter vector $\boldsymbol{\beta}'$ is obtained without having to specify $\lambda(t)$.

Breslow [1972] points out that if

$$\lambda(t) = \lambda_i \qquad (7.3.17)$$

for $t_{(i-1)} < t \leqslant t_{(i)}$, $i = 1,\ldots,k$, for distinct uncensored ordered failure times $t_{(1)} < t_{(2)} < \cdots < t_{(k)}$, $t_{(0)} = 0$, one obtains the Kaplan-Meier [1958] product-limit estimate for probability of survival at the point $t_{(i)}$ when $\boldsymbol{\beta}' = \mathbf{0}'$. That is,

$$\hat{S}(t_{(i)}) = \prod_{j=1}^{i} \frac{n_j - 1}{(n_j)}, \tag{7.3.18}$$

where $\hat{S}(t_{(i)})$ is the estimated probability of survival at the point $t_{(i)}$ and n_j is the number of failures in the interval $t_{(j-1)} < t \leqslant t_{(j)}$. We note that the Kaplan-Meier estimate $\hat{S}(t_{(i)})$ is used as an estimate for $S(t)$ in life tables. Breslow shows how the estimate $\hat{S}(t_{(i)})$ can be extended to the case when $\beta' \neq 0'$ and $\hat{\beta}'$ is the estimator.

Chapter 7 presents parametric and nonparametric methods for comparing two survival distributions with and without censoring. It should be noted that much research has been done on the comparison of two exponential distributions, and numerous references are listed in the bibliography. The results for the Weibull, gamma, and other distributions are more limited, which encourages the use of nonparametric techniques when the data are not exponentially distributed. Models such as Cox [1972] provide great generality and can be used to compare separate treatment groups, but at present many statisticians lack experience with such models. The descriptive techniques given in Chapter 2 should be used in conjunction with the formal tests of hypothesis to assist in interpolating experimental results. When comparing more than two treatment groups, the model of Cox can be used or nonparametric tests such as the Kruskal-Wallis test (Siegel [1956]) can be considered.

EXERCISES

1. Using (7.3.5) and (7.3.6), calculate $-2\log_e \Lambda$, the likelihood ratio for comparing two censored exponential densities.

2. Derive the likelihood ratio test for testing $H_0: \mu_1 = \mu_2 = \cdots = \mu_k$ against the alternative H_a: the means are not all equal when k treatments are to be compared, and survival in each treatment group is exponential with censoring.

3. Show that if there is no censoring and no tied observations—that is, if $m_1 = m_2 = \cdots = m_s = 1$, $l_1 = l_2 = \cdots = l_s = 0$, and $s = n_1 + n_2$ in (7.3.10)—then $\text{Var}(W|P, H_0) = \frac{1}{3} n_1 n_2 (n_1 + n_2 + 1)$, which is the variance of the Wilcoxon statistic.

4. In an experiment comparing the survival of 10 animals receiving an experimental treatment with that of 10 animals receiving a control the following data were obtained:

Survival times (days)

| Control: | 8, 8, 8, 9, 9, 10, 10, 10, 10, 11 |
| Experimental: | 8, 9, 10, 11, 14, 16, 17, 19, 30+, 30+ |

Where 30+ indicates survival at the end of the 30-day observation period. Using Gehan's generalization of the Wilcoxon test, test the hypothesis that control animals and experimental animals do not differ in their mean survival times at significance level $\alpha = 0.05$. Repeat the test using Cox's test, assuming that the data are exponentially distributed.

5. For the data in Exercise 4 perform a likelihood ratio test at the 0.01 level of significance.

6. Suppose the two surviving animals receiving the experimental treatment in Exercise 4 eventually died after 33 and 36 days, respectively. Using the Wilcoxon test, test the hypothesis that the control animals and experimental animals do not differ in their survival times at significance level $\alpha = 0.05$.

CHAPTER 8

Sample Size Determination in Clinical Trials

8.1 INTRODUCTORY REMARKS

How large a sample should be drawn from a population to make necessary inferences? A statistician is often called on to answer this fundamental question, particularly in clinical trials in which two (or more) samples must be chosen when two (or more) treatments are being compared with regard to lengthening survival or remission times of patients with an acute disease. Several reference books include extensive tables of sample sizes for tests for means, correlation coefficients, differences between proportions, and chi-square tests for contingency tables. One of the most complete is Cohen [1969].

In the straightforward laboratory experiment situation in which patients do not withdraw from the study, are not lost to follow-up, and all enter the study at the same point in time, the problem is seldom difficult. Furthermore, if the response is dichotomous—"cured" or "not cured," "dead" or "alive"—rather than continuous, the determination of appropriate sample sizes for treatment and control groups is quite elementary. For example, suppose a proposed treatment for patients with acute leukemia is to be compared against a standard treatment. The measure of success for each treatment is whether a patient who receives the treatment achieves complete remission.* If we let p_1 be the theoretical probability that a patient who receives the standard treatment achieves complete remission and p_2 be the corresponding probability for patients receiving the new treatment, we ask how large a sample is needed in each group (assuming patients neither

---

*Complete remission in acute leukemia means absences of all clinical and pathological symptoms of the disease.

withdraw nor are lost to follow-up) to detect a difference δ at significance level α. Since we would only replace the standard treatment by the new treatment if we conclude $p_2 > p_1$, we assume $\delta > 0$. In this clinical trial the distribution of patients is shown in Table 8.1, where $\hat{p}_1$ and $\hat{p}_1 + \hat{\delta}$ are proportions of patients achieving complete remission in the standard and new treatment groups, respectively, and $\hat{q}_1$ and $\hat{q}_1 - \hat{\delta}$ are the proportions of patients not achieving complete remission in the standard and new treatment groups, respectively. One statistic commonly used to test the significance of $\delta = p_2 - p_1$ is the chi-square statistic, with 1 degree of freedom corrected for continuity. Thus from Table 8.1 we have

$$\chi^2(1 - \alpha; 1) = \frac{n(\hat{\delta} - 1/n)^2}{2(\hat{p}_1 + \hat{\delta}/2)(\hat{q}_1 - \hat{\delta}/2)}, \tag{8.1.1}$$

Table 8.1 Proportions of leukemia patients achieving complete remission on standard and new therapies

Group	Complete Remission	Not Complete Remission	Total
		Outcome	
Standard treatment	$n\hat{p}_1$	$n\hat{q}_1$	n
New treatment	$n(\hat{p}_1 + \hat{\delta})$	$n(\hat{q}_1 - \hat{\delta})$	n
Total	$n(2\hat{p}_1 + \hat{\delta})$	$n(2\hat{q}_1 - \hat{\delta})$	$2n$

where $\chi^2(1 - \alpha; 1)$ is the $100(1 - \alpha)$ percentage point for the chi-square distribution with one degree of freedom. If we let A be the denominator of (8.1.1), the requisite sample size for determining significance of δ at level α is given by

$$n^* = \left[\begin{array}{ll} n & \text{if } n \text{ is an integer} \\ [n] + 1 & \text{otherwise,} \end{array} \right. \tag{8.1.2}$$

where

$$n = \frac{\hat{A}^2 \chi^2(1 - \alpha; 1)}{4(\hat{\delta})^2} \left[1 + \sqrt{1 + \frac{4\hat{\delta}}{\hat{A}^2 \chi^2(1 - \alpha; 1)}} \right]^2, \tag{8.1.3}$$

$[n]$ is the greatest integer less than or equal to n, and $\hat{A}^2$ is the denominator of (8.1.1). This development is given in Kramer and Greenhouse [1955].

Equation 8.1.2 accounts only for type I or α error; that is, the value n^* obtained is large enough to reject the null hypothesis $\delta = 0$, 100α percent of the time when in fact $\delta = 0$ is true. It is important to take into account also the type II or β error—that is, the probability of accepting the null hypothesis $\delta = 0$, when $\delta > 0$ is true. Kramer and Greenhouse obtain the following formula for n^*:

$$n^* = \left[\begin{array}{ll} n & \text{if } n \text{ is an integer} \\ [n]+1 & \text{otherwise,} \end{array} \right. \tag{8.1.4}$$

where

$$n = \left[\frac{A\chi(1-\alpha) + B\chi(1-\beta)}{d} \right]^2 , \tag{8.1.5}$$

where $\chi(1-\alpha)$ and $\chi(1-\beta)$ are the $100(1-\alpha)$ and $100(1-\beta)$ percentage points, respectively, of the chi distribution with 1 degree of freedom, A is the same as $\hat{A}$ without the "hat",

$$B = \sqrt{p_1 q_1 + (p_1 + \delta)(q_1 - \delta)} , \tag{8.1.6}$$

and $[n]$ is the greatest integer less than n. Kramer and Greenhouse give rather extensive tables for (8.1.4). To illustrate the use of these tables, Figures 8.1 and 8.2 are selected graphs of sample size estimates n (for each group) versus significant increase to be detected δ for fixed values of p_1, α, and β. For example, if $p_1 = 0.50$, the required sample size to detect an increase $\delta = 0.20$, with type I error, $\alpha = 0.05$, and type II error, $\beta = 0.10$, is $n = 101$ patients in each group, which can be read from Figure 8.2.

To detect a relatively small increase δ in the probability of treatment success, the required sample size for each treatment group is usually large. Hence this nonparametric method for determining sample sizes required to compare two treatment groups may result in obtaining a large number of patients in each group. We show in Section 8.2 how treatment group size can be reduced significantly in the presence of parametric procedures.

In most clinical trials, a continuous end point can be measured (survival time or remission time) rather than dichotomous. Furthermore, clinical

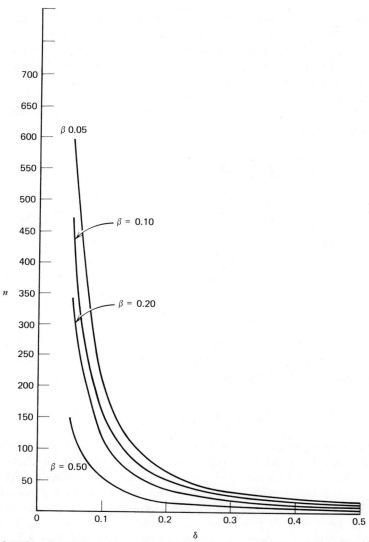

Figure 8.1 Number of samples n required in each of two groups to detect a significant increase of δ for $p_1 = 0.05$, $\alpha = 0.05$, $\beta = 0.05$, 0.10, 0.20, and 0.50. Source: Kramer and Greenhouse [1955].

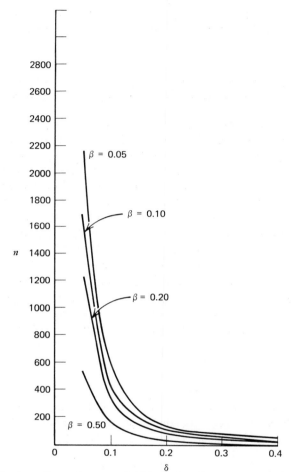

Figure 8.2 Number of samples n required in each of two groups to detect a significant increase of δ for $p_1 = 0.50$, $\alpha = 0.05$, $\beta = 0.05$, 0.10, 0.20, and 0.50. Source: Kramer and Greenhouse [1955].

trials are often completed before all patients die or relapse. Also, patients with different diseases follow different patterns of survival (i.e., different survival distributions). We discuss these problems in Section 8.2 for the exponential distribution and in Section 8.3 more generally. In Section 8.4 we present a brief discussion of other aspects of a clinical trial, including how one uses Bayesian methods in deciding which treatment is the better when the population size of individuals with the disease is known or a good estimate can be obtained. The reader may skip Section 8.4 at first reading without any loss of continuity. Section 8.5 takes up the problem of allocation of patients to the two treatment groups when cost of treatment varies between the groups.

8.2 SAMPLE SIZE ESTIMATES REQUIRED TO COMPARE SURVIVAL IN TWO EXPONENTIAL POPULATIONS

In a clinical trial designed to compare two treatments, it is assumed that survival time on each treatment follows the exponential distribution with hazard rates λ_1 for the standard treatment and λ_2 for the experimental treatment. In addition, we assume at the outset that the patients in each treatment group are followed until death and that no patient is lost to follow-up. Furthermore, even before a sample is taken, it is reasonable to expect that the sample size in each group will be large enough to permit the maximum likelihood estimators $\hat{\lambda}_1$ and $\hat{\lambda}_2$ each to be normally distributed, with means and variances λ_1 and λ_2 and λ_1^2/n and λ_2^2/n, respectively. Since λ_1 and λ_2 are unknown, we are unable to obtain estimates of $(\lambda_1-\lambda_2)$ and $\sigma^2_{\hat{\lambda}_1-\hat{\lambda}_2}$ directly. However, an estimate may be available for λ_1, the hazard rate of patients who receive the standard treatment. If the theoretical proportion of survivors after t_0 years is $S_1(t_0)\equiv p_1$ for patients who receive the standard treatment, the estimate for λ_1, λ_1^*, say, is

$$\lambda_1^* = \frac{-\log_e p_1(t_0)}{t_0}. \tag{8.2.1}$$

On the new treatment, furthermore, if a proportion $p_1 + \delta$, $\delta > 0$, of survivors must be observed for the new treatment to be deemed more effective than the standard, we estimate λ_2 as λ_2^*, where

$$\lambda_2^* = \frac{-\log_e(p_1+\delta)}{t_0}. \tag{8.2.2}$$

The acceptance region for a one-sided test of significance at level α is the region

$$(\hat{\lambda}_2-\hat{\lambda}_1) < \sqrt{\frac{\lambda_1^{*2}+\lambda_2^{*2}}{n}}\ Z(1-\alpha), \tag{8.2.3}$$

where under the null hypothesis $(\hat{\lambda}_2-\hat{\lambda}_1)$ is approximately a normal variable with mean zero and variance $(\lambda_1^{*2}+\lambda_2^{*2})/n$. If β is the probability of a type II error occurring—that is, the probability of accepting the null hypothesis of no treatment difference when actually $\lambda_2^* > \lambda_1^*, (\delta > 0)$, then

$$\beta = \text{pr}\left[\frac{\hat{\lambda}_2 - \hat{\lambda}_1}{\sqrt{(\lambda_1^{*2} + \lambda_2^{*2})/n}} < Z(1-\alpha) + \frac{\lambda_2^* - \lambda_1^*}{\sqrt{(\lambda_1^{*2} + \lambda_2^{*2})/n}}\right]. \quad (8.2.4)$$

Now

$$\beta = \int_{-\infty}^{Z(\beta)} \phi(t) \, dt, \quad (8.2.5)$$

where $\phi(t)$ is the standard normal density function. Thus we have

$$Z(1-\alpha) = Z(\beta) + \frac{\lambda_2^* - \lambda_1^*}{\sqrt{(\lambda_1^{*2} + \lambda_2^{*2})/n}}. \quad (8.2.6)$$

Solving (8.2.6) for n^*, we find

$$n^* = \left[\begin{array}{ll} n & \text{if } n \text{ is an integer} \\ [n] + 1 & \text{otherwise}, \end{array}\right. \quad (8.2.7)$$

where $n = (\lambda_1^{*2} + \lambda_2^{*2})(Z(1-\beta) + Z(1-\alpha))^2/(\lambda_2^* - \lambda_1^*)^2$ and $[n]$ is the greatest integer less than n. Thus n^* is the required sample size for each treatment group when the probability of falsely obtaining a difference $(\lambda_2^* - \lambda_1^*)$ is α and no difference exists, and when the probability of not observing a real difference $(\lambda_2^* - \lambda_1^*)$ is β.

In most clinical trial situations the experimenter is not concerned with detecting just a nonzero difference between the standard treatment and experimental treatment. Rather, he wants to increase survival times of his patients. Thus we do not consider the sample size problem for the two-sided alternative $\delta \neq 0$. It may be approximated for the two-sided case by using $Z(1 - \alpha/2)$ instead of $Z(1 - \alpha)$.

Assume now that patients are followed only for a fixed time interval of length T. Thus at the end of the proposed study the data may be censored. Patient survival time on each treatment is still assumed to be exponential, with hazard rates λ_1 for patients on the standard treatment and λ_2 for patients on the experimental treatment. We again assume that the sample size in each group is to be large enough (e.g., at least 25 in each group) to ensure that the maximum likelihood estimators of the censored data survival parameters in these groups have approximately normal distributions. As in the uncensored case, $\hat{\lambda}_1$ and $\hat{\lambda}_2$ are normally distributed with means λ_1 and λ_2, respectively. If we assume that patients enter the study on each treatment at random according to the uniform distribution on the interval $0 < x < T$, we recall from the theory developed in Section 3.4 that the variances $\sigma_{\hat{\lambda}_1}^2$ and $\sigma_{\hat{\lambda}_2}^2$ are given, approximately as

$$\sigma_{\hat{\lambda}_i}^2 = \frac{\lambda_i^2}{n}\left(1 - \frac{1 - \exp(-\lambda_i T)}{\lambda_i T}\right)^{-1}, \qquad i = 1, 2. \qquad (8.2.8)$$

[See (3.4.10).]

As with the uncensored case, λ_1 and λ_2 are unknown. However, if the theoretical proportion of survivors after t_0 years $(t_0 \leqslant T)$ is p_1 for patients receiving the standard treatment, the estimate for λ_1 is given as λ_1^* by (8.2.1). Furthermore, if the proportion of survivors must be $p_1 + \delta$, $\delta > 0$, for patients on the experimental treatment if that treatment is to be judged more effective than the standard, λ_2 is estimated by λ_2^*, which is given by (8.2.2).

Now following reasoning similar to that used in the uncensored case, we see that under the null hypothesis of no difference between treatments, $\hat{\lambda}_2 - \hat{\lambda}_1$ is approximately normally distributed, with mean zero and variance $[\lambda_1^{*2} L(\lambda_1^*, T) + \lambda_2^{*2} L(\lambda_2^*, T)]/n$, where

$$L(\lambda_i^*, T) = \left(1 - \frac{1 - \exp(-\lambda_i^* T)}{\lambda_i^* T}\right)^{-1}, \qquad i = 1, 2. \qquad (8.2.9)$$

When the probability of falsely obtaining a difference $(\lambda_2^* - \lambda_1^*)$ is α, and no difference exists, and when the probability of not observing a real difference $(\lambda_2^* - \lambda_1^*)$ is β, the required sample size for each treatment group n^* is

$$n^* = \begin{bmatrix} n & \text{if } n \text{ is an integer} \\ [n] + 1 & \text{otherwise,} \end{bmatrix} \qquad (8.2.10)$$

where $n = (\lambda_1^{*2} L(\lambda_1^*, T) + \lambda_2^{*2} L(\lambda_2^*, T))(Z(1 - \beta) + Z(1 - \alpha))^2 / (\lambda_2^* - \lambda_1^*)^2$ and $[n]$ is the greatest integer less than n. The details of this derivation, which are analogous to those for the uncensored case, are omitted.

Example 8.2.1. We assume that survival time after diagnosis of males with stage III prostate cancer (cancer that has spread beyond the prostate gland to nearby organs but not throughout the rest of the body) is exponentially distributed. Suppose we wish to design a clinical trial to compare two treatments for prostate cancer, in which approximately 60% of patients receiving the standard treatment are expected to be alive after 5 years. For the experimental therapy to be considered better than the standard treatment we consider two cases: (1) a 20% increase in patient survival and (2) a 10% increase in patient survival. Furthermore, $\alpha = 0.05$ and $\beta = 0.10$. That is, the size of the type I error (i.e., wrongly selecting the experimental treatment as the better) is 0.05, and the size of the type II error (i.e., wrongly not selecting the experimental treatment as the better) is 0.10.

Table 8.2 gives the required sample sizes for each group of patients for the two increases in survival percentage considered and for two censoring possibilities: no censoring at all, and censoring of all patients alive after 6 years of study. From Table 8.2 it follows that roughly four times as many patients will be needed for each treatment group if the study is to end 6 years after its beginning than would be necessary if all patients were followed until death. The experimenter must decide for himself which procedure is more economical on this basis.

Table 8.2 Number of patients required in the two treatment groups

Follow-up Period	$\delta = 0.10$	$\delta = 0.20$
Until death	140	33
6 years	587	149

Let us now abandon the assumption of exponential survival for the two treatment groups. Our measure becomes the proportion of patients surviving in each treatment group after 5 years. Thus $p_1 = 0.60$, $\delta = 0.20$ in the first case and $\delta = 0.10$ in the second case; $\alpha = 0.05$ and $\beta = 0.10$ for both cases. From the tables in Kramer and Greenhouse [1955] we find that the respective sample sizes are 89 and 388. Comparing these estimates of sample size to those based on exponentiality indicates roughly a 270% increase in using the dichotomous procedure. If the experimenter has reason to believe that survival follows an exponential distribution in his proposed clinical trial, economic reasons should dictate that he use the assumption in computing his sample size requirements. Finally, we note that the dichotomous procedure does not allow for censoring in that censored observations cannot be used, hence cannot be accounted for in estimating sample size. From the foregoing discussion we see that a great many more observations are needed if one treats a patient as surviving or not surviving versus measuring the length of time he survives and assuming that these survival times follow a given distribution.

In this section we have obtained sample sizes based on the assumption that survival times in the two treatment groups are exponentially distributed and that the estimated hazard rates $\hat{\lambda}_1$ and $\hat{\lambda}_2$ are normally distributed with means and variances λ_1, λ_2 and λ_1^2/n, λ_2^2/n respectively. Correspondingly, we could have looked at the estimates of mean survival time $\hat{\mu}_1$ and $\hat{\mu}_2$ for the two groups. Since $\hat{\mu}_1 = 1/\hat{\lambda}_1$ and $\hat{\mu}_2 = 1/\hat{\lambda}_2$, the resulting considerations would lead to exactly the same sample size estimate in (8.2.7).

EXERCISES

1. In a study of Hodgkin's disease, patient survival is approximately exponentially distributed. Among patients receiving the standard treatment, radiotherapy, 75% can expect to survive 5 years. In addition to radiotherapy, a new chemotherapeutic agent is given to patients which is expected to increase the survival proportion 10% for the 5-year period. Find the requisite number of patients for each treatment group if all patients are followed until death. Assume $\alpha = 0.10$ and $\beta = 0.05$.

2. If the study in Exercise 1 ceases 10 years after its inception, find the required number of patients for each treatment group.

3. Derive (8.2.10) for sample sizes in studies where survival is exponentially distributed in both treatment groups, although the data may be censored.

4. In a clinical trial, patients are first given a standard treatment for which their times in remission are recorded. After all patients have relapsed, they are given an experimental treatment; again their remission times are recorded. All patients are followed on the experimental treatment until they relapse again. Assuming knowledge of p_1, p_2, and t_0, as well as α and β, derive a formula for the required patient sample size.

5. In a clinical trial designed to compare two treatments, it is known that survival time on each treatment has a Weibull distribution with a known and common shape parameter α_0. The scale parameters λ_1 and λ_2 for the two treatment groups are unknown. Assuming knowledge of p_1, p_2, and t_0, as well as α and β, derive a formula for the required patient sample size estimate in each treatment group.

8.3 SAMPLE SIZE ESTIMATES REQUIRED TO COMPARE SURVIVAL IN SURVIVAL DISTRIBUTIONS DEPENDING ON TWO PARAMETERS

If the survival times of patients receiving the two treatments (standard and experimental) are not exponentially distributed, the problem of determining sample size estimates becomes considerably more complicated. In this section we consider survival times that are distributed according to a two-parameter survival distribution (e.g., a Weibull or gamma) and the concomitant problem of determining sample size estimates required in a clinical trial to compare survival times of patients who receive the standard treatment with those receiving the experimental treatment.

Let $S_i(t)$, $i = 1, 2$, be the survival distributions for patients who receive the standard and experimental therapies, respectively. We can then write

$$S_i(t) = \exp(-\Lambda(t; \lambda_i, \gamma_i)), \qquad i = 1, 2, \tag{8.3.1}$$

where

$$\Lambda(t) = \int_0^t \lambda(x) \, dx. \tag{8.3.2}$$

(See Section 3.6.)

Let $p_1(t_0)$ and $p_1(t_0')$ be the theoretical probabilities that a patient receiving the standard treatment is alive after t_0 and t_0' years, respectively, $t_0' > t_0$ without loss of generality. Furthermore, let $p_1(t_0) + \delta(t_0)$ and $p_1(t_0') + \delta(t_0'), \delta(t_0) > 0$, $\delta(t_0') > 0$ be the probabilities that a patient who receives experimental therapy is alive after t_0 and t_0' years, respectively. Estimates of the parameters λ_i and γ_i, $i = 1, 2$, are obtained by solving

$$p_i(t_0) = \exp(-\Lambda(t_0; \lambda_i^*, \gamma_i^*)) \tag{8.3.3}$$

and

$$p_i(t_0') = \exp(-\Lambda(t_0'; \lambda_i^*, \gamma_i^*)), \qquad i = 1, 2, \tag{8.3.4}$$

where $p_2(t_0)$ and $p_2(t_0')$ are, respectively, $p_1(t_0) + \delta(t_0)$ and $p_1(t_0') + \delta(t_0')$. Since we are testing for a longer survival of patients on the experimental treatment, we further assume $\delta(t_0) > 0$, and $\delta(t_0') > 0$.

As a rule, we cannot solve (8.3.3) and (8.3.4) in terms of λ_i^* and γ_i^*, $i = 1, 2$, explicitly. We discuss three special types of two-parameter survival distributions—the Weibull, the Rayleigh, and gamma—and the corresponding problems of estimating λ_i^* and γ_i^*, $i = 1, 2$, in each of them.

Returning to the general problem of sample size estimates for two-parameter survival distributions, let us assume that λ_i^* and γ_i^* are obtained from (8.3.3) and (8.3.4), $i = 1, 2$. Let μ_1, μ_2, σ_1^2, and σ_2^2 be the mean survival times and their variances for patients receiving the standard and experimental therapies, respectively. Then

$$\mu_i^* = \mu(\lambda_i^*, \gamma_i^*) \tag{8.3.5}$$

and

$$\sigma_i^{*2} = \sigma^2(\lambda_i^*, \gamma_i^*), \qquad i = 1, 2. \tag{8.3.6}$$

Before a sample is taken it is reasonable to expect that the sample size in each group will be large enough to ensure that $\overline{X}_i$, the sample mean survival time for each group, $i = 1$, 2, will be approximately normally distributed with mean μ_i^* and variance σ_i^{*2}/n.

Following the reasoning in Section 8.2, the requisite sample size for each treatment group is given by

$$n^* = \begin{cases} n & \text{if } n \text{ is an integer} \\ [n] + 1 & \text{otherwise,} \end{cases} \tag{8.3.7}$$

where $n = (\sigma_1^{*2} + \sigma_2^{*2})(Z(1-\alpha) + Z(1-\beta))^2/(\mu_2^* - \mu_1^*)^2$ and $Z(1-\alpha)$ and $Z(1-\beta)$ are the percentiles of the standard normal distribution. This general technique is illustrated using the Weibull, the Rayleigh, and the gamma survival distributions.

The Weibull Distribution

For the Weibull distribution we begin with

$$p_i(t_0) = \exp(-\lambda_i^* t_0^{\gamma_i^*}) \tag{8.3.8}$$

and

$$p_i(t_0') = \exp(-\lambda_i^* t_0'^{\gamma_i^*}), \qquad i = 1, 2. \tag{8.3.9}$$

Solving for λ_i^* and γ_i^*, we thus find

$$\gamma_i^* = \frac{\log_e \dfrac{\log_e p_i(t_0')}{\log_e p_i(t_0)}}{\log_e\left(\dfrac{t_0'}{t_0}\right)}, \tag{8.3.10}$$

and

$$\lambda_i^* = \frac{-\log_e p_i(t_0)}{t_0^{\gamma_i^*}}, \qquad i = 1, 2. \tag{8.3.11}$$

Recalling the formulas for the mean and variance of the Weibull variable —(4.1.8) and (4.1.9), respectively—we see that

$$\mu_i^* = \lambda_i^{*-1/\gamma_i^*}\Gamma\left(1+\frac{1}{\gamma_i^*}\right) \tag{8.3.12}$$

and

$$\sigma_i^{*2} = (\lambda_i^*)^{-2/\gamma_i^*}\left[\Gamma\left(1+\frac{2}{\gamma_i^*}\right) - \left\{\Gamma\left(1+\frac{1}{\gamma_i^*}\right)\right\}^2\right], \qquad i=1,2, \tag{8.3.13}$$

where $\Gamma(x)$ is the gamma function.

With knowledge of (8.3.12) and (8.3.13), which are obtained from (8.3.8) and (8.3.9) from knowledge of t_0, t_0', $p_i(t_0)$, and $p_i(t_0')$, which is usually supplied by the investigator, we can obtain sample size estimates in clinical trials where survival for both treatment groups follows the Weibull distribution for given levels of type I (α) and type II (β) error by means of (8.3.7), (8.3.12), and (8.3.13).

The Rayleigh Distribution

For the Rayleigh distribution,

$$p_i(t_0) = \exp\left(-\{\lambda_{0i}^* t_0 + \lambda_{1i}^* t_0^2\}\right) \tag{8.3.14}$$

and

$$p_i(t_0') = \exp\left(-\{\lambda_{0i}^* t_0' + \lambda_{1i}^* t_0'^2\}\right), \qquad i=1,2. \tag{8.3.15}$$

(Note: Here we use λ_{0i}^* and λ_{1i}^* instead of λ_i^* and γ_i^* in keeping with our earlier notation in Section 4.6.) Solving for λ_{0i}^* and λ_{1i}^*, we have

$$\lambda_{0i}^* = \frac{t_0'^2\log_e p_i(t_0') - t_0'^2\log_e p_i(t_0)}{(t_0'-t_0)t_0 t_0'} \tag{8.3.16}$$

and

$$\lambda_{1i}^* = \frac{t_0'\log_e p_i(t_0) - t_0\log_e p_i(t_0')}{(t_0'-t_0)t_0 t_0'}, \qquad i=1,2. \tag{8.3.17}$$

In Exercise 1, Section 4.8, the mean and variance of the Rayleigh distribution are derived. Using these results we have

$$\mu_i^* = \exp\left(\frac{\lambda_{0i}^{*2}}{4\lambda_{1i}^*}\right)\sqrt{\frac{\pi}{4\lambda_{1i}^*}}\left(1-I\left(\frac{\lambda_{0i}^{*2}}{4\lambda_{1i}^*},0.5\right)\right) - \frac{\lambda_{0i}^*}{2\lambda_{1i}^*} \tag{8.3.18}$$

and

$$\sigma^{*2}_i = \lambda^{*-1}_{1i}(1 - \lambda^*_{0i}\mu^*_i) - \mu^{*2}_i, \qquad i = 1, 2, \tag{8.3.19}$$

where

$$I(x,p) = \frac{\int_0^x t^p e^{-t}\,dt}{\Gamma(p+1)}, \tag{8.3.20}$$

or the incomplete gamma distribution, which has been tabled by Pearson [1922].

Using (8.3.7), requisite sample size estimates can be obtained in clinical trials where survival in both treatment groups follows the Rayleigh survival distribution.

The Gamma Distribution

For the gamma survival distribution,

$$p_i(t_0) = \frac{\int_{t_0}^\infty \lambda^*_i x^{\gamma^*_i - 1} \exp(-\lambda^*_i x)\,dx}{\Gamma(\gamma^*_i)} \tag{8.3.21}$$

and

$$p_i(t'_0) = \frac{\int_{t'_0}^\infty \lambda^*_i x^{\gamma^*_i - 1} \exp(-\lambda^*_i x)\,dx}{\Gamma(\gamma^*_i)}, \qquad i = 1, 2. \tag{8.3.22}$$

Solving for γ^*_i and λ^*_i generally requires numerical procedures such as discussed in Chapter 6. Before the statistician undertakes solving (8.3.21) and (8.3.22), he should see whether the Weibull survival distribution can be used instead of the gamma distribution (simply because of the reduction in numerical work that the former will bring). When he has obtained λ^*_i and γ^*_i from (8.3.21) and (8.3.22), $i = 1, 2$, the remainder of the task of estimating the requisite sample size for each of the two groups is not difficult.

Example 8.3.1. We assume that the survival time of patients with chronic lymphoma follows a Weibull distribution. Suppose we wish to design a clinical trial to compare two treatments for this disease. Of patients who receive the standard treatment, 60% are expected to survive 5 years and 30% are expected to attain a 10-year survival period. For the experimental therapy to be considered superior to the standard treatment we consider two cases: (1) a 20% increase in patient survival after 5 years and again at 10 years, and (2) a 10% increase in patient survival at the same two end points. The probabilities of wrongly choosing the experimental treatment to be superior when it is not and wrongly not choosing it to be superior when it is, are $\alpha = 0.05$ and $\beta = 0.10$, respectively. How large a sample is required in each treatment group? For case (1), using (8.3.7) with the aid of (8.3.10) and (8.3.11), we find $n = 76$ patients in each group, for case (2) $n = 249$ patients in each group. Comparison of these results with those in Example 8.2.1, should clearly indicate that sample size determination is quite sensitive to the distributional assumption of survival (or remission) made by the experimenter. Thus it is suggested that if sample size is to be determined on the basis of a particular survival distribution, a check of the validity of the distribution should be made, if possible. For certain diseases, such as lung cancer, angina pectoris, and tuberculosis, sufficient information on survival is available to permit investigators to make a preliminary estimate concerning the parametric form of the survival distribution.

Discussion

As we have seen in Sections 8.2 and 8.3, sample size estimates when comparing two treatments depend on a number of prior considerations such as the form of the survival distributions, the size of the type I and type II errors involved, whether the trial is allowed to continue until all patients on both treatments have died or relapsed or whether the trial is ended after some finite period regardless of the patient status, and estimates of the proportions of patients alive or in remission after some fixed period or periods. Often an experimenter can supply the statistician with these prior quantities, and the statistician in turn can provide the experimenter the requisite sample size estimates for the two treatment groups. It should be noted that these sample size estimates are just that—estimates. For example, an experimental treatment may be 20% more effective than the standard treatment, whereas the experimenter first thought it might only be 10% more effective. In such a case, the required sample size estimates would be overestimates of what is actually needed. On the other hand, the experimental treatment might only be 5% more effective than the control. Here the experimenter may want an additional sample in each group to detect this 5% difference or he may feel that this difference is not large enough to warrant an additional sample to choose the experimental treatment in preference to the standard treatment. Moreover, the sample size estimates for each of the two treatment groups tend to be smaller when a specific survival distribution is assumed than when only the proportion of survivors is known after a given period. Thus it is to the experimenter's advantage to be able to specify the form of the survival distribution to the statistician. Other practical problems in determination of sample size are listed and discussed briefly:

1. Ideally, the experimenter should be able to provide sensible levels of type I and type II errors to the statistician. Although investigators understand the concept of type I and type II errors, in practice few can simultaneously decide on proper levels of these quantities because they do not have their definitions firmly in mind. Minor shifts in levels of type I and/or type II errors have a decided effect on sample size obtained from the formulas.

2. Often investigators are not able to specify clearly the percentage or proportion of patients who survive or who are in remission after a given time period t_0. For two-parameter survival distributions, two such time periods t_0 and t_0' are required.

3. It is difficult to decide on the percentage difference in survival to be required between the two groups of patients. That is, what value of $p_2 - p_1 \equiv \delta$, or $p_2(t_0) - p_1(t_0) \equiv \delta(t_0)$ and $p_2(t_0') - p_1(t_0') \equiv \delta(t_0')$ should an investigator choose in a real-life situation?

4. The equations developed for sample size determination of the various distributions are precise formulas; we fill them in with guesses, and then expect to obtain a precise result. We should keep clearly in mind that these formulas provide estimates of sample sizes to be used by investigators conducting a clinical trial before the trial has begun. They may be modified after information on the clinical trial becomes available.

5. The problem is complicated further because in most clinical trials several alternative proportions of patients surviving both the standard and experimental treatments are considered. Also, other concomitant variables —side effects of the treatments, the efficacy of the treatments in reducing morbidity, and other patient differences due to the different therapies— result in the simultaneous consideration of several variables, including survival. Such consideration may lead to untenable sample sizes. However, experienced experimenters are often dubious about studies when the sample sizes are too small.

6. The sample size estimation formulas assume simple random sampling from the same population. This condition is not often met, since it is possible that not all individuals with a particular serious illness (for which a clinical trial is being devised to test the differences between two treatments) have the disease in exactly the same way. For example, if we are testing patients' responses to lung cancer, it is likely that not all lung cancer patients available for the study have exactly the same form of the disease.

7. Sample sizes to this point have assumed that both treatment groups are allocated the same number of patients. Such patient allocation is not always best from an economical standpoint. That is, there may be a fixed total cost allowed for the clinical trial, whereas the cost of patients in each treatment group is not the same. In such situations equal patient allocation seldom provides the most economical trial. This point is discussed further in Section 8.5 and a complete treatment can be found in Nam [1973].

Most experimenters find sample size estimates helpful in computing a lower limit to the required sample size in a given study, and this is their principal use. The sample size estimate obtained from one of the given formulas is then compared with the realistic sample size the experimenter can obtain, given the limitations imposed by cost, facilities, staff, and/or incidence of disease. If the sample size estimate is less than the realistic sample size, the study can begin. If not, the study must be dropped or so altered that the estimated and realistic sample sizes come together. Performing a medical or clinical study with an insufficient sample is not recommended because the experimental drug under consideration may not undergo a fair trial. That is, even though a significant improvement in survival or remission of patients may exist, there may be too few patients on study to detect it. In medical studies the problem of obtaining an adequate sample is often acute because patients are not eligible for the study unless they have certain characteristics (e.g., age limitations, previous disease history, and previous exposure to treatment). Furthermore, if patients are followed for a long time, some are likely to drop out of the study or die of another cause. The problem of patient dropout will result in a further reduction of patient sample size. Deterministic models for obtaining sample size estimates in the two-sample binomial response case with patient dropouts are obtained by Schork and Remington [1967], and Halperin et al. [1968].

EXERCISES

1. Derive the equation for sample size determination, (8.3.7), for a general two-parameter survival distribution.

2. Suppose n_1 is the sample size for each treatment group in a proposed clinical trial when the survival distribution is exponential, and n_2 is the sample size when survival follows the gamma distribution. For fixed λ_1^* and λ_2^* as well as fixed type I and type II errors, prove that $n_1 \geqslant n_2$ if $\max(\gamma_1^*, \gamma_2^*) \leqslant \min(\gamma_1^{*2}, \gamma_2^{*2})$ and $n_1 \leqslant n_2$ if $\min(\gamma_1^*, \gamma_2^*) \geqslant \max(\gamma_1^{*2}, \gamma_2^{*2})$. Under what circumstances do we have $n_1 = n_2$ in both cases?

3. Assume that patients who suffer from chronic lymphoma survive according to the Weibull distribution and that a clinical trial has been proposed to test a

new chemotherapeutic agent against a standard treatment for which at most 50 patients in each treatment group can be obtained. If 60% of patients receiving the standard treatment survive 5 years and 30% survive 10 years, and the type I and type II errors are $\alpha = 0.05$ and $\beta = 0.10$, respectively, what increase in survival in the treatment group would have to be observed after 10 years if after 5 years a 25% increase is observed?

4. Show that for sample sizes based on the gamma survival distribution, the sample size estimate is independent of the scale parameter λ if it is known prior to the study that $\lambda_1 = \lambda_2 = \lambda$. Do similar results hold for sample sizes based on the Weibull and linear hazard rate survival distributions? Explain your answer.

8.4 SAMPLE SIZE ESTIMATES WHEN THE POPULATION SIZE IS KNOWN

Suppose there is a population of N patients with a given disease for which some reasonable estimate of N can be obtained. For certain relatively rare diseases (e.g., malignant melanoma) such an estimate from incidence rates (new cases) and prevalent rates (old cases) can be obtained. To determine with an adequate degree of assurance which of two treatments is the better, a clinical trial is performed on a sample of $2n$ patients, n in each treatment group. The treatment that is the better one in the clinical trial is then selected for the remaining N-$2n$ patients in the population of patients who have the disease. Our problem here is arriving at the size of n. That is, how large a sample is needed for each treatment group? Since we are dealing with a relatively rare disease, determination of n by the formulas developed in Sections 8.2 and 8.3 may lead to an unacceptably large sample size.

A word of caution is in order at this point. Choice of treatment is not an eternal decision. At some future time there may be justification for ceasing to use the "better" treatment. For example, a new, more efficacious treatment may be found, or the treatment itself may produce dangerous side effects that were dormant initially. Thus N will be dependent on the length of time the treatment judged superior is used in the future, which may be hard to predict.

In this section we discuss rather briefly three procedures for estimating n based on knowledge of N. The procedures were put forth by Colton [1963], Cornfield, Halperin, and Greenhouse [1969], and Zelen [1969]. The first two involve a minimax and a Bayes procedure, respectively, whereas Zelen's play-the-winner rule uses a Markov chain approach. These procedures have not been widely used in actual studies. Readers unfamiliar with loss and risk functions may find it helpful to refer to Wasan [1970].

Colton's Procedure

Let x_1 and x_2 be the individual responses to treatments one and two respectively, where treatment one is the standard treatment and treatment two is the experimental treatment. We assume further that $x_i \sim N(\mu_i, \sigma^2)$, $i = 1$, 2, where σ^2 is the common known variance. A clinical trial is performed on $2n$ patients, n assigned to each treatment group. The remaining N-$2n$ patients receive the treatment judged as the better at the end of the trial. If we let $\delta = \mu_2 - \mu_1$, we then wish to choose treatment two (one), when $\delta > 0 (\delta < 0)$.

If the wrong treatment is selected, it is assumed a loss l proportional to δ will be incurred. The expected loss $E(l)$ for all N patients is

$$E(l) = C\delta \left[n + (N - 2n)\Phi\left(-\frac{\delta\sqrt{n}}{\sigma\sqrt{2}} \right) \right], \tag{8.4.1}$$

where

$$\Phi(u) = \int_{-\infty}^{u} \phi(t)\, dt, \tag{8.4.2}$$

$\phi(t) = 1/\sqrt{2\pi} \exp(-t^2/2)$ is the standard normal density function, and C is a constant of proportionality. Equation 8.4.1 follows, since a loss $C\delta n$ must occur because n patients are allocated to the inferior treatment, and if the inferior treatment is chosen, a loss $(N - 2n)C\delta$ occurs; the inferior treatment is chosen with probability $\Phi(-\delta\sqrt{n}/\sigma\sqrt{2})$.

Letting $p = n/N$, the problem of determining an optimum value of p is easier than determining an optimum n value directly. Colton [1963] and Maurice [1959] discuss extensively the mathematical problems incurred in obtaining the optimum value of p. We merely point out that the optimum value of p can only be obtained when the loss δ is restricted. This optimum value is in the form of a minimax solution for p.

Another approach is to maximize the minimum expected net gain. We assume that when a patient receives the superior treatment, there is a gain proportional to δ, whereas the patient receiving the inferior treatment loses in direct proportion to δ. The expected net gain $E(g_{net})$ for all N patients is

$$E(g_{net}) = G\delta N - \frac{2G}{C}E(l), \tag{8.4.3}$$

where G is a constant of proportionality. The straightforward derivation of (8.4.3) is left as an exercise to the reader.

If we maximize the minimum expected net gain, we note that expected net gain is zero for $\delta=0$. Then maximizing $E(g_{net}/N)$ for $\tilde{x}=0$ (which is equivalent to $\tilde{\delta}=0$, where $x=\delta\sqrt{Np}/\sigma\sqrt{2}$), we find that a candidate value $\tilde{p}$ is the solution to

$$\frac{1-2\tilde{p}}{2\tilde{p}} = \lim_{x\to 0}\left[\frac{2\Phi(x)-1}{x\phi(x)}\right]. \tag{8.4.4}$$

Applying L'Hôpital's rule to the right-hand side of (8.4.4), we find

$$\lim_{x\to 0}\left[\frac{2\Phi(x)-1}{x\phi(x)}\right] = 2\lim_{x\to 0}\left\{\frac{\phi(x)}{x\frac{d\phi(x)}{dx}+\phi(x)}\right\} = 2. \tag{8.4.5}$$

Then $\tilde{p}=\frac{1}{6}$ is the maximum solution for p which is verified by considering the signs of the second derivatives at $\tilde{\delta}=0$, $\tilde{p}=\frac{1}{6}$. Thus maximizing the minimum expected net gain when it is proportional to the expected treatment difference is equivalent to placing $N/3$ patients in the first stage of the trial, $N/6$ on each treatment, and the remaining $2N/3$ patients in the second stage on the better treatment.

An Adaptive Procedure

Cornfield, Halperin, and Greenhouse [1969] generalize Colton's model in three directions. First, the unknown treatment difference δ has a prior normal distribution with mean d_0 and variance σ_0^2. Second, instead of a 50–50 patient split, an arbitrary proportion of patients is assigned in the testing stage to each of the two treatments. This proportion along with the other constants is determined to minimize cost. Finally, this two-stage optimum allocation can be generalized to a multistage allocation procedure, which although not optimal leads to lower costs than the optimal two-stage allocation procedure. Here we consider only the two-stage procedure, referring the interested reader to Cornfield, Halperin, and Greenhouse [1969] for details concerning multistaging.

As before we let μ_i be the true treatment effect for the ith treatment, $i=1, 2$, and $\delta=\mu_1-\mu_2$ be the true treatment difference; $\delta>0(<0)$ implies preference for treatment one (two). Following the notation of Cornfield, Halperin, and Greenhouse, the N patients to be treated in the two stages are assigned to stage and treatment as follows: $2pN$ patients are assigned to the first stage $(0\leqslant p\leqslant\frac{1}{2})$. Of these, $2pN\theta_1$ are assigned to the first treatment and $2pN(1-\theta_1)$ are assigned the second treatment. The $(1-2p)N$ patients in the second stage are assigned to the two treatments in the proportion θ_2 to $1-\theta_2$ for treatments one and two, respectively, with θ_2 depending on the outcome of the first stage.* As in the previous section the cost of assigning a patient to the inferior treatment is assumed directly proportional to $|\delta|$.

Loss for fixed δ is given by

$$L = \begin{cases} [2p(1-\theta_1)+(1-2p)(1-\theta_2)]N\delta, & \delta > 0 \\ -[2p\theta_1+(1-2p)\theta_2]N\delta, & \delta < 0, \end{cases} \qquad (8.4.6)$$

where the constant of proportionality is omitted because it plays no role in the optimization procedure. Equation 8.4.6 follows directly, under the assumption that a loss proportional to δ is incurred if the wrong treatment is used. The computation of $E(L)$ is somewhat complicated, involving the joint distribution of $d = \bar{x}_2 - \bar{x}_1$ and $\delta = \mu_2 - \mu_1$. That is, δ is assumed to have a prior normal distribution with mean d_0 and variance σ_0^2, whereas the conditional distribution of d given δ is assumed normal with mean δ and variance $\sigma^2/n(\sigma^2$ and σ_0^2 are assumed known). After a considerable amount of algebra, it can be shown that

$$E(L) = \frac{N}{\sigma_0} \int_0^\infty \delta\phi\left(\frac{\delta-d_0}{\sigma_0}\right) d\delta - \frac{2p\theta_1 N}{\sigma_0} \int_{-\infty}^\infty \delta\phi\left(\frac{\delta-d_0}{\sigma_0}\right) d\delta$$

$$-(1-2p)N \int_{-\infty}^\infty \int_{-\infty}^\infty \theta_2\delta f(\delta,d_1) d\delta\, dd_1, \qquad (8.4.7)$$

where $\phi(x)$ is the standard normal density and $f(d_1,\delta)$ is the joint density function of d_1 and δ. Minimization of $E(L)$ is a rather complex problem and is not discussed here. However, it can be shown from (8.4.7) that if $d_0 = 0$ as in Colton's model, the optimum rule is to place all patients in stage II of the clinical trial on the treatment that is found to be superior in the first stage.

Note that in these models both the population variance and N are assumed known.

Zelen's Modified Play-the-Winner Rule

Zelen [1969] describes yet a third method of designing a two-stage clinical trial for selecting the (apparently) better of two treatments in the first stage

---

*In actual practice $2pN\theta_1$ and $2pN(1-\theta_1)$ as well as $(1-2pN)\theta_2$ and $(1-2pN)(1-\theta_2)$ must be integers. However, the solution given here depends on p, θ_1, and θ_2 being continuous. Thus, in practice, the closest integer values are used.

for use in the second stage. He suggests the following procedure, the modified play-the-winner rule (MPWR). The first patient entering the trial is assigned to one of the two treatments with probability $\frac{1}{2}$ (i.e., random assignment of the first patient is made). Label the two treatments 0 and 1, and allow patients to enter the first stage of the trial sequentially. If the $(j-1)$th patient is on treatment i, $i=0$, 1, the jth patient is placed on i, provided the $(j-1)$th patient does not fail. Otherwise the jth patient is placed on treatment $l \neq i$, $l=0$, 1, $j \geqslant 2$. Thus a patient is assigned to the treatment that was successful on the previous trial. This requires a dichotomous response (success or failure) that can be obtained before the start of the next patient. This is an example of the "two-armed bandit" problem in which one plays the arm of a slot machine again if one wins on it. The problem in using MPWR with survival data is that a length of time T is set and an outcome is called a success if the patient survives longer than T. This implies all patients who live less than T are failures regardless of their survival times. Thus the procedure discards information. Also, this only would work when the length of survival is short and/or the intake of patients is very slow, ensuring that the outcome is known before the next patient enters. If the outcome of the $(j-1)$th treatment is not known prior to treating the jth patient, the procedure is to randomly select either treatment 0 or 1 with the probability of selection proportional to the number of known prior successes of the two treatments. If treatment 0 had 10 successes and treatment 1 had 5, the probability of the next treatment being treatment 0 should be $\frac{2}{3}$. If nothing is known, choose with equal probability. The latter procedure is called play-the-winner rule (PWR).

Suppose in the first stage there are n patients who are subject to MPWR. We define the random variable Z_j as follows:

$$Z_j = \begin{cases} 1 & \text{if} \quad \text{the } j\text{th patient receives treatment 1} \\ 0 & \text{if} \quad \text{the } j\text{th patient receives treatment 0,} \end{cases} \tag{8.4.8}$$

$j=1, 2, \ldots, n$; N and n are assumed to be large.

Let p_i be defined as

$$p_i = Pr\{Z_j = i | Z_{j-1} = i\}, \qquad i=0, 1, j=2, \ldots, n. \tag{8.4.9}$$

That is, p_0 and p_1 are treatment success probabilities for treatments 0 and 1, respectively. Note that $\Pr\{Z_1 = 0\} = \Pr\{Z_1 = 1\} = \frac{1}{2}$. Let $q_i = 1 - p_i$, $i=0$, 1.

The MPWR is then equivalent to the two-state Markov chain whose matrix M of transition probabilities is

$$M = \begin{bmatrix} p_0 & q_0 \\ q_1 & p_1 \end{bmatrix}. \tag{8.4.10}$$

If N_0 is the number of successes in the first stage for patients receiving treatment 0 (say), it follows from the asymptotic theory of Markov chains (see Kemeny and Snell [1960] for a complete discussion) that N_0 has an asymptotically normal distribution with mean $nq_1/(q_0+q_1)$ and variance $nq_0q_1(p_0+p_1)/(q_0+q_1)^3$.

The approximate expected numbers of successes in the two stages of testing for the first and second stages, respectively, are

$$p_0 E(N_0) + p_1 E(N_1) \doteq \frac{n(p_0 q_1 + p_1 q_0)}{q_0 + q_1} \tag{8.4.11}$$

and

$$p_0 \Pr\{N_0 > N_1\} + p_1 \Pr\{N_0 < N_1\} \doteq p_0 \Phi(x) + p_1(1 - \Phi(x)), \tag{8.4.12}$$

where since N_0 and N_1 have approximate normal distributions, it follows that $\Pr\{N_0 > N_1\} \doteq \Phi(x)$, where

$$x = \frac{\sqrt{n}\,(\tfrac{1}{2} - q_1/(q_0+q_1))}{\sqrt{q_0 q_1(p_0+p_1)/(q_0+q_1)^3}},$$

and $\Phi(x)$ is the cumulative standard normal distribution. We now make the following definitions:

$$\bar{p} = \frac{p_0 q_1 + p_1 q_0}{q_0 + q_1}, \tag{8.4.13}$$

$$\bar{\bar{p}} = p_0 \Phi(x) + p_1(1 - \Phi(x)), \tag{8.4.14}$$

and

$$w = \frac{n}{N}. \tag{8.4.15}$$

The expected proportion of successes or the utility for the two stages of experimentation U is then approximately

$$U \doteq w\bar{p} + (1-w)\bar{\bar{p}}. \tag{8.4.16}$$

Thus the utility U is the weighted average of the proportion of successes in the two stages.

Let us define n_0 to be that value of n which maximizes $U \equiv U(n, N)$ for fixed N. That is,

$$U(n_0, N) = \max_n U(n, N). \tag{8.4.17}$$

It can then be shown that n_0 is approximately

$$n_0 \doteq \frac{N}{3} + \frac{2}{9}\rho_\alpha^2 \pm \frac{2}{9}\rho_\alpha\sqrt{3N + \rho_\alpha^2}, \tag{8.4.18}$$

where

$$\rho_\alpha = Z(\alpha)\left[\frac{\alpha(1-\alpha)(p_0 + p_1)}{(\alpha - \frac{1}{2})^2(q_0 + q_1)}\right]^{1/2} \tag{8.4.19}$$

and $Z(\alpha)$ is a percentage point of the standard normal density such that $\Phi(Z(\alpha)) = \alpha$ and α is $q_1/(q_0 + q_1)$. The value ρ_α remains fairly constant in the interval defined by (8.4.18). The ambiguity of sign in (8.4.18) is a result of solving a quadratic equation. Any value of n contained in the interval obtained by utilizing the plus and minus sign would be satisfactory as a value of n_0.

Finally we see that as $N \to \infty$, $w_0 = n_0/N \to \frac{1}{3}$. That is, for large values of N, the optimum allocation to stage for the clinical trial using the play-the-winner formulation is the same allocation that Colton [1963] found, assigning an equal number of patients to each of the two treatments during the first stage of testing. In both cases it was found that $N/3$ patients should be allocated to the first stage.

Example 8.4.1. Suppose in a clinical trial a modified play-the-winner rule is used for which p_0 (the probability of success for patients who receive treatment 0, the standard) is 0.5. Assume further that the total number of patients for both stages of the clinical trial is $N = 100$. We want to find n_0, the optimum number of patients at the first stage, and compute $U(n_0, N)$ and $U(N/3, N)$, the optimum value of the utility function and the value of the utility function when the number of patients at the first stage is $N/3$, respectively. We do this for the set of values $p_1 = 0.1$, 0.2, 0.3, 0.4, 0.5, 0.6, 0.7, 0.8, and 0.9. That is, we compare a fixed probability of treatment success of 0.5 for the standard treatment, as opposed to various success probabilities ranging from 0.1 to 0.9 for the experimental treatment. The values of n_0 are found using Table 8.3 and (8.4.18) and (8.4.19), whereas $U(n, N)$ for $n = n_0$, and $n = N/3$ is found by applying (8.4.16) with the aid of (8.4.13) through (8.4.15).

Table 8.3 Comparison of $U(n, N)$ for
$n = n_0$ and $n = N/3, p_0 = 0.5$ and
various values of p_1

p_1	n_0	$U(n_0, N)$	$U(N/3, N)$
0.1	30	0.455	0.451
0.2	30	0.457	0.455
0.3	30	0.457	0.457
0.4	39	0.466	0.466
0.5	—	—	0.500
0.6	39	0.569	0.569
0.7	28	0.660	0.660
0.8	28	0.769	0.768
0.9	28	0.881	0.878

As a sample calculation we compute $U(n_0, N)$ and $U(N/3, N)$ for $p_0 = 0.5$ and $p_1 = 0.2$. First, $\alpha = 0.5/(0.5 + 0.8) = 0.385$. Thus $Z(\alpha)$, the percentage point of the standard normal density corresponding to an area of 0.385 above $Z(\alpha)$, is 0.292. Now ρ_α, which is given by (8.4.19), is 0.907. Hence n_0, given by (8.4.18), has two possible values, 29 or 37. We find, however, that $U(29, N) > U(37, N)$; thus $n_0 = 29$. A final application of (8.4.16) shows $U(N/3, N) = 0.455$. As Zelen [1969] points out, use of $U(N/3, N)$ in place of $U(n_0, N)$ does not change the optimum value of the utility function by a significant amount.

In general Zelen's method (MPWR) has two advantages over the Colton-Cornfield-Halperin-Greenhouse (CCHG) approach. Zelen does not assume a particular loss function; instead he defines his utility function in a very natural way as the expected proportion of successes in the entire experiment. Furthermore, if a treatment shows a high probability of success in the first stage, its usage is correspondingly high. Similarly, a treatment with a low success probability will not be used extensively in the first stage unless the other treatment also has a low success probability. In the latter case each of the two poor treatments is used roughly 50% of the time in the first stage.

On the other hand, the CCHG considers a continuous rather than a dichotomous treatment response. Since the allocation of patients in the first stage is not dependent on the response to treatment of previously treated patients in the first stage, total patient allocation to the two treatments in the first stage can be made prior to completion of the first stage.

A great deal of work has been done on sequential procedures for use in clinical trials, and a good summary is contained in Wetherill [1966]. Armitage [1960] gives some examples of use of sequential procedures in which there is a fixed upper limit to the sample size. With these sequential procedures, the sample size does not need to be determined in advance; sampling is continued until the chance of making an α (type I) or β (type II) error has reached the desired size. The procedures have been attractive to medical investigations because the trial is stopped as soon as a decision can be made and patients are not put on the poorer treatment unnecessarily. To take advantage of the benefits, however, two features are necessary:

1. The length of time to obtain the response from the patients must be short. For example, if length of survival is the response variable and on the average patients survive about 5 years, a sequential procedure that depends on previous outcomes is not practical because these outcomes will be mostly unknown.

2. The investigator must be convinced that he will not decide to look at subgroups of his sample after the trial is completed. Frequently after a clinical trial is completed investigators decide to compare subgroups (men and women, different age groups, etc.) that were not obvious candidates for subgrouping at the start of the trial. A procedure that stops sampling as soon as a minimum size is reached will not allow for later divisions of the sample.

Thus many medical investigators iike to have generous sample sizes taken, and there is some practical merit to this viewpoint. A statistician can help reduce this needed sample size by encouraging the investigator to take the following steps:

1. Measure the survival time instead of simply determining whether a person has survived a particular period.

2. Measure possible covariates along with the length of survival. An example of using concomitant information is given by Feigl and Zelen [1965] in Section 3.5 and Cox [1972].

3. Use techniques such as those mentioned in this last section, particularly for rare diseases.

EXERCISES

1. Derive the formula

$$E(g_{net}) = G\delta N - \frac{2G}{C} E(l).$$

2. Suppose that δ has a prior distribution that is normal, with mean zero and variance σ_0^2. Letting $\overline{E(g_{net})}$ denote the result of integrating $E(g_{net})$ over the probability distribution for δ, show that

$$\frac{\overline{E(g_{net})}}{N} = \frac{2G\sigma_0}{(2\pi)^{1/2}} (1 - 2p) \left(\frac{Rp}{1 + Rp} \right)^{1/2},$$

where $R = N\sigma_0^2/\sigma^2$.

3. Show that the value of p, p_0, that maximizes $\overline{E(g_{net})}/N$ in Exercise 3, is $(3 + \sqrt{q + R})^{-1}$.

4. If $R = 0$, then p_0 and $\tilde{p}$, the maximum solution, coincide at $\frac{1}{6}$. If we adhere to the rule of using $p = \frac{1}{6}$ no matter what the circumstances, verify Table 8.4 (Colton [1963]) of efficiencies of net gains when comparing the net gain of p_0 to the net gain of $\tilde{p} = \frac{1}{6}$.

Table 8.4

R	p_0	Efficiency of $\tilde{p} = \frac{1}{6}$
0	0.167	100.0
0.5	0.158	99.9
1	0.151	99.7
2	0.140	99.0
4	0.125	97.4
5	0.119	96.6
10	0.100	93.2
20	0.080	88.7
50	0.057	82.6
100	0.043	78.7
∞	0	66.7

5. Let $U(n_0, N)$ be that value of n which maximizes the utility function $U(n, N)$. That is, let $U(n_0, N)$ be given by (8.4.17). Show that

$$\lim_{N \to \infty} U(n_0, N) = \frac{\bar{p}}{3} + \frac{2}{3} \max(p_0, p_1)$$

approximately.

6. Assume that $n \geq \rho_\alpha^2$ in the first stage of testing and is fixed. Show that if $N_2 > N_1$, $U(n, N_2) \geq U(n, N_1)$. That is, the utility function increases as the number of patients in the second stage increases as long as n, the number of patients in the first stage, is at least as large as ρ_α^2.
7. Verify the computations in Table 8.3.

8.5 SAMPLE SIZE ALLOCATION TO THE TWO TREATMENT GROUPS

Occasionally in determining sample sizes in a clinical trial there is justification for not allocating an equal number of patients to each of the two treatment groups. As Nam [1973] points out, equal allocation is generally not optimal if the treatment costs are different in the two groups. Optimality is based on the power of the test used to distinguish between the two treatments. That is, we investigate the problem of maximizing the power of the test for a given total experimental cost or equivalently minimizing cost for a specified power of the test.

Following Nam's procedure, suppose there are n_1 and n_2 patients in the two treatment groups. Each of the n_1 patients who receives the standard treatment incurs a cost of c_1 dollars, and correspondingly each of the n_2 patients receiving the experimental therapy has a cost of c_2 dollars to bear. Thus

$$C = c_1 n_1 + c_2 n_2 \qquad (8.5.1)$$

is the total cost of the clinical trial. We assume here that the survival time of the jth patient in the ith treatment group, $t_{ij}, j = 1, 2, \ldots, n_i$, is normally distributed with mean μ_i and variance σ^2, $i = 1, 2$, for the two treatment groups, respectively.

The null hypothesis is $H_0: \mu_1 = \mu_2$. The alternative hypothesis is $H_a: \mu_1 \neq \mu_2$. That is, under H_0 the survival times for patients in the two treatment groups have the same normal distribution, whereas these distributions differ under H_a. When H_a is true the test statistic

$$t = \frac{\bar{x}_1 - \bar{x}_2}{s_p \sqrt{1/n_1 + 1/n_2}}, \qquad (8.5.2)$$

has a noncentral t distribution with $n_1 + n_2 - 2$ degrees of freedom, where

$$s_p^2 = \frac{\sum_{i=1}^{2} \sum_{j=1}^{n_i} (t_{ij} - \bar{t}_i)^2}{n_1 + n_2 - 2}. \qquad (8.5.3)$$

The noncentrality parameter is

$$\delta = \frac{\mu_2 - \mu_1}{\sigma \sqrt{1/n_1 + 1/n_2}}. \tag{8.5.4}$$

Furthermore, the power function (i.e., the probability of rejecting H_0 when $\delta \neq 0$ is the true value of the noncentrality parameter) is an increasing function of $|\delta|$. Thus for a given fixed cost C [see 8.5.1] the power is maximized when $|\delta|$ is maximized. Rewriting (8.5.4) we see that

$$|\delta| = \frac{a}{\sqrt{1/n_1 + 1/n_2}}, \tag{8.5.5}$$

where $a = |\mu_2 - \mu_1|/\sigma$. The problem then is to find the values of n_1 and n_2 that maximize (8.5.5) subject to the constraint given by (8.5.1)—that is, subject to a fixed cost of the clinical trial. This is accomplished by means of a Lagrange multiplier. Let

$$LM = \frac{a}{\sqrt{1/n_1 + 1/n_2}} + \xi(n_1 c_1 + n_2 c_2). \tag{8.5.6}$$

It then follows that

$$\frac{\partial LM}{\partial n_i} = \frac{a}{2} \left(\frac{1}{n_1} + \frac{1}{n_2} \right)^{-3/2} \cdot \frac{1}{n_i^2} + \xi c_i, \qquad i = 1, 2. \tag{8.5.7}$$

Setting $\partial LM / \partial n_i = 0$, $i = 1, 2$, we have

$$\frac{T}{n_i^2} = -\xi c_i, \qquad i = 1, 2, \tag{8.5.8}$$

where $T = a/2(1/n_1 + 1/n_2)^{-3/2}$. Hence the optimal ratio of the sizes of the two samples as a function of the costs c_1 and c_2 is

$$\frac{n_2}{n_1} = \left(\frac{c_1}{c_2} \right)^{1/2}. \tag{8.5.9}$$

Substituting (8.5.9) into (8.5.1) we thus find that

$$n_i = \frac{C}{\left\{ c_i + (c_1 c_2)^{1/2} \right\}}, \qquad i = 1, 2, \tag{8.5.10}$$

are the requisite optimal sizes of the two samples. It is now clear that equal sample size allocation to the two treatment groups is not optimal in the sense of maximizing power with respect to a fixed cost of the clinical trial, unless $c_1 = c_2$.

Most clinical trials are designed to put an equal number of patients (n, say) in each treatment group. If in addition we require the same power of the test under equal patient and optimal patient assignments, it is not difficult to show that the ratio of the costs of the two trials (R_c, say) is given by

$$R_c = \frac{2(1+s)}{(1+s^{1/2})^2},$$ (8.5.11)

where $s = c_2/c_1$ and it can be assumed without loss of generality that $s \geqslant 1$.

The R_c is a monotone increasing function of s, and $1 \leqslant R_c \leqslant 2$ for all values $s \geqslant 1$. Figure 8.3 is a graph of R_c as a function s. This analysis has shown that although equal sample allocation is more expensive than optimal sample allocation, it is never more than twice as expensive. Finally, the intuitive approach in most clinical trials (placing equal numbers of patients in the two treatment groups) is actually near optimal from the standpoint of maximizing power of the test for a fixed cost, unless the costs of experimentation differ markedly in the two treatment groups. This of course assumes the sample survival times in the two treatment groups are normally distributed.

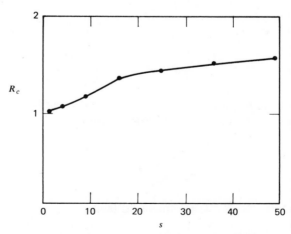

Figure 8.3 Ratio of cost of equal sample allocation to optimal sample allocation with identical powers of the t test for various control group and experimental group costs.

Nam also considers what happens when the treatment costs per individual are calculated on a daily basis. That is, patient costs are dependent on the length of time a patient survives. The analysis is again based on the normality assumption concerning the survival times in the respective treatment groups.

When the survival times are not normally distributed, it may be possible to make a transformation on the survival times to ensure that the transformed data are approximately normally distributed. Thus the results can be used to determine at least roughly patient allocations to the two treatment groups.

In Chapter 8 we discuss the determination of required sample sizes in survival studies. This problem is vexing to statisticians involved in studies with human patients because ethical considerations dictate using no more patients than necessary. However, the statistician must also consider the dropout of patients over time, the possibility of the clinician desiring to examine results for subgroups of patients, and perhaps even changes in the disease condition itself occurring over time. Because patients are entered into the study gradually over time, those entering late in the study may have characteristics somewhat different from those of early subjects, despite well-written study entrance protocols. Because government contracts are written for particular time periods and personnel must be hired for these periods, it is frequently difficult to estimate the length of time needed to intake these patients. If a uniform entry rate can be assumed over a given period, the number of patients entering can be estimated utilizing the cumulative of the Poisson distribution for that entry rate.

If the treatment effect can be determined quickly after treatment, sequential procedures may be utilized. If it takes a long time to intake the requisite patients, an additional complication is presented in the case of sequential analysis because there is no idea of how long the trial will take unless a closed design is used. Such considerations have led some investigators to consider double sampling procedures similar to those developed many years ago in quality control work. Fortunately the sample size problems in animal experimental work tend to be simpler: if small animals are used, they can all be started on the treatment at the same time, and their characteristics are better known.

EXERCISES

1. Suppose the total cost of a clinical trial is $C = \$10,000$, with a cost $c_1 = \$25$ for each patient in the control group and a cost $c_2 = \$50$ for each patient in the experimental group. Find the requisite sample sizes in the two treatment groups under optimal sample allocation. Assuming equal patient allocation to the two treatment groups, how much more would the clinical trial cost?

2. Show that R_c (the ratio of the costs of the clinical trial under equal patient allocation to that under optimal patient allocation) is given by (8.5.11).

3. Show that $2(1+s)/(1+s^{1/2})^2$, $s \geqslant 1$, is an increasing function of s and that $\lim_{s\to\infty} 2(1+s)/(1+s^{1/2})^2 = 2$.

Appendix

Table 1 Shape parameter estimates of the gamma distribution for uncensored samples. The estimate $\hat{\gamma}$ is given for values of $(1 - R)^{-1}$ where R is the ratio of the geometric to the arithmetic sample mean.

Reproduced by permission of the editor of Biometrika from M. B. Wilk et al., Estimation of Parameters of the Gamma Distribution Using Order Statistics, *Biometrika*, Vol. 49, pp. 525–545 (1962).

$(1-R)^{-1}$	$\hat{\gamma}$	$(1-R)^{-1}$	$\hat{\gamma}$	$(1-R)^{-1}$	$\hat{\gamma}$
1·000	0·00000	1·11	0·29854	2·4	1·06335
1·001	0·11539	1·12	0·30749	2·5	1·11596
1·002	0·12663	1·13	0·31616	2·6	1·16833
1·003	0·13434	1·14	0·32461	2·7	1·22050
1·004	0·14043	1·15	0·33285	2·8	1·27248
1·005	0·14556	1·16	0·34090	2·9	1·32430
1·006	0·15005	1·18	0·35654	3·0	1·37599
1·007	0·15408	1·20	0·37165	3·2	1·47899
1·008	0·15775	1·22	0·38631	3·4	1·58158
1·009	0·16115	1·24	0·40060	3·6	1·68385
1·010	0·16432	1·26	0·41457	3·8	1·78585
1·012	0·17011	1·28	0·42825	4·0	1·88763
1·014	0·17535	1·30	0·44170	4·2	1·98921
1·016	0·18017	1·32	0·45492	4·4	2·09063
1·018	0·18464	1·34	0·46795	4·6	2·19191
1·020	0·18884	1·36	0·48081	4·8	2·29308
1·022	0·19282	1·38	0·49351	5·0	2·39414
1·024	0·19659	1·40	0·50608	5·5	2·64643
1·026	0·20020	1·45	0·53694	6·0	2·89830
1·028	0·20366	1·50	0·56715	6·5	3·14984
1·030	0·20700	1·55	0·59682	7·0	3·40115
1·035	0·21486	1·60	0·62604	8·0	3·90325
1·040	0·22217	1·65	0·65488	9·0	4·40482
1·045	0·22904	1·70	0·68340	10·0	4·90608
1·050	0·23554	1·75	0·71163	12·0	5·90790
1·055	0·24175	1·80	0·73961	14·0	6·90914
1·060	0·24771	1·85	0·76737	16·0	7·91005
1·065	0·25345	1·90	0·79494	18·0	8·91073
1·070	0·25899	1·95	0·82233	20·0	9·91125
1·075	0·26437	2·00	0·84957	30·0	14·91301
1·080	0·26959	2·10	0·90364	40·0	19·91401
1·090	0·27966	2·20	0·95724	50·0	24·91457
1·100	0·28928	2·30	1·01046		

APPENDIX

Table 2 Shape and a function of the scale parameter estimates of the gamma distribution for censored samples.[a]

[a] In using this table, consult the following key: $n/d = K/M$, $G = P$, $A = S$, $\gamma = \eta$, $\lambda = \lambda$, $\mu = \mu$. Each page contains a specific value of K/M ranging from 1.0 to 3.0; P is the horizontal listing on each page, and S is the vertical listing. For $P \leqslant .52$ read S from the left-hand margin, and for $P \geqslant .56$ read S from the right-hand margin. Note that the numbers in region 2 are printed in bold roman type and enclosed in a solid line; the numbers in region 3 are in bold italic type and enclosed in a dashed line; the numbers in region 1 are in a lighter type and are located in the lower left-hand and upper right-hand corners of the table.

Reproduced by permission of the editor of Biometrika from M. B. Wilk et al., Estimation of Parameters of the Gamma Distribution Using Order Statistics, *Biometrika*, Vol. 49, 525–545 (1962).

S		0·96	0·92	0·88	0·84	0·80	0·76	0·72	0·68	0·64	0·60	0·56			P/S
0·08	η	0·850		12·412	6·158	4·070	3·024	2·394	1·972	1·669	1·440	1·261	1·116	0·996	1·00
	μ	0·080		1·000	1·000	1·000	1·000	1·000	1·000	1·000	1·000	1·000	1·000	1·000	
0·12	η	0·567	1·376		11·912	5·908	3·903	2·898	2·293	1·888	1·596	1·376	1·203	1·063	0·96
	μ	0·120	0·120		0·960	0·960	0·960	0·960	0·960	0·960	0·960	0·960	0·960	0·960	
0·16	η	0·464	0·850	1·887		11·412	5·657	3·736	2·772	2·192	1·803	1·523	1·311	1·145	0·92
	μ	0·160	0·160	0·160		0·920	0·920	0·920	0·920	0·920	0·920	0·920	0·920	0·920	
0·20	η	0·408	0·664	1·116	2·394		10·912	5·407	3·569	2·646	2·091	1·718	1·450	1·247	0·88
	μ	0·200	0·200	0·200	0·200		0·880	0·880	0·880	0·880	0·880	0·880	0·880	0·880	
0·24	η	0·372	0·567	0·850	1·376	2·898		10·411	5·157	3·401	2·520	1·989	1·633	1·376	0·84
	μ	0·240	0·240	0·240	0·240	0·240		0·840	0·840	0·840	0·840	0·840	0·840	0·840	
0·28	η	0·346	0·506	0·712	1·028	1·633	3·401		9·911	4·906	3·234	2·394	1·888	1·547	0·80
	μ	0·280	0·280	0·280	0·280	0·280	0·280		0·800	0·800	0·800	0·800	0·800	0·800	
0·32	η	0·327	0·464	0·626	0·850	1·203	1·888	3·903		9·411	4·655	3·066	2·268	1·786	0·76
	μ	0·320	0·320	0·320	0·320	0·320	0·320	0·320		0·760	0·760	0·760	0·760	0·760	
0·36	η	0·312	0·432	0·567	0·740	0·984	1·376	2·141	4·405		8·911	4·405	2·898	2·141	0·72
	μ	0·360	0·360	0·360	0·360	0·360	0·360	0·360	0·360		0·720	0·720	0·720	0·720	
0·40	η	0·300	0·408	0·524	0·664	0·850	1·116	1·547	2·394	4·906		8·410	4·154	2·730	0·68
	μ	0·400	0·400	0·400	0·400	0·400	0·400	0·400	0·400	0·400		0·680	0·680	0·680	
0·44	η	0·289	0·388	0·490	0·609	0·758	0·957	1·247	1·718	2·646	5·407		7·910	3·903	0·64
	μ	0·440	0·440	0·440	0·440	0·440	0·440	0·440	0·440	0·440	0·440		0·640	0·640	
0·48	η	0·281	0·372	0·464	0·567	0·691	0·850	1·063	1·376	1·888	2·898	5·908		7·409	0·60
	μ	0·480	0·480	0·480	0·480	0·480	0·480	0·480	0·480	0·480	0·480	0·480		0·600	
0·52	η	0·273	0·358	0·442	0·534	0·641	0·771	0·939	1·168	1·505	2·057	3·150	6·408		
	μ	0·520	0·520	0·520	0·520	0·520	0·520	0·520	0·520	0·520	0·520	0·520	0·520		
0·56	η	0·266	0·346	0·423	0·506	0·600	0·712	0·850	1·028	1·272	1·633	2·226	3·401	6·909	
	μ	0·560	0·560	0·560	0·560	0·560	0·560	0·560	0·560	0·560	0·560	0·560	0·560	0·560	
0·60	η	0·261	0·336	0·408	0·483	0·567	0·664	0·781	0·927	1·116	1·376	1·760	2·394	3·652	
	μ	0·600	0·600	0·600	0·600	0·600	0·600	0·600	0·600	0·600	0·600	0·600	0·600	0·600	
0·64	η	0·255	0·327	0·394	0·464	0·540	0·626	0·727	0·850	1·003	1·203	1·479	1·888	2·562	
	μ	0·640	0·640	0·640	0·640	0·640	0·640	0·640	0·640	0·640	0·640	0·640	0·640	0·640	
0·68	η	0·251	0·319	0·382	0·447	0·516	0·594	0·683	0·789	0·917	1·078	1·290	1·582	2·015	
	μ	0·680	0·680	0·680	0·680	0·680	0·680	0·680	0·680	0·680	0·680	0·680	0·680	0·680	
0·72	η	0·246	0·312	0·372	0·432	0·496	0·567	0·647	0·740	0·850	0·984	1·153	1·376	1·684	
	μ	0·720	0·720	0·720	0·720	0·720	0·720	0·720	0·720	0·720	0·720	0·720	0·720	0·720	
0·76	η	0·242	0·305	0·362	0·419	0·479	0·544	0·616	0·699	0·795	0·910	1·050	1·228	1·462	
	μ	0·760	0·760	0·760	0·760	0·760	0·760	0·760	0·760	0·760	0·760	0·760	0·760	0·760	
0·80	η	0·239	0·300	0·354	0·408	0·464	0·524	0·590	0·664	0·750	0·850	0·969	1·116	1·302	
	μ	0·800	0·800	0·800	0·800	0·800	0·800	0·800	0·800	0·800	0·800	0·800	0·800	0·800	
0·84	η	0·236	0·294	0·346	0·397	0·450	0·506	0·567	0·635	0·712	0·800	0·904	1·028	1·181	
	μ	0·840	0·840	0·840	0·840	0·840	0·840	0·840	0·840	0·840	0·840	0·840	0·840	0·840	
0·88	η	0·232	0·289	0·339	0·388	0·438	0·490	0·547	0·609	0·679	0·758	0·850	0·957	1·087	
	μ	0·880	0·880	0·880	0·880	0·880	0·880	0·880	0·880	0·880	0·880	0·880	0·880	0·880	
0·92	η	0·230	0·285	0·333	0·379	0·427	0·476	0·529	0·587	0·651	0·722	0·804	0·899	1·010	
	μ	0·920	0·920	0·920	0·920	0·920	0·920	0·920	0·920	0·920	0·920	0·920	0·920	0·920	
0·96	η	0·227	0·281	0·327	0·372	0·417	0·464	0·513	0·567	0·626	0·691	0·765	0·850	0·948	
	μ	0·960	0·960	0·960	0·960	0·960	0·960	0·960	0·960	0·960	0·960	0·960	0·960	0·960	
1·00	η	0·225	0·277	0·322	0·365	0·408	0·452	0·499	0·549	0·604	0·664	0·732	0·808	0·895	
	μ	1·000	1·000	1·000	1·000	1·000	1·000	1·000	1·000	1·000	1·000	1·000	1·000	1·000	
S/P		0·04	0·08	0·12	0·16	0·20	0·24	0·28	0·32	0·36	0·40	0·44	0·48	0·52	

Table 2 (*Continued*)

		1.00	0.96	0.92	0.88	0.84	0.80	0.76	0.72	0.68	0.64	0.60	0.56	P/S
0.08	η	0.501		12.505	6.224	4.124	3.070	2.435	2.009	1.703	1.471	1.290	1.143	1.021	1.00
	μ	0.195		1.023	1.034	1.043	1.051	1.058	1.065	1.072	1.079	1.085	1.092	1.099	
0.12	η	0.438	0.759		11.866	5.927	3.932	2.929	2.323	1.916	1.624	1.402	1.228	1.087	0.96
	μ	0.243	0.226		0.986	0.996	1.005	1.013	1.021	1.028	1.035	1.041	1.048	1.055	
0.16	η	0.398	0.634	1.029		11.212	5.628	3.740	2.787	2.211	1.824	1.544	1.332	1.166	0.92
	μ	0.291	0.272	0.259		0.948	0.959	0.968	0.976	0.983	0.990	0.997	1.004	1.011	
0.20	η	0.370	0.558	0.827	1.321		10.546	5.326	3.547	2.646	2.099	1.731	1.465	1.263	0.88
	μ	0.338	0.318	0.305	0.294		0.911	0.922	0.931	0.939	0.946	0.953	0.960	0.967	
0.24	η	0.349	0.507	0.711	1.025	1.639		9.872	5.023	3.354	2.504	1.987	1.638	1.385	0.84
	μ	0.386	0.364	0.351	0.339	0.329		0.874	0.885	0.893	0.901	0.909	0.916	0.923	
0.28	η	0.333	0.470	0.634	0.863	1.232	1.986		9.193	4.721	3.161	2.363	1.875	1.545	0.80
	μ	0.433	0.411	0.396	0.385	0.374	0.365		0.837	0.847	0.856	0.864	0.872	0.879	
0.32	η	0.319	0.442	0.580	0.759	1.018	1.449	2.365		8.516	4.419	2.968	2.222	1.764	0.76
	μ	0.480	0.457	0.442	0.430	0.419	0.410	0.400		0.800	0.810	0.819	0.827	0.835	
0.36	η	0.308	0.419	0.539	0.687	0.884	1.177	1.677	2.776		7.846	4.119	2.777	2.082	0.72
	μ	0.528	0.503	0.487	0.475	0.464	0.455	0.446	0.436		0.763	0.774	0.782	0.790	
0.40	η	0.299	0.401	0.507	0.633	0.792	1.010	1.340	1.914	3.220		7.186	3.821	2.588	0.68
	μ	0.575	0.549	0.533	0.520	0.509	0.500	0.491	0.482	0.472		0.727	0.737	0.745	
0.44	η	0.291	0.385	0.481	0.591	0.724	0.897	1.138	1.507	2.162	3.698		6.543	3.529	0.64
	μ	0.622	0.595	0.578	0.565	0.554	0.544	0.535	0.527	0.518	0.508		0.690	0.700	
0.48	η	0.284	0.372	0.460	0.557	0.672	0.815	1.002	1.267	1.678	2.419	4.208		5.920	0.60
	μ	0.669	0.641	0.623	0.610	0.599	0.589	0.580	0.571	0.563	0.554	0.544		0.653	
0.52	η	0.277	0.361	0.442	0.529	0.630	0.752	0.905	1.109	1.399	1.854	2.685	4.750		
	μ	0.716	0.687	0.669	0.655	0.644	0.634	0.624	0.616	0.607	0.599	0.590	0.581		
0.56	η	0.272	0.351	0.426	0.506	0.596	0.702	0.831	0.996	1.216	1.532	2.033	2.959	5.321	
	μ	0.763	0.733	0.714	0.700	0.688	0.678	0.669	0.660	0.652	0.644	0.635	0.627	0.617	
0.60	η	0.267	0.342	0.412	0.486	0.568	0.662	0.773	0.910	1.086	1.324	1.668	2.215	3.241	
	μ	0.810	0.779	0.759	0.745	0.733	0.723	0.713	0.705	0.696	0.688	0.680	0.672	0.663	
0.64	η	0.262	0.334	0.401	0.469	0.544	0.628	0.726	0.844	0.989	1.178	1.433	1.804	2.400	
	μ	0.856	0.824	0.805	0.790	0.778	0.767	0.758	0.749	0.741	0.733	0.725	0.717	0.709	
0.68	η	0.258	0.327	0.390	0.454	0.523	0.600	0.687	0.790	0.914	1.068	1.269	1.543	1.942	
	μ	0.903	0.870	0.850	0.835	0.822	0.811	0.802	0.793	0.785	0.777	0.769	0.761	0.754	
0.72	η	0.254	0.320	0.380	0.441	0.505	0.575	0.654	0.745	0.853	0.984	1.147	1.361	1.653	
	μ	0.950	0.916	0.895	0.879	0.867	0.856	0.846	0.837	0.829	0.821	0.813	0.806	0.798	
0.76	η	0.251	0.314	0.372	0.429	0.489	0.554	0.626	0.708	0.803	0.916	1.054	1.227	1.453	
	μ	0.996	0.962	0.940	0.924	0.911	0.900	0.890	0.881	0.873	0.865	0.857	0.850	0.842	
0.80	η	0.247	0.309	0.364	0.418	0.475	0.535	0.602	0.676	0.761	0.860	0.979	1.123	1.306	
	μ	1.043	1.007	0.985	0.969	0.956	0.944	0.934	0.925	0.917	0.909	0.901	0.894	0.887	
0.84	η	0.244	0.304	0.357	0.409	0.462	0.519	0.580	0.648	0.725	0.814	0.917	1.041	1.193	
	μ	1.089	1.053	1.030	1.014	1.000	0.989	0.978	0.969	0.961	0.953	0.945	0.938	0.931	
0.88	η	0.241	0.300	0.350	0.400	0.451	0.504	0.561	0.624	0.695	0.774	0.866	0.974	1.103	
	μ	1.136	1.098	1.075	1.058	1.045	1.033	1.023	1.013	1.005	0.997	0.989	0.982	0.974	
0.92	η	0.239	0.295	0.344	0.392	0.440	0.491	0.544	0.603	0.668	0.740	0.822	0.918	1.030	
	μ	1.182	1.144	1.120	1.103	1.089	1.077	1.067	1.057	1.049	1.040	1.033	1.025	1.018	
0.96	η	0.236	0.291	0.339	0.385	0.431	0.479	0.529	0.584	0.644	0.710	0.785	0.870	0.970	
	μ	1.229	1.189	1.165	1.148	1.134	1.121	1.111	1.101	1.092	1.084	1.076	1.069	1.062	
1.00	η	0.234	0.288	0.334	0.378	0.422	0.467	0.515	0.567	0.623	0.684	0.752	0.830	0.918	
	μ	1.275	1.235	1.210	1.192	1.178	1.166	1.155	1.145	1.136	1.128	1.120	1.113	1.106	
S/P		0.04	0.08	0.12	0.16	0.20	0.24	0.28	0.32	0.36	0.40	0.44	0.48	0.52	

Table 2 (*Continued*)

		0·96	0·92	0·88	0·84	0·80	0·76	0·72	0·68	0·64	0·60	0·56			P/S
0·08	η	0·423		12·587	6·283	4·172	3·111	2·471	2·042	1·733	1·499	1·315	1·167	1·043	1·00
	μ	0·313		1·044	1·065	1·082	1·098	1·112	1·126	1·140	1·153	1·167	1·180	1·194	
0·12	η	0·393	0·616		11·827	5·945	3·958	2·956	2·350	1·942	1·648	1·425	1·249	1·107	0·96
	μ	0·400	0·352		1·009	1·030	1·047	1·063	1·078	1·092	1·106	1·119	1·133	1·147	
0·16	η	0·370	0·554	0·816		11·045	5·603	3·744	2·801	2·228	1·842	1·563	1·351	1·184	0·92
	μ	0·456	0·405	0·373		0·974	0·995	1·013	1·028	1·043	1·058	1·072	1·086	1·100	
0·20	η	0·352	0·510	0·713	1·033		10·248	5·259	3·529	2·646	2·107	1·742	1·478	1·277	0·88
	μ	0·512	0·458	0·424	0·398		0·940	0·961	0·978	0·994	1·009	1·023	1·038	1·052	
0·24	η	0·337	0·476	0·643	0·879	1·274		9·447	4·916	3·315	2·491	1·986	1·643	1·394	0·84
	μ	0·567	0·511	0·476	0·449	0·426		0·906	0·926	0·944	0·960	0·975	0·990	1·004	
0·28	η	0·325	0·450	0·592	0·779	1·056	1·542		8·651	4·574	3·103	2·338	1·866	1·545	0·80
	μ	0·623	0·564	0·527	0·500	0·476	0·455		0·872	0·892	0·910	0·926	0·941	0·956	
0·32	η	0·315	0·429	0·553	0·708	0·919	1·245	1·841		7·871	4·237	2·893	2·186	1·747	0·76
	μ	0·678	0·616	0·579	0·550	0·526	0·505	0·485		0·838	0·858	0·876	0·892	0·908	
0·36	η	0·307	0·412	0·522	0·654	0·825	1·066	1·447	2·175		7·116	3·906	2·685	2·036	0·72
	μ	0·733	0·669	0·630	0·601	0·577	0·555	0·535	0·515		0·805	0·825	0·842	0·859	
0·40	η	0·299	0·397	0·497	0·612	0·755	0·945	1·219	1·664	2·547		*6·393*	3·583	2·482	0·68
	μ	0·788	0·721	0·681	0·651	0·626	0·605	0·585	0·566	0·546		*0·772*	0·791	0·809	
0·44	η	0·292	0·384	0·476	0·579	0·702	0·858	1·069	1·379	1·895	*2·961*		*5·708*	3·270	0·64
	μ	0·842	0·774	0·732	0·701	0·676	0·654	0·634	0·616	0·597	*0·578*		*0·739*	0·758	
0·48	η	0·286	0·373	0·458	0·551	0·659	0·791	0·962	1·196	1·546	*2·142*	*3·418*		*5·066*	0·60
	μ	0·897	0·826	0·783	0·751	0·726	0·704	0·684	0·665	0·647	*0·629*	*0·609*		*0·706*	
0·52	η	0·281	0·363	0·442	0·527	0·624	0·739	0·881	1·068	1·327	*1·721*	*2·403*	*3·920*		
	μ	0·951	0·878	0·834	0·802	0·775	0·753	0·733	0·714	0·696	*0·678*	*0·661*	*0·641*		
0·56	η	0·276	0·355	0·429	0·507	0·594	0·696	0·819	0·973	1·176	*1·463*	*1·902*	*2·679*	*4·470*	
	μ	1·005	0·930	0·885	0·852	0·825	0·892	0·782	0·763	0·745	0·728	*0·711*	*0·693*	*0·675*	
0·60	η	0·272	0·347	0·417	0·490	0·570	0·661	0·768	0·899	1·065	1·287	*1·601*	*2·090*	*2·968*	
	μ	1·059	0·982	0·935	0·901	0·874	0·851	0·830	0·811	0·794	0·777	0·760	*0·743*	*0·725*	
0·64	η	0·268	0·340	0·406	0·474	0·548	0·631	0·726	0·840	0·980	1·159	1·399	*1·743*	*2·283*	
	μ	1·113	1·033	0·986	0·951	0·923	0·900	0·879	0·860	0·842	0·825	0·899	0·793	0·776	
0·68	η	0·264	0·333	0·397	0·461	0·529	0·605	0·691	0·791	0·912	1·061	1·254	1·514	1·888	
	μ	1·167	1·085	1·037	1·001	0·973	0·949	0·927	0·908	0·890	0·874	0·857	0·841	0·825	
0·72	η	0·261	0·328	0·388	0·449	0·513	0·583	0·661	0·751	0·856	0·984	1·144	1·350	1·630	
	μ	1·221	1·137	1·087	1·051	1·022	0·997	0·976	0·957	0·939	0·922	0·906	0·890	0·874	
0·76	η	0·257	0·322	0·380	0·438	0·498	0·563	0·635	0·716	0·810	0·922	1·057	1·226	1·447	
	μ	1·274	1·188	1·137	1·100	1·071	1·046	1·024	1·005	0·987	0·970	0·954	0·938	0·923	
0·80	η	0·254	0·317	0·373	0·428	0·485	0·545	0·612	0·686	0·771	0·870	0·987	1·130	1·310	
	μ	1·328	1·240	1·188	1·150	1·120	1·095	1·073	1·053	1·035	1·018	1·002	0·986	0·971	
0·84	η	0·252	0·313	0·366	0·419	0·473	0·530	0·592	0·660	0·737	0·826	0·929	1·052	1·203	
	μ	1·381	1·291	1·238	1·200	1·169	1·143	1·121	1·101	1·082	1·065	1·049	1·034	1·019	
0·88	η	0·249	0·308	0·360	0·411	0·462	0·516	0·574	0·637	0·708	0·788	0·880	0·988	1·118	
	μ	1·434	1·343	1·288	1·249	1·218	1·192	1·169	1·149	1·130	1·113	1·097	1·081	1·067	
0·92	η	0·247	0·304	0·355	0·403	0·452	0·503	0·558	0·617	0·682	0·755	0·839	0·935	1·048	
	μ	1·487	1·394	1·338	1·298	1·267	1·240	1·217	1·197	1·178	1·161	1·144	1·129	1·114	
0·96	η	0·244	0·301	0·349	0·396	0·443	0·492	0·543	0·599	0·659	0·727	0·802	0·889	0·989	
	μ	1·540	1·445	1·389	1·348	1·316	1·289	1·265	1·245	1·226	1·208	1·192	1·176	1·161	
1·00	η	0·242	0·297	0·344	0·389	0·434	0·481	0·530	0·582	0·639	0·701	0·771	0·849	0·939	
	μ	1·593	1·496	1·439	1·397	1·364	1·337	1·313	1·292	1·273	1·256	1·239	1·224	1·209	
S/P		0·04	0·08	0·12	0·16	0·20	0·24	0·28	0·32	0·36	0·40	0·44	0·48	0·52	

Table 2 (*Continued*)

		1·00	0·96	0·92	0·88	0·84	0·80	0·76	0·72	0·68	0·64	0·60	0·56		P/S
0·08	η	0·368		12·728	6·382	4·253	3·181	2·533	2·097	1·783	1·545	1·358	1·207	1·081	1·00
	μ	0·760		1·081	1·120	1·153	1·183	1·211	1·239	1·266	1·293	1·321	1·349	1·378	
0·12	η	0·355	0·514		11·765	5·975	4·002	3·002	2·395	1·985	1·689	1·463	1·286	1·142	0·96
	μ	0·835	0·671		1·051	1·090	1·123	1·154	1·182	1·211	1·238	1·266	1·295	1·324	
0·16	η	0·343	0·486	0·661		10·779	5·565	3·752	2·824	2·257	1·873	1·594	1·382	1·214	0·92
	μ	0·909	0·739	0·643		1·021	1·061	1·094	1·125	1·154	1·183	1·211	1·240	1·269	
0·20	η	0·334	0·463	0·614	0·820		9·789	5·156	3·503	2·647	2·121	1·762	1·501	1·301	0·88
	μ	0·982	0·806	0·708	0·637		0·992	1·032	1·065	1·097	1·126	1·156	1·185	1·214	
0·24	η	0·325	0·445	0·578	0·749	0·996		8·818	4·753	3·258	2·473	1·986	1·653	1·409	0·84
	μ	1·055	0·873	0·772	0·699	0·641		0·964	1·003	1·037	1·069	1·099	1·129	1·159	
0·28	η	0·318	0·429	0·548	0·695	0·893	1·194		7·884	4·358	3·017	2·302	1·854	1·545	0·80
	μ	1·128	0·940	0·836	0·761	0·702	0·651		0·936	0·976	1·010	1·042	1·073	1·103	
0·32	η	0·311	0·415	0·524	0·652	0·817	1·049	1·417		7·000	3·977	2·782	2·134	1·724	0·76
	μ	1·200	1·007	0·899	0·823	0·763	0·711	0·664		0·909	0·948	0·983	1·015	1·047	
0·36	η	0·306	0·403	0·503	0·617	0·759	0·947	1·220	1·671		6·177	3·611	2·554	1·971	0·72
	μ	1·271	1·073	0·962	0·884	0·823	0·771	0·725	0·681		0·883	0·922	0·957	0·990	
0·40	η	0·300	0·393	0·486	0·589	0·713	0·870	1·085	1·406	1·960		5·423	3·263	2·335	0·68
	μ	1·342	1·138	1·025	0·946	0·883	0·831	0·784	0·741	0·699		0·857	0·896	0·931	
0·44	η	0·296	0·384	0·470	0·565	0·675	0·810	0·987	1·233	1·611	2·290		4·738	2·935	0·64
	μ	1·413	1·204	1·088	1·007	0·943	0·890	0·843	0·800	0·759	0·719		0·832	0·871	
0·48	η	0·291	0·376	0·457	0·544	0·643	0·762	0·911	1·109	1·390	1·834	2·665		4·124	0·60
	μ	1·483	1·269	1·150	1·067	1·002	0·948	0·901	0·859	0·818	0·779	0·740		0·808	
0·52	η	0·287	0·368	0·445	0·526	0·616	0·722	0·851	1·016	1·237	1·558	2·077	3·092		
	μ	1·553	1·334	1·213	1·128	1·062	1·007	0·959	0·916	0·877	0·839	0·801	0·761		
0·56	η	0·283	0·361	0·434	0·510	0·593	0·689	0·803	0·943	1·124	1·372	1·737	2·342	3·577	
	μ	1·623	1·399	1·275	1·188	1·121	1·065	1·017	0·974	0·931	0·897	0·860	0·823	9·781	
0·60	η	0·280	0·355	0·424	0·496	0·573	0·660	0·762	0·884	1·037	1·237	1·513	1·926	2·627	
	μ	1·692	1·463	1·336	1·248	1·180	1·123	1·075	1·031	0·991	0·954	0·918	0·883	0·847	
0·64	η	0·277	0·349	0·416	0·483	0·555	0·636	0·727	0·836	0·968	1·134	1·353	1·660	2·126	
	μ	1·762	1·528	1·398	1·308	1·238	1·181	1·132	1·088	1·048	1·011	0·976	0·941	0·907	
0·68	η	0·274	0·344	0·408	0·472	0·539	0·614	0·698	0·795	0·911	1·053	1·233	1·473	1·813	
	μ	1·830	1·592	1·460	1·368	1·297	1·239	1·189	1·145	1·105	1·067	1·032	0·998	0·965	
0·72	η	0·271	0·339	0·400	0·461	0·525	0·595	0·672	0·760	0·863	0·987	1·140	1·335	1·597	
	μ	1·899	1·656	1·521	1·428	1·355	1·296	1·246	1·201	1·161	1·123	1·088	1·055	1·022	
0·76	η	0·268	0·335	0·394	0·452	0·512	0·578	0·649	0·730	0·823	0·932	1·064	1·228	1·439	
	μ	1·967	1·719	1·582	1·487	1·414	1·354	1·302	1·257	1·217	1·179	1·144	1·111	1·079	
0·80	η	0·266	0·330	0·387	0·443	0·501	0·562	0·629	0·703	0·788	0·886	1·002	1·142	1·318	
	μ	2·036	1·783	1·643	1·546	1·472	1·411	1·359	1·313	1·272	1·235	1·200	1·166	1·135	
0·84	η	0·263	0·326	0·382	0·435	0·490	0·548	0·611	0·680	0·757	0·846	0·949	1·072	1·221	
	μ	2·104	1·846	1·704	1·606	1·530	1·468	1·415	1·369	1·328	1·290	1·255	1·222	1·190	
0·88	η	0·261	0·323	0·376	0·428	0·480	0·535	0·595	0·659	0·731	0·812	0·904	1·013	1·143	
	μ	2·171	1·910	1·765	1·665	1·588	1·525	1·472	1·425	1·383	1·345	1·310	1·276	1·245	
0·92	η	0·259	0·319	0·371	0·421	0·471	0·524	0·580	0·640	0·707	0·781	0·866	0·963	1·077	
	μ	2·239	1·973	1·826	1·724	1·646	1·582	1·528	1·481	1·438	1·400	1·364	1·331	1·299	
0·96	η	0·257	0·316	0·366	0·415	0·463	0·513	0·566	0·623	0·685	0·754	0·832	0·920	1·022	
	μ	2·306	2·036	1·886	1·783	1·704	1·639	1·584	1·536	1·494	1·455	1·419	1·385	1·354	
1·00	η	0·255	0·313	0·362	0·409	0·455	0·502	0·554	0·608	0·666	0·730	0·802	0·882	0·974	
	μ	2·374	2·099	1·947	1·842	1·761	1·696	1·640	1·592	1·548	1·509	1·473	1·440	1·408	
S/P		0·04	0·08	0·12	0·16	0·20	0·24	0·28	0·32	0·36	0·40	0·44	0·48	0·52	

Table 2 (*Continued*)

		1·00	0·96	0·92	0·88	0·84	0·80	0·76	0·72	0·68	0·64	0·60	0·56		P/S
0·08	η	0·346		12·845	6·465	4·320	3·238	2·583	2·143	1·825	1·584	1·394	1·239	1·111	1·00
	μ	1·366		1·113	1·168	1·215	1·259	1·300	1·341	1·381	1·422	1·464	1·507	1·552	
0·12	η	0·338	0·473		11·716	6·000	4·039	3·040	2·432	2·020	1·722	1·495	1·316	1·170	0·96
	μ	1·460	1·089		1·087	1·143	1·191	1·235	1·277	1·319	1·361	1·403	1·446	1·491	
0·16	η	0·331	0·456	0·598		10·575	5·536	3·760	2·844	2·281	1·899	1·620	1·408	1·239	0·92
	μ	1·553	1·172	0·975		1·062	1·119	1·167	1·212	1·255	1·298	1·341	1·385	1·430	
0·20	η	0·325	0·441	0·570	0·732		9·451	5·079	3·485	2·650	2·133	1·779	1·520	1·321	0·88
	μ	1·645	1·255	1·053	0·916		1·038	1·095	1·144	1·190	1·234	1·278	1·323	1·368	
0·24	η	0·320	0·429	0·546	0·688	0·880		8·373	4·633	3·216	2·461	1·988	1·662	1·421	0·84
	μ	1·736	1·336	1·130	0·991	0·884		1·016	1·073	1·123	1·169	1·215	1·260	1·306	
0·28	η	0·315	0·418	0·526	0·652	0·816	1·046		7·363	4·205	2·956	2·277	1·846	1·547	0·80
	μ	1·827	1·418	1·206	1·064	0·956	0·866		0·995	1·052	1·102	1·150	1·196	1·243	
0·32	η	0·310	0·408	0·508	0·623	0·764	0·954	1·234		6·435	3·798	2·705	2·098	1·709	0·76
	μ	1·916	1·498	1·282	1·137	1·027	0·936	0·857		0·975	1·032	1·083	1·131	1·180	
0·36	η	0·306	0·400	0·493	0·597	0·723	0·884	1·107	1·448		5·597	3·414	2·466	1·927	0·72
	μ	2·006	1·578	1·358	1·210	1·098	1·006	0·927	0·854		0·956	1·013	1·065	1·115	
0·40	η	0·302	0·392	0·480	0·576	0·689	0·828	1·012	1·275	1·694		*4·850*	3·057	2·239	0·68
	μ	2·094	1·658	1·433	1·282	1·168	1·075	0·996	0·924	0·856		*0·938*	0·995	1·048	
0·44	η	0·298	0·385	0·468	0·557	0·660	0·783	0·940	1·151	1·461	*1·976*		*4·192*	2·727	0·64
	μ	2·182	1·737	1·507	1·354	1·238	1·144	1·064	0·992	0·926	*0·862*		*0·921*	0·979	
0·48	η	0·295	0·378	0·457	0·541	0·635	0·745	0·881	1·058	1·301	*1·667*	*2·302*		*3·615*	0·60
	μ	2·269	1·816	1·582	1·425	1·307	1·212	1·131	1·060	0·994	*0·932*	*0·870*		*0·906*	
0·52	η	0·292	0·372	0·447	0·526	0·613	0·713	0·834	0·984	1·183	*1·462*	*1·895*	*2·679*		
	μ	2·356	1·894	1·656	1·496	1·376	1·280	1·198	1·126	1·061	*1·000*	*0·940*	*0·880*		
0·56	η	0·289	0·367	0·438	0·513	0·594	0·685	0·793	0·925	1·092	*1·316*	*1·637*	*2·146*	*3·114*	
	μ	2·442	1·972	1·729	1·567	1·445	1·347	1·265	1·192	1·127	*1·066*	*1·008*	*0·951*	*0·893*	
0·60	η	0·286	0·361	0·430	0·501	0·577	0·661	0·759	0·876	1·020	1·205	*1·457*	*1·824*	*2·423*	
	μ	2·528	2·050	1·802	1·637	1·514	1·414	1·331	1·258	1·192	1·132	1·075	*1·019*	*0·964*	
0·64	η	0·284	0·357	0·423	0·490	0·561	0·640	0·730	0·835	0·961	1·118	1·324	1·606	2·025	
	μ	2·613	2·127	1·876	1·708	1·582	1·481	1·396	1·323	1·257	1·196	1·140	1·086	1·033	
0·68	η	0·281	0·352	0·416	0·480	0·548	0·621	0·704	0·799	0·911	1·048	1·221	1·447	1·762	
	μ	2·699	2·204	1·948	1·778	1·650	1·548	1·462	1·387	1·321	1·261	1·204	1·151	1·099	
0·72	η	0·279	0·348	0·410	0·471	0·535	0·604	0·681	0·768	0·869	0·990	1·138	1·326	1·576	
	μ	2·783	2·281	2·021	1·847	1·718	1·614	1·527	1·452	1·385	1·324	1·268	1·215	1·164	
0·76	η	0·277	0·344	0·404	0·463	0·524	0·589	0·661	0·741	0·833	0·941	1·070	1·230	1·435	
	μ	2·868	2·358	2·093	1·917	1·785	1·680	1·592	1·516	1·449	1·388	1·331	1·279	1·228	
0·80	η	0·274	0·341	0·399	0·455	0·514	0·575	0·643	0·717	0·802	0·899	1·014	1·152	1·325	
	μ	2·951	2·434	2·165	1·986	1·852	1·746	1·657	1·580	1·512	1·450	1·394	1·341	1·291	
0·84	η	0·272	0·337	0·393	0·448	0·504	0·563	0·626	0·696	0·774	0·862	0·966	1·088	1·236	
	μ	3·035	2·510	2·237	2·055	1·920	1·811	1·721	1·643	1·575	1·513	1·456	1·404	1·354	
0·88	η	0·271	0·334	0·389	0·441	0·495	0·551	0·611	0·676	0·749	0·830	0·924	1·033	1·163	
	μ	3·118	2·585	2·308	2·124	1·987	1·877	1·785	1·707	1·638	1·575	1·518	1·465	1·416	
0·92	η	0·269	0·331	0·384	0·435	0·487	0·540	0·597	0·659	0·727	0·802	0·888	0·986	1·101	
	μ	3·201	2·661	2·380	2·193	2·053	1·942	1·850	1·770	1·700	1·637	1·580	1·527	1·477	
0·96	η	0·267	0·328	0·380	0·429	0·479	0·531	0·585	0·643	0·707	0·777	0·856	0·945	1·048	
	μ	3·284	2·736	2·451	2·261	2·120	2·007	1·914	1·833	1·762	1·699	1·641	1·588	1·538	
1·00	η	0·265	0·325	0·376	0·424	0·472	0·521	0·573	0·629	0·688	0·754	0·827	0·909	1·003	
	μ	3·367	2·811	2·522	2·330	2·186	2·072	1·977	1·896	1·825	1·760	1·702	1·649	1·599	
S/P		0·04	0·08	0·12	0·16	0·20	0·24	0·28	0·32	0·36	0·40	0·44	0·48	0·52	

Table 2 (*Continued*)

S/P		1·00	0·96	0·92	0·88	0·84	0·80	0·76	0·72	0·68	0·64	0·60	0·56		P/S
0·08	η	0·335		12·945	6·535	4·377	3·287	2·626	2·181	1·860	1·616	1·423	1·267	1·137	1·00
	μ	2·196		1·140	1·211	1·271	1·327	1·381	1·434	1·487	1·542	1·598	1·656	1·717	
0·12	η	0·330	0·451		11·677	6·023	4·070	3·072	2·463	2·050	1·750	1·522	1·341	1·194	0·96
	μ	2·310	1·613		1·119	1·190	1·251	1·309	1·364	1·419	1·474	1·531	1·589	1·650	
0·16	η	0·325	0·439	0·564		10·412	5·515	3·767	2·861	2·302	1·920	1·642	1·429	1·260	0·92
	μ	2·422	1·712	1·370		1·099	1·171	1·233	1·292	1·349	1·405	1·463	1·522	1·583	
0·20	η	0·321	0·429	0·544	0·685		9·188	5·019	3·471	2·654	2·144	1·794	1·536	1·338	0·88
	μ	2·534	1·810	1·461	1·237		1·080	1·153	1·217	1·276	1·335	1·394	1·453	1·515	
0·24	η	0·317	0·420	0·527	0·654	0·817		8·039	4·512	3·185	2·453	1·990	1·670	1·432	0·84
	μ	2·644	1·907	1·551	1·323	1·153		1·064	1·137	1·201	1·263	1·323	1·384	1·446	
0·28	η	0·313	0·412	0·512	0·627	0·771	0·965		6·984	4·090	2·910	2·259	1·842	1·549	0·80
	μ	2·753	2·003	1·641	1·408	1·236	1·098		1·049	1·122	1·188	1·251	1·313	1·377	
0·32	η	0·310	0·404	0·499	0·605	0·733	0·899	1·132		6·038	3·667	2·648	2·072	1·698	0·76
	μ	2·861	2·098	1·729	1·492	1·318	1·179	1·061		1·036	1·109	1·176	1·241	1·306	
0·36	η	0·306	0·398	0·487	0·585	0·701	0·845	1·039	1·322		5·201	3·274	2·401	1·895	0·72
	μ	2·968	2·192	1·817	1·576	1·399	1·259	1·140	1·035		1·024	1·098	1·167	1·234	
0·40	η	0·303	0·391	0·476	0·568	0·674	0·802	0·967	1·195	1·541		*4·471*	2·913	2·170	0·68
	μ	3·075	2·286	1·904	1·659	1·480	1·337	1·218	1·114	1·018		*1·015*	1·089	1·160	
0·44	η	0·301	0·386	0·467	0·553	0·650	0·765	0·909	1·099	1·369	*1·795*		*3·839*	2·585	0·64
	μ	3·181	2·380	1·991	1·742	1·559	1·416	1·295	1·191	1·096	*1·007*		*1·007*	1·083	
0·48	η	0·298	0·380	0·458	0·539	0·629	0·734	0·862	1·024	1·243	*1·562*	*2·088*		*3·296*	0·60
	μ	3·286	2·472	2·077	1·824	1·639	1·493	1·372	1·267	1·173	*1·085*	*1·000*		*1·001*	
0·52	η	0·295	0·375	0·450	0·527	0·611	0·707	0·822	0·963	1·147	*1·399*	*1·778*	*2·431*		
	μ	3·390	2·564	2·163	1·905	1·718	1·570	1·447	1·342	1·248	*1·161*	*1·079*	*0·998*		
0·56	η	0·293	0·371	0·442	0·515	0·595	0·684	0·788	0·913	1·071	1·278	*1·569*	*2·019*	*2·830*	
	μ	3·494	2·656	2·248	1·987	1·796	1·646	1·522	1·416	1·322	1·236	*1·155*	*1·077*	*0·998*	
0·60	η	0·291	0·366	0·435	0·505	0·580	0·663	0·758	0·871	1·009	1·184	1·418	*1·754*	*2·287*	
	μ	3·597	2·747	2·333	2·067	1·874	1·722	1·597	1·490	1·395	1·309	1·229	*1·153*	*1·078*	
0·64	η	0·289	0·362	0·429	0·496	0·567	0·644	0·732	0·834	0·957	1·108	1·303	1·568	1·954	
	μ	3·699	2·838	2·418	2·148	1·952	1·798	1·671	1·563	1·468	1·382	1·302	1·227	1·155	
0·68	η	0·287	0·359	0·423	0·487	0·555	0·628	0·709	0·803	0·913	1·046	1·212	1·429	1·727	
	μ	3·802	2·928	2·502	2·228	2·029	1·873	1·745	1·635	1·539	1·453	1·374	1·300	1·229	
0·72	η	0·285	0·355	0·418	0·479	0·544	0·613	0·689	0·775	0·875	0·993	1·138	1·320	1·560	
	μ	3·903	3·018	2·586	2·308	2·106	1·948	1·818	1·708	1·611	1·524	1·445	1·371	1·301	
0·76	η	0·283	0·352	0·412	0·472	0·533	0·599	0·670	0·751	0·842	0·949	1·076	1·233	1·433	
	μ	4·004	3·108	2·669	2·387	2·183	2·022	1·891	1·779	1·682	1·595	1·515	1·442	1·372	
0·80	η	0·281	0·348	0·408	0·465	0·524	0·586	0·654	0·729	0·813	0·910	1·024	1·161	1·331	
	μ	4·105	3·197	2·752	2·467	2·259	2·097	1·964	1·851	1·752	1·665	1·585	1·511	1·442	
0·84	η	0·279	0·345	0·403	0·459	0·515	0·575	0·639	0·709	0·787	0·876	0·979	1·101	1·249	
	μ	4·205	3·286	2·835	2·545	2·335	2·171	2·036	1·922	1·823	1·734	1·654	1·580	1·512	
0·88	η	0·278	0·343	0·399	0·452	0·507	0·564	0·625	0·691	0·764	0·846	0·940	1·050	1·180	
	μ	4·305	3·374	2·918	2·624	2·411	2·245	2·108	1·993	1·892	1·803	1·723	1·649	1·580	
0·92	η	0·276	0·340	0·394	0·449	0·499	0·554	0·612	0·674	0·743	0·819	0·906	1·005	1·121	
	μ	4·404	3·462	3·000	2·703	2·487	2·318	2·180	2·063	1·962	1·872	1·791	1·717	1·648	
0·96	η	0·275	0·337	0·391	0·441	0·492	0·545	0·600	0·660	0·724	0·795	0·875	0·966	1·071	
	μ	4·503	3·550	3·082	2·781	2·562	2·391	2·252	2·134	2·031	1·941	1·859	1·784	1·715	
1·00	η	0·273	0·335	0·387	0·436	0·486	0·536	0·589	0·646	0·707	0·774	0·848	0·932	1·027	
	μ	4·601	3·638	3·164	2·859	2·637	2·465	2·323	2·204	2·100	2·009	1·926	1·851	1·782	
S/P		0·04	0·08	0·12	0·16	0·20	0·24	0·28	0·32	0·36	0·40	0·44	0·48	0·52	

Table 2 (*Continued*)

S/P		1·00	0·96	0·92	0·88	0·84	0·80	0·76	0·72	0·68	0·64	0·60	0·56		P/S
0·08	η	0·319	13·237	6·740	4·542	3·428	2·751	2·293	1·961	1·708	1·509	1·346	1·210		1·00
	μ	8·478	1·226	1·344	1·449	1·548	1·647	1·746	1·848	1·955	2·067	2·186	2·315		
0·12	η	0·317	0·418	11·575	6·089	4·160	3·165	2·553	2·135	1·831	1·598	1·412	1·261		0·96
	μ	8·670	4·915	1·219	1·340	1·448	1·552	1·654	1·758	1·866	1·978	2·098	2·226		
0·16	η	0·315	0·413	0·512	9·987	5·461	3·792	2·911	2·361	1·983	1·704	1·490	1·318		0·92
	μ	8·861	5·075	3·600	1·216	1·340	1·451	1·559	1·666	1·775	1·889	2·009	2·137		
0·20	η	0·314	0·409	0·504	0·610	8·532	4·869	3·441	2·667	2·177	1·836	1·582	1·386		0·88
	μ	9·049	5·233	3·743	2·906	1·217	1·344	1·458	1·570	1·682	1·797	1·918	2·047		
0·24	η	0·312	0·405	0·496	0·596	0·716	7·241	4·320	3·110	2·436	2·001	1·694	1·465		0·84
	μ	9·237	5·390	3·884	3·038	2·477	1·222	1·351	1·470	1·586	1·704	1·826	1·955		
0·28	η	0·311	0·402	0·489	0·584	0·695	0·833	6·124	3·818	2·802	2·218	1·834	1·560		0·80
	μ	9·423	5·545	4·023	3·168	2·601	2·185	1·232	1·363	1·486	1·607	1·731	1·862		
0·32	η	0·310	0·398	0·483	0·574	0·677	0·803	0·964	5·174	3·366	2·518	2·014	1·677		0·76
	μ	9·607	5·699	4·162	3·297	2·723	2·303	1·975	1·246	1·380	1·507	1·635	1·767		
0·36	η	0·308	0·395	0·477	0·564	0·661	0·776	0·921	1·113	4·377	2·963	2·257	1·825		0·72
	μ	9·790	5·852	4·299	3·424	2·844	2·420	2·089	1·816	1·265	1·402	1·535	1·669		
0·40	η	0·307	0·392	0·471	0·554	0·646	0·753	0·884	1·052	1·284	3·711	2·605	2·019		0·68
	μ	9·972	6·004	4·435	3·551	2·964	2·535	2·201	1·928	1·693	1·289	1·430	1·569		
0·44	η	0·306	0·389	0·466	0·546	0·633	0·733	0·852	1·002	1·200	1·483	3·157	2·289		0·64
	μ	10·153	6·154	4·570	3·676	3·082	2·649	2·312	2·037	1·803	1·595	1·318	1·465		
0·48	η	0·304	0·386	0·461	0·538	0·621	0·715	0·825	0·959	1·132	1·366	1·714	2·696		0·60
	μ	10·333	6·304	4·705	3·800	3·200	2·762	2·422	2·145	1·910	1·704	1·516	1·355		
0·52	η	0·303	0·384	0·457	0·531	0·610	0·698	0·800	0·923	1·075	1·275	1·554	1·987		
	μ	10·512	6·453	4·838	3·924	3·317	2·874	2·530	2·251	2·015	1·810	1·625	1·452		
0·56	η	0·302	0·381	0·452	0·524	0·600	0·683	0·779	0·891	1·028	1·201	1·433	1·768	2·311	
	μ	10·689	6·600	4·970	4·046	3·433	2·985	2·638	2·356	2·119	1·913	1·729	1·560	1·399	
0·60	η	0·301	0·379	0·448	0·517	0·590	0·670	0·759	0·863	0·987	1·140	1·338	1·609	2·012	
	μ	10·866	6·747	5·102	4·168	3·548	3·095	2·744	2·460	2·221	2·015	1·831	1·664	1·508	
0·64	η	0·300	0·376	0·444	0·511	0·581	0·657	0·741	0·838	0·951	1·088	1·261	1·487	1·804	
	μ	11·041	6·893	5·233	4·289	3·662	3·204	2·850	2·563	2·322	2·115	1·932	1·766	1·612	
0·68	η	0·299	0·374	0·440	0·506	0·573	0·645	0·725	0·815	0·920	1·044	1·196	1·390	1·649	
	μ	11·216	7·038	5·363	4·410	3·776	3·313	2·955	2·665	2·422	2·214	2·030	1·865	1·712	
0·72	η	0·298	0·372	0·437	0·500	0·565	0·635	0·710	0·795	0·892	1·005	1·142	1·311	1·529	
	μ	11·390	7·183	5·492	4·529	3·888	3·421	3·059	2·766	2·522	2·312	2·127	1·962	1·810	
0·76	η	0·297	0·370	0·433	0·495	0·558	0·625	0·697	0·777	0·867	0·971	1·095	1·244	1·432	
	μ	11·563	7·326	5·621	4·649	4·001	3·528	3·163	2·867	2·620	2·408	2·223	2·057	1·906	
0·80	η	0·296	0·368	0·430	0·490	0·551	0·615	0·684	0·760	0·845	0·941	1·054	1·188	1·352	
	μ	11·735	7·469	5·749	4·767	4·113	3·635	3·265	2·967	2·718	2·504	2·318	2·151	2·001	
0·84	η	0·295	0·366	0·427	0·486	0·545	0·607	0·673	0·744	0·824	0·914	1·018	1·139	1·285	
	μ	11·906	7·611	5·876	4·885	4·224	3·741	3·368	3·066	2·815	2·599	2·412	2·244	2·093	
0·88	η	0·294	0·364	0·424	0·481	0·539	0·598	0·662	0·730	0·805	0·890	0·986	1·096	1·227	
	μ	12·076	7·753	6·003	5·002	4·335	3·847	3·470	3·165	2·911	2·694	2·505	2·337	2·185	
0·92	η	0·293	0·362	0·421	0·477	0·533	0·591	0·652	0·717	0·788	0·868	0·957	1·059	1·177	
	μ	12·246	7·894	6·129	5·119	4·445	3·952	3·571	3·263	3·007	2·788	2·597	2·428	2·276	
0·96	η	0·292	0·361	0·418	0·473	0·528	0·583	0·642	0·705	0·773	0·847	0·931	1·025	1·134	
	μ	12·415	8·034	6·255	5·236	4·555	4·057	3·672	3·361	3·102	2·881	2·689	2·518	2·366	
1·00	η	0·291	0·359	0·416	0·469	0·522	0·577	0·633	0·693	0·758	0·829	0·907	0·995	1·095	
	μ	12·583	8·174	6·380	5·352	4·664	4·161	3·772	3·458	3·197	2·974	2·780	2·608	2·455	
S/P		0·04	0·08	0·12	0·16	0·20	0·24	0·28	0·32	0·36	0·40	0·44	0·48	0·52	

Table 2 (*Continued*)

		1·00	0·96	0·92	0·88	0·84	0·80	0·76	0·72	0·68	0·64	0·60	0·56	P/S
0·08	η	0·316		13·343	6·815	4·602	3·479	2·795	2·333	1·997	1·742	1·539	1·374	1·236	1·00
	μ	13·912		1·258	1·396	1·520	1·638	1·756	1·876	2·001	2·132	2·271	2·421	2·584	
0·12	η	0·315	0·411		11·540	6·113	4·193	3·199	2·585	2·166	1·860	1·625	1·438	1·284	0·96
	μ	14·142	7·370		1·258	1·400	1·528	1·651	1·775	1·902	2·034	2·174	2·324	2·486	
0·16	η	0·314	0·408	0·502		9·850	5·445	3·802	2·929	2·383	2·005	1·727	1·511	1·339	0·92
	μ	14·371	7·560	5·113		1·263	1·409	1·541	1·670	1·800	1·934	2·075	2·225	2·387	
0·20	η	0·313	0·405	0·496	0·595		8·328	4·822	3·433	2·674	2·189	1·851	1·599	1·402	0·88
	μ	14·598	7·747	5·280	3·966		1·273	1·422	1·560	1·694	1·831	1·974	2·124	2·286	
0·24	η	0·312	0·403	0·490	0·585	0·695		7·003	4·252	3·088	2·433	2·006	1·704	1·477	0·84
	μ	14·823	7·933	5·446	4·119	3·273		1·288	1·441	1·584	1·725	1·870	2·022	2·185	
0·28	η	0·311	0·400	0·485	0·575	0·679	0·805		5·878	3·738	2·771	2·207	1·834	1·565	0·80
	μ	15·047	8·118	5·610	4·270	3·416	2·810		1·309	1·466	1·615	1·764	1·918	2·081	
0·32	η	0·310	0·397	0·480	0·567	0·665	0·782	0·928		4·938	3·281	2·480	1·999	1·672	0·76
	μ	15·270	8·301	5·773	4·420	3·557	2·946	2·480		1·336	1·498	1·653	1·811	1·976	
0·36	η	0·309	0·395	0·475	0·559	0·652	0·761	0·894	1·068		4·160	2·877	2·217	1·806	0·72
	μ	15·491	8·483	5·934	4·569	3·697	3·080	2·611	2·234		1·370	1·536	1·699	1·868	
0·40	η	0·308	0·393	0·471	0·552	0·640	0·742	0·865	1·020	1·228		3·518	2·522	1·978	0·68
	μ	15·711	8·663	6·094	4·716	3·835	3·212	2·739	2·360	2·044		1·411	1·583	1·756	
0·44	η	0·307	0·390	0·467	0·545	0·630	0·726	0·839	0·979	1·161	1·413		2·989	2·212	0·64
	μ	15·929	8·842	6·253	4·862	3·973	3·343	2·866	2·485	2·167	1·893		1·460	1·640	
0·48	η	0·306	0·388	0·463	0·538	0·620	0·711	0·816	0·944	1·105	1·320	1·630		2·552	0·60
	μ	16·147	9·020	6·410	5·007	4·108	3·472	2·991	2·607	2·289	2·015	1·771		1·518	
0·52	η	0·305	0·386	0·459	0·532	0·610	0·697	0·796	0·913	1·058	1·244	1·500	1·886		
	μ	16·363	9·197	6·567	5·150	4·243	3·601	3·114	2·727	2·407	2·134	1·892	1·671		
0·56	η	0·304	0·384	0·455	0·527	0·602	0·684	0·777	0·886	1·017	1·182	1·399	1·706	2·189	
	μ	16·578	9·372	6·722	5·293	4·377	3·728	3·237	2·847	2·525	2·250	2·009	1·792	1·588	
0·60	η	0·304	0·382	0·452	0·521	0·594	0·672	0·760	0·862	0·982	1·129	1·318	1·572	1·942	
	μ	16·792	9·547	6·877	5·435	4·510	3·854	3·358	2·964	2·640	2·364	2·124	1·908	1·709	
0·64	η	0·303	0·380	0·449	0·516	0·586	0·661	0·745	0·840	0·951	1·084	1·250	1·466	1·764	
	μ	17·005	9·721	7·031	5·575	4·641	3·979	3·479	3·081	2·754	2·477	2·236	2·021	1·825	
0·68	η	0·302	0·379	0·446	0·511	0·579	0·651	0·731	0·820	0·923	1·045	1·193	1·380	1·628	
	μ	17·217	9·893	7·183	5·715	4·772	4·104	3·598	3·197	2·867	2·588	2·346	2·132	1·937	
0·72	η	0·301	0·377	0·443	0·507	0·572	0·642	0·718	0·802	0·898	1·010	1·144	1·309	1·521	
	μ	17·428	10·065	7·335	5·855	4·903	4·228	3·717	3·312	2·979	2·698	2·455	2·240	2·046	
0·76	η	0·300	0·375	0·440	0·502	0·566	0·633	0·706	0·786	0·876	0·979	1·102	1·249	1·433	
	μ	17·638	10·236	7·486	5·993	5·032	4·350	3·834	3·425	3·090	2·807	2·562	2·347	2·153	
0·80	η	0·300	0·373	0·437	0·498	0·560	0·625	0·694	0·770	0·855	0·952	1·064	1·197	1·360	
	μ	17·847	10·406	7·636	6·131	5·161	4·473	3·951	3·539	3·200	2·914	2·668	2·452	2·258	
0·84	η	0·299	0·372	0·434	0·494	0·554	0·617	0·684	0·756	0·837	0·927	1·031	1·152	1·298	
	μ	18·055	10·575	7·786	6·268	5·289	4·594	4·068	3·651	3·309	3·021	2·773	2·556	2·361	
0·88	η	0·298	0·370	0·432	0·490	0·549	0·610	0·674	0·743	0·820	0·905	1·001	1·113	1·244	
	μ	18·262	10·744	7·935	6·404	5·417	4·715	4·184	3·763	3·418	3·127	2·877	2·658	2·463	
0·92	η	0·298	0·369	0·429	0·487	0·544	0·603	0·665	0·731	0·804	0·884	0·974	1·077	1·197	
	μ	18·468	10·912	8·083	6·540	5·543	4·835	4·299	3·874	3·525	3·232	2·981	2·760	2·564	
0·96	η	0·297	0·367	0·427	0·483	0·539	0·596	0·656	0·720	0·789	0·865	0·950	1·046	1·156	
	μ	18·674	11·078	8·230	6·675	5·670	4·955	4·413	3·984	3·633	3·337	3·083	2·861	2·663	
1·00	η	0·296	0·366	0·425	0·480	0·534	0·590	0·648	0·709	0·776	0·848	0·928	1·017	1·119	
	μ	18·878	11·245	8·377	6·809	5·795	5·074	4·527	4·094	3·739	3·441	3·185	2·961	2·762	
S/P		0·04	0·08	0·12	0·16	0·20	0·24	0·28	0·32	0·36	0·40	0·44	0·48	0·52	

Table 3 Percentage of the gamma for values of the shape parameter, $\gamma = \eta$. Reproduced by permission of the editor of *Technometrics* from M. B. Wilk et al., Probability Plots for the Gamma Distribution, *Technometrics*, Vol. 4, pp. 1–20 (1962).

PER CENT	ETA= 0.1	ETA= 0.2	ETA= 0.3	ETA= 0.4	ETA= 0.5
0.1	6.0730398E-31	6.5254914E-16	9.7227015E-11	2.3449403E-08	7.8539858E-07
0.5	5.9307050E-24	2.0392159E-12	1.4903943E-08	1.3108620E-06	1.9635211E-05
0.7	3.4199421E-22	1.5485293E-11	5.7580040E-08	3.6123166E-06	4.4179955E-05
1.0	6.0730419E-21	6.5254908E-11	1.5022232E-07	7.4153871E-06	7.8543941E-05
1.5	3.50202C8E-19	4.9552939E-10	5.8037043E-07	2.0434596E-05	1.7673543E-04
2.0	6.2187947E-18	2.0881566E-09	1.5141467E-06	4.1948807E-05	3.1422512E-04
2.5	5.7917027E-17	6.3725492E-09	3.1856788E-06	7.3283179E-05	4.9103468E-04
3.0	3.5860676E-16	1.5856933E-08	5.8497902E-06	1.1560341E-04	7.0719174E-04
4.0	6.3680429E-15	6.6821009E-08	1.5261804E-05	2.3733141E-04	1.2576912E-03
5.0	5.93C7026E-14	2.0392152E-07	3.2110344E-05	4.1465349E-04	1.9660703E-03
7.5	3.4199416E-12	1.5485298E-06	1.2406425E-04	1.1432451E-03	4.4309273E-03
10.0	6.0730380E-11	6.5255202E-06	3.2372462E-04	2.3488758E-03	7.8953891E-03
15.0	3.5020194E-09	4.9554952E-05	1.2515738E-03	6.4918979E-03	1.7882892E-02
20.0	6.2187893E-08	2.0985191E-04	3.2703397E-03	1.3392227E-02	3.2092386E-02
25.0	5.7917008E-07	6.3759328E-04	6.8998047E-03	2.3564933E-02	5.0765538E-02
30.0	3.5860783E-06	1.5877920E-03	1.2726660E-02	3.7541870E-02	7.4235954E-02
40.0	6.3684073E-05	6.7195739E-03	3.3739803E-02	7.9361846E-02	1.3749799E-01
50.0	5.9339001E-04	2.0746364E-02	7.3131159E-02	1.4507806E-01	2.2746834E-01
60.0	3.6844445E-03	5.3010665E-02	1.4125254E-01	2.4475218E-01	3.5416332E-01
70.0	1.7427737E-02	1.2103773E-01	2.5656503E-01	3.9725703E-01	5.3709736E-01
75.0	3.5306306E-02	1.7885940E-01	3.4289970E-01	5.0480585E-01	6.6165223E-01
80.0	6.9389746E-02	2.6354398E-01	4.6007422E-01	6.4557067E-01	8.2118779E-01
85.0	1.3466307E-01	3.9239831E-01	6.2662798E-01	8.3910112E-01	1.0361264E 00
90.0	2.6615398E-01	6.0490358E-01	8.8481154E-01	1.1298418E 00	1.3527737E 00
92.5	3.8439176E-01	7.7335898E-01	1.0811301E 00	1.3461192E 00	1.5850286E 00
95.0	5.8043370E-01	1.0305303E 00	1.3723524E 00	1.6619615E 00	1.9207334E 00
97.5	9.7790323E-01	1.5111122E 00	1.9002707E 00	2.2247126E 00	2.5119509E 00
98.0	1.1190291E 00	1.6744477E 00	2.0765711E 00	2.4107092E 00	2.7059585E 00
99.0	1.5884692E 00	2.2023303E 00	2.6394268E 00	3.0000888E 00	3.3174701E 00
99.5	2.0945469E 00	2.7547526E 00	3.2206054E 00	3.6035306E 00	3.9397731E 00
99.9	3.3636821E 00	4.1C23043E 00	4.6191710E 00	5.0425654E 00	5.4140809E 00

PER CENT	ETA= 0.6	ETA= 0.8	ETA= 1.0	ETA= 1.2	ETA= 1.4
0.1	8.2890664E-06	1.6272338E-04	1.0005003E-03	3.4337143E-03	8.4321789E-03
0.5	1.2119545E-04	1.2173541E-03	5.0125419E-03	1.3187497E-02	2.6824049E-02
0.7	2.3823386E-04	2.0217386E-03	7.5282661E-03	1.8533525E-02	3.5971068E-02
1.0	3.8483497E-04	2.8980758E-03	1.0050336E-02	2.3608727E-02	4.4330284E-02
1.5	7.5659043E-04	4.8159973E-03	1.5113637E-02	3.3243716E-02	5.9597079E-02
2.0	1.2224146E-03	6.9081841E-03	2.0202707E-02	4.2425602E-02	7.3617880E-02
2.5	1.7737191E-03	9.1419638E-03	2.5317809E-02	5.1301112E-02	8.6809725E-02
3.0	2.4044995E-03	1.1496962E-02	3.0459208E-02	5.9952244E-02	9.9398144E-02
4.0	3.8873902E-03	1.6518275E-02	4.0821993E-02	7.6773139E-02	1.2327700E-01
5.0	5.6448357E-03	2.1897531E-02	5.1293299E-02	9.3145212E-02	1.4592649E-01
7.5	1.1133291E-02	3.6647758E-02	7.7961543E-02	1.3293248E-01	1.8922959E-01
10.0	1.8060443E-02	5.2981821E-02	1.0536052E-01	1.7189840E-01	2.4973644E-01
15.0	3.5894110E-02	8.9739645E-02	1.6251895E-01	2.4937285E-01	3.4689076E-01
20.0	5.8803372E-02	1.3152850E-01	2.2314357E-01	3.2797601E-01	4.4246886E-01
25.0	8.6771754E-02	1.7827535E-01	2.8768212E-01	4.0903381E-01	5.3886871E-01
30.0	1.1998889E-01	2.3019176E-01	3.5667497E-01	4.9357830E-01	6.3769279E-01
40.0	2.0382268E-01	3.5143359E-01	5.1082570E-01	6.7712659E-01	8.4793257E-01
50.0	3.1570220E-01	5.0135124E-01	6.9314729E-01	8.8793657E-01	1.0843713E 00
60.0	4.6590956E-01	6.9127123E-01	9.1629100E-01	1.1400349E 00	1.3623938E 00
70.0	6.7474850E-01	9.4321526E-01	1.2039732E 00	1.4587471E 00	1.7088393E 00
75.0	8.1365435E-01	1.1058806E 00	1.3862949E 00	1.6581306E 00	1.9234758E 00
80.0	9.8899279E-01	1.3073711E 00	1.6094384E 00	1.9000906E 00	2.1822733E 00
85.0	1.2219616E 00	1.5702344E 00	1.8971206E 00	2.2093879E 00	2.5109470E 00
90.0	1.5605061E 00	1.9452591E 00	2.3025857E 00	2.6414636E 00	2.9669431E 00
92.5	1.8063390E 00	2.2138442E 00	2.5902694E 00	2.9459091E 00	3.2864993E 00
95.0	2.1590177E 00	2.5951444E 00	2.9957345E 00	3.3726729E 00	3.7325141E 00
97.5	2.7747879E 00	3.2526998E 00	3.6888885E 00	4.0973864E 00	4.4858795E 00
98.0	2.9757511E 00	3.4656007E 00	3.9120296E 00	4.3296577E 00	4.7264682E 00
99.0	3.6066221E 00	4.1298904E 00	4.6051941E 00	5.0486563E 00	5.4690573E 00
99.5	4.2455626E 00	4.7977945E 00	5.2983655E 0C	5.7645890E 00	6.2058056E 00
99.9	5.7512864E 00	6.3586366E 00	6.9080574E 00	7.4187547E 00	7.9005137E 00

Table 3 (*Continued*)

PER CENT	ETA= 1.6	ETA= 1.8	ETA= 2.0	ETA= 2.2	ETA= 2.4
0.1	1.6780863E-02	2.9006379E-02	4.5402019E-02	6.6083795E-02	9.1045250E-02
0.5	4.6409061E-02	7.2C19480E-02	1.0349455E-01	1.4055196E-01	1.8285795E-01
0.7	6.0108387E-02	9.0817968E-02	1.2777440E-01	1.7057086E-01	2.1878191E-01
1.0	7.2283818E-02	1.0717552E-01	1.4855474E-01	1.9592441E-01	2.4879515E-01
1.5	9.39C2475E-02	1.3560738E-01	1.8407833E-01	2.3869266E-01	2.9887579E-01
2.0	1.1322857E-01	1.6050999E-01	2.1469912E-01	2.7508877E-01	3.4104964E-01
2.5	1.3105810E-01	1.8314475E-01	2.4220929E-01	3.0748374E-01	3.7830102E-01
3.0	1.4781163E-01	2.0416656E-01	2.6752685E-01	3.3707901E-01	4.1212894E-01
4.0	1.7902662E-01	2.4280678E-01	3.1357260E-01	3.9044885E-01	4.7270738E-01
5.0	2.0808401E-01	2.7826796E-01	3.5536154E-01	4.3845224E-01	5.2679405E-01
7.5	2.7489323E-01	3.5837546E-01	4.4846799E-01	5.4422134E-01	6.4488062E-01
10.0	3.3669067E-01	4.3112658E-01	5.3181166E-01	6.3780034E-01	7.4835367E-01
15.0	4.5272915E-01	5.6524168E-01	6.8323869E-01	8.0583392E-01	9.3234974E-01
20.0	5.6432575E-01	6.9199598E-01	8.2438848E-01	9.6070601E-01	1.1003482E 00
25.0	6.7504999E-01	8.1617533E-01	9.6127882E-01	1.1096652E 00	1.2608163E 00
30.0	7.8710858E-01	9.4C60314E-01	1.0973494E 00	1.2567599E 00	1.4184017E 00
40.0	1.0219104E C0	1.1982486E 00	1.3764216E 00	1.5560676E 00	1.7369269E 00
50.0	1.2817967E 00	1.4798565E 00	1.6783474E 00	1.8771417E 00	2.0761571E 00
60.0	1.5834684E 00	1.8C33957E 00	2.0223137E 00	2.2403417E 00	2.4575810E 00
70.0	1.9551916E 00	2.2198482C7E 00	2.4392173E 00	2.6777864E 00	2.9144873E 00
75.0	2.1837358E 00	2.4398846E 00	2.6926356E 00	2.9425202E 00	3.1899446E 00
80.0	2.4579001E 00	2.7282669E 00	2.9943091E 00	3.2567191E 00	3.5160193E 00
85.0	2.8042824E 00	3.0910513E 00	3.3724436E 00	3.6493274E 00	3.9223586E 00
90.0	3.2821878E 00	3.5892844E 00	3.8897218E 00	4.1845905E 00	4.4746965E 00
92.5	3.6156165E 00	3.9356059E 00	4.2481450E 00	4.5544488E 00	4.8554263E 00
95.0	4.0793485E 00	4.4158303E 00	4.7438694E 00	5.0648451E 00	5.3797783E 00
97.5	4.8591827E 00	5.2203591E 00	5.5716554E 00	5.9146804E 00	6.2506155E 00
98.0	5.1074690E 00	5.4758451E 00	5.8339376E 00	6.1834139E 00	6.5255030E 00
99.0	5.8719870E 00	6.2608809E 00	6.6383883E 00	7.0063273E 00	7.3660060E 00
99.5	6.6282540E 00	7.0353775E 00	7.4301966E 00	7.8146312E 00	8.1900520E 00
99.9	8.3615569E 00	8.8C45225E 00	9.2336768E 00	9.6511943E 00	1.0057676E 01

PER CENT	ETA= 2.6	ETA= 2.8	ETA= 3.0	ETA= 3.2	ETA= 3.4
0.1	1.2020084E-01	1.5341695E-01	1.9053338E-01	2.3137730E-01	2.7577227E-01
0.5	2.3006679E-01	2.8184123E-01	3.3786340E-01	3.9783849E-01	4.6149604E-01
0.7	2.7199522E-01	3.2982536E-01	3.9191958E-01	4.5795804E-01	5.2765206E-01
1.0	3.0670981E-01	3.6925137E-01	4.3604521E-01	5.0675595E-01	5.8108456E-01
1.5	3.6411288E-01	4.3394811E-01	5.0798807E-01	5.8585852E-01	6.6727104E-01
2.0	4.1203167E-01	4.8755657E-01	5.6720963E-01	6.5063010E-01	7.3750284E-01
2.5	4.5408805E-01	5.3435255E-01	6.1867216E-01	7.0668188E-01	7.9806521E-01
3.0	4.9203797E-C1	5.7645527E-01	6.6480418E-01	7.5676816E-01	8.5203115E-01
4.0	5.5974186E 00	6.5104351E-01	7.4618340E-01	8.4479663E-01	9.4657009E-01
5.0	6.1977374E-01	7.1688152E-01	8.1769154E-01	9.2184297E-01	1.0290283E 00
7.5	7.4983816E-01	8.5859696E-01	9.7074668E-01	1.0859143E 00	1.2038973E 00
10.0	8.6288491E-01	9.8091840E-01	1.1020654E 00	1.2260021E 00	1.3524560E 00
15.0	1.0622576E 00	1.1951340E 00	1.3306367E 00	1.4684834E 00	1.6084377E 00
20.0	1.2428518E 00	1.3878487E 00	1.5350443E 00	1.6841965E 00	1.8351031E 00
25.0	1.4143367E 00	1.5699138E 00	1.7272996E 00	1.8862906E 00	2.0467182E 00
30.0	1.5819476E 00	1.7471400E 00	1.9137761E 00	2.0816902E 00	2.2507454E 00
40.0	1.9188084E 00	2.1015632E 00	2.2850772E 00	2.4692581E 00	2.6540295E 00
50.0	2.2753401E 00	2.4746498E 00	2.6740610E 00	2.8735522E 00	3.0731070E 00
60.0	2.6741217E 00	2.8900326E 00	3.1053791E 00	3.3202146E 00	3.5345823E 00
70.0	3.1495635E 00	3.3832035E 00	3.6155689E 00	3.8467910E 00	4.0769786E 00
75.0	3.4352389E 00	3.6786528E 00	3.9204034E 00	4.1606653E 00	4.3995793E 00
80.0	3.7726309E 00	4.0268778E 00	4.2790313E 00	4.5293148E 00	4.7779093E 00
85.0	4.1920675E 00	4.4588519E 00	4.7230354E 00	4.9849489E 00	5.2447592E 00
90.0	4.7607076E 00	5.0431069E 00	5.3223249E 00	5.5986977E 00	5.8724992E 00
92.5	5.1518242E 00	5.4441808E .00	5.7329780E 00	6.0185937E 00	6.3013375E 00
95.0	5.6895199E 00	5.9946764E 00	6.2958025E 00	6.5933195E 00	6.8875671E 00
97.5	6.5804864E 00	6.9049583E 00	7.2247059E 00	7.5402121E 00	7.8518687E 00
98.0	6.8612753E 00	7.1914166E 00	7.5166242E 00	7.8374205E 00	8.1542076E 00
99.0	7.7186885E 00	8.0650784E 00	8.4059950E 00	8.7419776E 00	9.0734380E 00
99.5	8.5579089E 00	8.9188460E 00	9.2738498E 00	9.6235664E 00	9.9682106E 00
99.9	1.0456185E C1	1.0846008E 01	1.1229337E 01	1.1606503E 01	1.1977438E 01

Table 3 (*Continued*)

PER CENT	ETA= 3.6	ETA= 3.8	ETA= 4.0	ETA= 4.2	ETA= 4.4
0.1	3.2354432E-01	3.7452469E-01	4.2855244E-01	4.8547497E-01	5.4514865E-01
0.5	5.2858958E-01	5.9889438E-01	6.7220659E-01	7.4834049E-01	8.2712702E-01
0.7	6.0074238E-01	6.7699502E-01	7.5619978E-01	8.3816664E-01	9.2272364E-01
1.0	6.587645CE-01	7.3955699E-01	8.2324872E-01	9.0964804E-01	9.9858204E-01
1.5	7.5194496E-01	8.3963767E-01	9.3013401E-01	1.0232418E 00	1.1187889E 00
2.0	8.2755259E-01	9.2053593E-01	1.0162386E 00	1.1144695E 00	1.2150586E 00
2.5	8.9254687E-C1	9.8988446E-01	1.0898654E 00	1.1923011E 00	1.2970235E 00
3.0	9.5031885E-01	1.0513911E 00	1.1550374E 00	1.2610717E 00	1.3693284E 00
4.0	1.0512345E 00	1.1585543E 00	1.2683247E 00	1.3803650E 00	1.4945144E 00
5.0	1.1389841E 00	1.2514805E 00	1.3663186E 00	1.4833227E 00	1.6023380E 00
7.5	1.3243603E C0	1.4471184E 00	1.5719868E 00	1.6988037E 00	1.8274260E 00
10.0	1.4811965E 00	1.6120247E 00	1.7447698E 00	1.8792817E 00	2.0154290E 00
15.0	1.7503018E 00	1.8939042E 00	2.0390997E 00	2.1857612E 00	2.3337768E 00
20.0	1.9875967E C0	2.1415323E 00	2.2967870E 00	2.4532543E 00	2.6108400E 00
25.0	2.2084427E 00	2.3713439E 00	2.5353205E 00	2.7002840E 00	2.8661566E 00
30.0	2.4208299E C0	2.5918456E 00	2.7637115E 00	2.9363569E 00	3.1097192E 00
40.0	2.8393317E 00	3.0251106E 00	3.2113231E 00	3.3979324E 00	3.5849041E 00
50.0	3.2727174E 00	3.4723707E 00	3.6720615E 00	3.8717843E 00	4.0715334E 00
60.0	3.7485254E C0	3.9620748E 00	4.1752641E 00	4.3881181E 00	4.6006591E 00
70.0	4.3062314E 00	4.5346232E 00	4.7622305E 00	4.9891121E 00	5.2153195E 00
75.0	4.6372802E 00	4.8738642E 00	5.1094297E 00	5.3440557E 00	5.5778067E 00
80.0	5.0249804E 00	5.2706529E 00	5.5150478E 00	5.7582676E 00	6.0003931E 00
85.0	5.5026938E 00	5.7589013E 00	6.0135390E 00	6.2667284E 00	6.5185704E 00
90.0	6.1439911E 00	6.4133576E 00	6.6807875E 00	6.9464370E 00	7.2104183E 00
92.5	6.5814911E 00	6.8592682E 00	7.1348786E 00	7.4084946E 00	7.6802465E 00
95.0	7.1788929E 00	7.4675052E 00	7.7536630E 00	8.0375611E 00	8.3193413E 00
97.5	8.1601093E 00	8.4651448E 00	8.7672890E 00	9.0667806E 00	9.3637615E 00
98.0	8.4674107E 00	8.7772766E 00	9.0841264E 00	9.3882200E 00	9.6896919E 00
99.0	9.4009742E 00	9.7247509E 00	1.0045182E 01	1.0362505E 01	1.0676853E 01
99.5	1.0308698E 01	1.0644946E 01	1.0977605E 01	1.1306876E 01	1.1632821E 01
99.9	1.2344066E 01	1.2705508E 01	1.3062754E 01	1.3416387E 01	1.3765564E 01

PER CENT	ETA= 4.6	ETA= 4.8	ETA= 5.0	ETA= 5.5	ETA= 6.0
0.1	6.0743883E-01	6.7221912E-01	7.3937174E-01	9.1692643E-01	1.1071047E 00
0.5	9.0841244E-01	9.9205569E-01	1.0779283E 00	1.3016111E 00	1.5369119E 00
0.7	1.0097155E C0	1.0590001E 00	1.1904491E 00	1.4277901E 00	1.6763566E 00
1.0	1.0898952E 00	1.1834459E 00	1.2791062E 00	1.5267421E 00	1.7852847E 00
1.5	1.2166211E 00	1.3165983E 00	1.4185948E 00	1.6816918E 00	1.9551829E 00
2.0	1.3178539E 00	1.4227173E 00	1.5295258E 00	1.8043437E 00	2.0891438E 00
2.5	1.4038829E 00	1.5127442E 00	1.6234866E 00	1.9078744E 00	2.2018944E 00
3.0	1.4796606E 00	1.5919354E 00	1.7060345E 00	1.9985808E 00	2.3004521E 00
4.0	1.6106316E 00	1.7285883E 00	1.8482710E 00	2.1543736E 00	2.4692747E 00
5.0	1.7232276E 00	1.8458680E 00	1.9701498E 00	2.2874067E 00	2.6130151E 00
7.5	1.9577285E 00	2.0895982E 00	2.2229352E 00	2.5621264E 00	2.9085574E 00
10.0	2.1530969E 00	2.2921812E 00	2.4325913E 00	2.7888928E 00	3.1518984E 00
15.0	2.4830503E 00	2.6334934E 00	2.7850300E 00	3.1682177E 00	3.5569183E 00
20.0	2.7694638E 00	2.9290513E 00	3.0895401E 00	3.4943373E 00	3.9036642E 00
25.0	3.0328729E C0	3.2003714E 00	3.3686008E 00	3.7920720E 00	4.2192098E 00
30.0	3.2837474E 00	3.4583909E 00	3.6336099E 00	4.0739347E 00	4.5171390E 00
40.0	3.7722123E 00	3.9598298E 00	4.1477370E 00	4.6186435E 00	5.0909867E 00
50.0	4.2713065E 00	4.4710989E 00	4.6709096E 00	5.1705011E 00	5.6701630E 00
60.0	4.8129125E 00	5.0248932E 00	5.2362202E 00	5.7649193E 00	6.2919204E 00
70.0	5.4409055E 00	5.6659053E 00	5.8903626E 00	6.4493365E 00	7.0055531E 00
75.0	5.8107550E 00	6.0429444E 00	6.2744328E 00	6.8503502E 00	7.4227052E 00
80.0	6.2415134E 00	6.4816867E 00	6.7209831E 00	7.3157148E 00	7.9059964E 00
85.0	6.7691775E 00	7.0186161E 00	7.2669716E 00	7.8835530E 00	8.4946588E 00
90.0	7.4728816E C0	7.7339032E 00	7.9935977E 00	8.6375158E 00	9.2746798E 00
92.5	7.9503019E 00	8.2187408E 00	8.4856923E 00	9.1471189E 00	9.8009979E 00
95.0	8.5991930E 00	8.8772030E 00	9.1535279E 00	9.8375876E 00	1.0513056E 01
97.5	9.6584998E 00	9.9510543E 00	1.0241620E 01	1.0960063E 01	1.1668369E 01
98.0	9.9888175E C0	1.0285654E 01	1.0580429E 01	1.1309022E 01	1.2027013E 01
99.0	1.0988642E 01	1.1297808E 01	1.1604693E 01	1.2362570E 01	1.3108568E 01
99.5	1.1956081E C1	1.2276368E 01	1.2594225E 01	1.3378652E 01	1.4150016E 01
99.9	1.4112402E 01	1.4454964E 01	1.4794938E 01	1.5632967E 01	1.6455929E 01

Table 3 (*Continued*)

PER CENT	ETA= 6.5	ETA= 7.0	ETA= 7.5	ETA= 8.0	ETA= 8.5
0.1	1.3086092E 00	1.5203364E 00	1.7413424E 00	1.9708140E 00	2.2080464E 00
0.5	1.7825174E 00	2.0373377E 00	2.3004580E 00	2.5711030E 00	2.8486089E 00
0.7	1.9348533E 00	2.2022113E 00	2.4775370E 00	2.7600756E 00	3.0491823E 00
1.0	2.0534580E 00	2.3302127E 00	2.6146747E 00	2.9061064E 00	3.2038803E 00
1.5	2.2378305E 00	2.5286208E 00	2.8267128E 00	3.1313983E 00	3.4420770E 00
2.0	2.3827228E 00	2.6840991E 00	2.9924585E 00	3.3071189E 00	3.6275021E 00
2.5	2.5043755E 00	2.8143633E 00	3.1310691E 00	3.4538326E 00	3.7820935E 00
3.0	2.6105067E 00	2.9278158E 00	3.2516130E 00	3.5812568E 00	3.9162049E 00
4.0	2.7918850E 00	3.1213215E 00	3.4568573E 00	3.7978841E 00	4.1438895E 00
5.0	2.9459324E 00	3.2853161E 00	3.6304726E 00	3.9808231E 00	4.3358805E 00
7.5	3.2618875E 00	3.6207577E 00	3.9847463E 00	4.3533348E 00	4.7260877E 00
10.0	3.5207531E 00	3.8947672E 00	4.2733786E 00	4.6561186E 00	5.0425941E 00
15.0	3.9504190E 00	4.3481487E 00	4.7496417E 00	5.1545107E 00	5.5624310E 00
20.0	4.3169312E 00	4.7336645E 00	5.1534806E 00	5.5760590E 00	6.0011336E 00
25.0	4.6495332E 00	5.0826579E 00	5.5182703E 00	5.9561104E 00	6.3959645E 00
30.0	4.9628425E 00	5.4107397E 00	5.8605856E 00	6.3121755E 00	6.7653399E 00
40.0	5.5645716E 00	6.0392421E 00	6.5148761E 00	6.9913707E 00	7.4686382E 00
50.0	6.1698797E 00	6.6696393E 00	7.1694322E 00	7.6692510E 00	8.1690941E 00
60.0	6.8177888E 00	7.3426500E 00	7.8666144E 00	8.3897712E 00	8.9121968E 00
70.0	7.5593646E 00	8.1110528E 00	8.6608512E 00	9.2089498E 00	9.7555171E 00
75.0	7.9919587E 00	8.5584704E 00	9.1225476E 00	9.6844350E 00	1.0244344E 01
80.0	8.4924042E 00	9.0753887E 00	9.6553365E 00	1.0232545E 01	1.0807289E 01
85.0	9.1009954E 00	9.7031254E 00	1.0301515E 01	1.0896538E 01	1.1488524E 01
90.0	9.9059781E 00	1.0532086E 01	1.1153581E 01	1.1770923E 01	1.2384529E 01
92.5	1.0448295E 01	1.1089785E 01	1.1726100E 01	1.2357771E 01	1.2985264E 01
95.0	1.1181042E 01	1.1842418E 01	1.2497918E 01	1.3148141E 01	1.3793597E 01
97.5	1.2367854E 01	1.3059525E 01	1.3744270E 01	1.4422725E 01	1.5095555E 01
98.0	1.2735805E 01	1.3436437E 01	1.4129837E 01	1.4816652E 01	1.5497598E 01
99.0	1.3844243E 01	1.4570737E 01	1.5289087E 01	1.6000054E 01	1.6704526E 01
99.5	1.4909999E 01	1.5659867E 01	1.6401025E 01	1.7133823E 01	1.7859475E 01
99.9	1.7265614E 01	1.8062641E 01	1.8849757E 01	1.9627376E 01	2.0397113E 01

PER CENT	ETA= 9.0	ETA= 9.5	ETA=10.0	ETA=11.0	ETA=12.0
0.1	2.4524246E 00	2.7034082E 00	2.9605206E 00	3.4914847E 00	4.0424412E 00
0.5	3.1324024E 00	3.4219864E 00	3.7169226E 00	4.3213587E 00	4.9431173E 00
0.7	3.3443029E 00	3.6449553E 00	3.9507180E 00	4.5761350E 00	5.2180602E 00
1.0	3.5074561E 00	3.8163654E 00	4.1301997E 00	4.7712465E 00	5.4281815E 00
1.5	3.7582328E 00	4.0794198E 00	4.4052488E 00	5.0695062E 00	5.7487117E 00
2.0	3.9531113E 00	4.2835185E 00	4.6183496E 00	5.3000156E 00	5.9959114E 00
2.5	4.1153736E 00	4.4532587E 00	4.7953894E 00	5.4911615E 00	6.2005758E 00
3.0	4.2559932E 00	4.6002209E 00	4.9485406E 00	5.6562685E 00	6.3771362E 00
4.0	4.4944343E 00	4.8491418E 00	5.2076828E 00	5.9351506E 00	6.6749172E 00
5.0	4.6952280E 00	5.0585075E 00	5.4254066E 00	6.1690078E 00	6.9242137E 00
7.5	5.1026338E 00	5.4876530E 00	5.8658682E 00	6.6409487E 00	7.4262587E 00
10.0	5.4324691E 00	5.8254560E 00	6.2213050E 00	7.0207483E 00	7.8293430E 00
15.0	5.9731265E 00	6.3863619E 00	6.8019311E 00	7.6393782E 00	8.4842802E 00
20.0	6.4284776E 00	6.8578969E 00	7.2892214E 00	8.1570214E 00	9.0309031E 00
25.0	6.8376465E 00	7.2810003E 00	7.7258626E 00	8.6198114E 00	9.5186276E 00
30.0	7.2199330E 00	7.6758316E 00	8.1329297E 00	9.0503625E 00	9.9716172E 00
40.0	7.9466078E 00	8.4252195E 00	8.9044168E 00	9.8643986E 00	1.0826247E 01
50.0	8.6689538E 00	9.1688287E 00	9.6687176E 00	1.0668527E 01	1.1668365E 01
60.0	9.4339553E 00	9.9551048E 00	1.0475689E 01	1.1515335E 01	1.2553178E 01
70.0	1.0300684E 01	1.0844570E 01	1.1387277E 01	1.2469514E 01	1.3547990E 01
75.0	1.0802450E 01	1.1358912E 01	1.1913852E 01	1.3019639E 01	1.4120585E 01
80.0	1.1379781E 01	1.1950216E 01	1.2518760E 01	1.3650738E 01	1.4776670E 01
85.0	1.2077742E 01	1.2664436E 01	1.3248803E 01	1.4411246E 01	1.5566245E 01
90.0	1.2994726E 01	1.3601808E 01	1.4206015E 01	1.5406661E 01	1.6598145E 01
92.5	1.3608924E 01	1.4229087E 01	1.4846038E 01	1.6071234E 01	1.7286267E 01
95.0	1.4434687E 01	1.5071797E 01	1.5705250E 01	1.6962263E 01	1.8207551E 01
97.5	1.5763251E 01	1.6426253E 01	1.7084491E 01	1.8390440E 01	1.9682139E 01
98.0	1.6173158E 01	1.6843826E 01	1.7509872E 01	1.8829848E 01	2.0135297E 01
99.0	1.7402804E 01	1.8095583E 01	1.8783283E 01	2.0144920E 01	2.1490142E 01
99.5	1.8578447E 01	1.9291487E 01	1.9998780E 01	2.1398395E 01	2.2779675E 01
99.9	2.1158045E 01	2.1912130E 01	2.2658903E 01	2.4136056E 01	2.5590844E 01

Table 3 (*Continued*)

PER CENT	ETA=13.0	ETA=14.0	ETA=15.0	ETA=16.0	ETA=17.0
0.1	4.6110638E 00	5.1954403E 00	5.7939761E 00	6.4053278E 00	7.0283506E 00
0.5	5.5801195E 00	6.2306684E 00	6.8933615E 00	7.5670171E 00	8.2506377E 00
0.7	5.8744963E 00	6.5438208E 00	7.2246892E 00	7.9159749E 00	8.6167239E 00
1.0	6.0990738E 00	6.7823563E 00	7.4767294E 00	8.1811088E 00	8.8945751E 00
1.5	6.4410392E 00	7.1450050E 00	7.8593836E 00	8.5831507E 00	9.3154399E 00
2.0	6.7042936E 00	7.4237416E 00	8.1530889E 00	8.8913561E 00	9.6377208E 00
2.5	6.9219536E 00	7.6539314E 00	8.3953875E 00	9.1453834E 00	9.9031279E 00
3.0	7.1095212E 00	7.8521072E 00	8.6038142E 00	9.3637351E 00	1.0131110E 01
4.0	7.4254577E 00	8.1855372E 00	8.9541411E 00	9.7304205E 00	1.0513662E 01
5.0	7.6895794E 00	8.4639391E 00	9.2463311E 00	1.0035958E 01	1.0832143E 01
7.5	8.2205200E 00	9.0226990E 00	9.8319496E 00	1.0647562E 01	1.1468939E 01
10.0	8.6459442E 00	9.4696220E 00	1.0299619E 01	1.1135299E 01	1.1976129E 01
15.0	9.3356943E 00	1.0192867E 01	1.1055175E 01	1.1922099E 01	1.2793207E 01
20.0	9.9100994E 00	1.0793988E 01	1.1682060E 01	1.2573894E 01	1.3469138E 01
25.0	1.0421718E 01	1.1328581E 01	1.2238806E 01	1.3152057E 01	1.4068045E 01
30.0	1.0896202E 01	1.1823732E 01	1.2753882E 01	1.3686390E 01	1.4621032E 01
40.0	1.1789720E 01	1.2754630E 01	1.3720816E 01	1.4688148E 01	1.5656520E 01
50.0	1.2668234E 01	1.3668121E 01	1.4668023E 01	1.5667934E 01	1.6667863E 01
60.0	1.3589448E 01	1.4624319E 01	1.5657936E 01	1.6690443E 01	1.7721924E 01
70.0	1.4623175E 01	1.5695444E 01	1.6765128E 01	1.7832464E 01	1.8897699E 01
75.0	1.5217729E 01	1.6310262E 01	1.7399882E 01	1.8486499E 01	1.9570403E 01
80.0	1.5897320E 01	1.7013296E 01	1.8125106E 01	1.9233168E 01	2.0337850E 01
85.0	1.6714752E 01	1.7857509E 01	1.8995142E 01	2.0128173E 01	2.1257023E 01
90.0	1.7731609E 01	1.8957984E 01	2.0180054E 01	2.1292423E 01	2.2451631E 01
92.5	1.8492109E 01	1.9690108E 01	2.0881007E 01	2.2065500E 01	2.3244234E 01
95.0	1.9442629E 01	2.0668632E 01	2.1886569E 01	2.3097160E 01	2.4301234E 01
97.5	2.0961681E 01	2.2230526E 01	2.3489773E 01	2.4740289E 01	2.5983167E 01
98.0	2.1428066E 01	2.2709562E 01	2.3981039E 01	2.5243490E 01	2.6497842E 01
99.0	2.2821100E 01	2.4139324E 01	2.5446398E 01	2.6743183E 01	2.8030844E 01
99.5	2.4145321E 01	2.5497245E 01	2.6836654E 01	2.8166678E 01	2.9482640E 01
99.9	2.7028923E 01	2.8449401E 01	2.9854549E 01	3.1246408E 01	3.2626712E 01

PER CENT	ETA=18.0	ETA=19.0	ETA=20.0	ETA=21.0	ETA=22.0
0.1	7.6620569E 00	8.3055943E 00	8.9582144E 00	9.6192597E 00	1.0288149E 01
0.5	8.9433638E 00	9.6444577E 00	1.0353269E 01	1.1069233E 01	1.1791848E 01
0.7	9.3261169E 00	1.0043450E 01	1.0768106E 01	1.1499544E 01	1.2237289E 01
1.0	9.6163391E 00	1.0345723E 01	1.1082133E 01	1.1825048E 01	1.2574014E 01
1.5	1.0055510E 01	1.0802721E 01	1.1556513E 01	1.2316403E 01	1.3081958E 01
2.0	1.0391474E 01	1.1152007E 01	1.1918788E 01	1.2691355E 01	1.3469300E 01
2.5	1.0667943E 01	1.1439243E 01	1.2216521E 01	1.2999333E 01	1.3787286E 01
3.0	1.0905281E 01	1.1685690E 01	1.2471847E 01	1.3263322E 01	1.4059740E 01
4.0	1.1303252E 01	1.2098665E 01	1.2899443E 01	1.3705187E 01	1.4515540E 01
5.0	1.1634307E 01	1.2441954E 01	1.3254656E 01	1.4072029E 01	1.4893742E 01
7.5	1.2295574E 01	1.3127027E 01	1.3962919E 01	1.4802916E 01	1.5646724E 01
10.0	1.2821652E 01	1.3671478E 01	1.4525265E 01	1.5382713E 01	1.6243567E 01
15.0	1.3668128E 01	1.4546540E 01	1.5428164E 01	1.6312768E 01	1.7200125E 01
20.0	1.4367484E 01	1.5268673E 01	1.6172481E 01	1.7078708E 01	1.7987179E 01
25.0	1.4986524E 01	1.5907287E 01	1.6830154E 01	1.7754998E 01	1.8681570E 01
30.0	1.5557613E 01	1.6495978E 01	1.7435974E 01	1.8377436E 01	1.9320403E 01
40.0	1.6625837E 01	1.7596011E 01	1.8566986E 01	1.9538694E 01	2.0511088E 01
50.0	1.7667794E 01	1.8667731E 01	1.9667683E 01	2.0667618E 01	2.1667592E 01
60.0	1.8752478E 01	1.9782188E 01	2.0811008E 01	2.1839308E 01	2.2866829E 01
70.0	1.9961006E 01	2.1022545E 01	2.2082456E 01	2.3140855E 01	2.4197863E 01
75.0	2.0651832E 01	2.1730978E 01	2.2808022E 01	2.3883145E 01	2.4956471E 01
80.0	2.1439427E 01	2.2538165E 01	2.3634289E 01	2.4728014E 01	2.5819494E 01
85.0	2.2382071E 01	2.3503613E 01	2.4621950E 01	2.5737329E 01	2.6849966E 01
90.0	2.3606131E 01	2.4756340E 01	2.5902573E 01	2.7045162E 01	2.8184337E 01
92.5	2.4417700E 01	2.5586345E 01	2.6750551E 01	2.7910633E 01	2.9066888E 01
95.0	2.5499299E 01	2.6691864E 01	2.7879356E 01	2.9062158E 01	3.0240542E 01
97.5	2.7218828E 01	2.8447969E 01	2.9671091E 01	3.0888593E 01	3.2100914E 01
98.0	2.7744701E 01	2.8984669E 01	3.0218336E 01	3.1446061E 01	3.2668638E 01
99.0	2.9310028E 01	3.0581402E 01	3.1845937E 01	3.3103699E 01	3.4355160E 01
99.5	3.0791157E 01	3.2091351E 01	3.3383925E 01	3.4668762E 01	3.5946929E 01
99.9	3.3995731E 01	3.5355368E 01	3.6703910E 01	3.8046123E 01	3.9379733E 01

Table 4 Percentage points for testing the equality of the Weibull shape parameters.

Reproduced by permission of the editor of *Technometrics* from D. R. Thoman and L. J. Bain, Two Sample Test in the Weibull Distribution, *Technometrics*, Vol. 11, pp. 805–815 (1969).

N \ $1-\alpha$	.60	.70	.75	.80	.85	.90	.95	.98
5	1.158	1.351	1.478	1.636	1.848	2.152	2.725	3.550
6	1.135	1.318	1.418	1.573	1.727	1.987	2.465	3.146
7	1.127	1.283	1.370	1.502	1.638	1.869	2.246	2.755
8	1.119	1.256	1.338	1.450	1.573	1.780	2.093	2.509
9	1.111	1.236	1.311	1.410	1.524	1.711	1.982	2.339
10	1.104	1.220	1.290	1.380	1.486	1.655	1.897	2.213
11	1.098	1.206	1.273	1.355	1.454	1.609	1.829	2.115
12	1.093	1.195	1.258	1.334	1.428	1.571	1.774	2.036
13	1.088	1.186	1.245	1.317	1.406	1.538	1.727	1.922
14	1.084	1.177	1.233	1.301	1.386	1.509	1.688	1.917
15	1.081	1.170	1.224	1.288	1.369	1.485	1.654	1.870
16	1.077	1.164	1.215	1.277	1.355	1.463	1.624	1.829
17	1.075	1.158	1.207	1.266	1.341	1.444	1.598	1.793
18	1.072	1.153	1.200	1.257	1.329	1.426	1.574	1.762
19	1.070	1.148	1.194	1.249	1.318	1.411	1.553	1.733
20	1.068	1.144	1.188	1.241	1.308	1.396	1.534	1.709
22	1.064	1.136	1.178	1.227	1.291	1.372	1.501	1.663
24	1.061	1.129	1.169	1.216	1.276	1.351	1.473	1.625
26	1.058	1.124	1.162	1.206	1.263	1.333	1.449	1.593
28	1.055	1.119	1.155	1.197	1.252	1.318	1.428	1.566
30	1.053	1.114	1.149	1.190	1.242	1.304	1.409	1.541
32	1.051	1.110	1.144	1.183	1.233	1.292	1.393	1.520
34	1.049	1.107	1.139	1.176	1.224	1.281	1.378	1.500
36	1.047	1.103	1.135	1.171	1.217	1.272	1.365	1.483
38	1.046	1.100	1.131	1.166	1.210	1.263	1.353	1.467
40	1.045	1.098	1.127	1.161	1.204	1.255	1.342	1.453
42	1.043	1.095	1.124	1.156	1.198	1.248	1.332	1.439
44	1.042	1.093	1.121	1.152	1.193	1.241	1.323	1.427
46	1.041	1.091	1.118	1.149	1.188	1.235	1.314	1.416
48	1.040	1.088	1.115	1.145	1.184	1.229	1.306	1.405
50	1.039	1.087	1.113	1.142	1.179	1.224	1.299	1.396
52	1.038	1.085	1.111	1.139	1.175	1.219	1.292	1.387
54	1.037	1.083	1.108	1.136	1.172	1.215	1.285	1.378
56	1.036	1.081	1.106	1.133	1.168	1.210	1.279	1.370
58	1.036	1.080	1.104	1.131	1.165	1.206	1.274	1.363
60	1.035	1.078	1.102	1.128	1.162	1.203	1.268	1.355
62	1.034	1.077	1.101	1.126	1.159	1.199	1.263	1.349
64	1.034	1.076	1.099	1.124	1.156	1.196	1.258	1.342
66	1.033	1.075	1.097	1.122	1.153	1.192	1.253	1.336
68	1.032	1.073	1.096	1.120	1.151	1.189	1.249	1.331
70	1.032	1.072	1.094	1.118	1.148	1.186	1.245	1.325
72	1.031	1.071	1.093	1.116	1.146	1.184	1.241	1.320
74	1.031	1.070	1.091	1.114	1.143	1.181	1.237	1.315
76	1.030	1.069	1.090	1.112	1.141	1.179	1.233	1.310
78	1.030	1.068	1.089	1.111	1.139	1.176	1.230	1.306
80	1.030	1.067	1.088	1.109	1.137	1.174	1.227	1.301
90	1.028	1.063	1.082	1.102	1.128	1.164	1.212	1.282
100	1.026	1.060	1.078	1.097	1.121	1.155	1.199	1.266
120	1.023	1.054	1.071	1.087	1.109	1.142	1.180	1.240

Table 5 Percentage points for testing the equality of the Weibull scale parameters

Reproduced by permission of the editor of *Technometrics* from D. R. Thoman and L. J. Bain, Two Sample Test in the Weibull Distribution, *Technometrics*, Vol. 11, pp. 805–815 (1969).

N \ 1-α	.60	.70	.75	.80	.85	.90	.95	.98
5	.228	.476	.608	.777	.960	1.226	1.670	2.242
6	.190	.397	.522	.642	.821	1.050	1.404	1.840
7	.164	.351	.461	.573	.726	.918	1.215	1.592
8	.148	.320	.415	.521	.658	.825	1.086	1.421
9	.136	.296	.383	.481	.605	.757	.992	1.294
10	.127	.277	.356	.449	.563	.704	.918	1.195
11	.120	.261	.336	.423	.528	.661	.860	1.115
12	.115	.248	.318	.401	.499	.625	.811	1.049
13	.110	.237	.303	.383	.474	.594	.770	.993
14	.106	.227	.290	.366	.453	.567	.734	.945
15	.103	.218	.279	.352	.434	.544	.704	.904
16	.099	.210	.269	.339	.417	.523	.676	.867
17	.096	.203	.260	.328	.403	.505	.654	.834
18	.094	.197	.251	.317	.389	.488	.631	.805
19	.091	.191	.244	.308	.377	.473	.611	.779
20	.089	.186	.237	.299	.366	.459	.593	.755
22	.085	.176	.225	.284	.347	.435	.561	.712
24	.082	.168	.215	.271	.330	.414	.534	.677
26	.079	.161	.206	.259	.316	.396	.510	.646
28	.076	.154	.198	.249	.303	.380	.490	.619
30	.073	.149	.191	.240	.292	.366	.472	.595
32	.071	.144	.185	.232	.282	.354	.455	.574
34	.069	.139	.179	.225	.273	.342	.441	.555
36	.067	.135	.174	.218	.265	.332	.427	.537
38	.065	.131	.169	.212	.258	.323	.415	.522
40	.064	.127	.165	.206	.251	.314	.404	.507
42	.062	.124	.160	.201	.245	.306	.394	.494
44	.061	.121	.157	.196	.239	.298	.384	.482
46	.059	.118	.153	.192	.234	.292	.376	.470
48	.058	.115	.150	.188	.229	.285	.367	.460
50	.057	.113	.147	.184	.224	.279	.360	.450
52	.056	.110	.144	.180	.220	.273	.353	.440
54	.055	.108	.141	.176	.215	.268	.346	.432
56	.054	.106	.138	.173	.212	.263	.340	.423
58	.053	.104	.136	.170	.208	.258	.334	.416
60	.052	.102	.134	.167	.204	.254	.328	.408
62	.051	.100	.131	.164	.201	.250	.323	.402
64	.050	.099	.129	.162	.198	.246	.317	.395
66	.049	.097	.127	.159	.195	.242	.313	.389
68	.049	.095	.125	.157	.192	.238	.308	.383
70	.048	.094	.123	.154	.190	.235	.304	.377
72	.047	.092	.122	.152	.187	.231	.299	.372
74	.046	.091	.120	.150	.184	.228	.295	.366
76	.046	.090	.118	.148	.182	.225	.291	.361
78	.045	.089	.117	.146	.180	.222	.288	.357
80	.045	.087	.115	.144	.178	.219	.284	.352
90	.042	.082	.109	.136	.168	.207	.268	.332
100	.040	.077	.103	.128	.160	.196	.255	.315
120	.036	.070	.094	.117	.147	.179	.233	.287

Annotated Bibliography

Afifi, A. A. and S. P. Azen (1972), *Statistical Analysis: A Computer Oriented Approach*, Academic Press, New York.
The authors give a description of the use of BMDX85 program in solving a nonlinear regression equation. This procedure can be used to maximize a likelihood function as shown in Chapter 6.

Aitken, A. C. (1935), On Least-squares and Linear Combination of Observations, *Proceedings of the Royal Society of Edinburgh*, Vol. 55, pp. 42–48.
The author in this classic paper develops a generalized least squares theory that can be used when the variance–covariance matrix of the errors is not scalar (i.e. when there is nonzero correlation between errors and where the error variances are not homoscedastic.)

Anderson, T. W. (1964), Sequential Analysis with Delayed Observations, *Journal of the American Statistical Association*, Vol. 59, pp. 1006–1015.
The author utilizes Wald's sequential probability ratio test as a stopping rule but considers the case when delayed observations are available some time after stopping. The formulas are applied to the normal distribution with σ^2 known.

Armitage, P. (1959), The Comparison of Survival Curves, *Journal of the Royal Statistical Society*, Vol. A122, pp. 279–300.
The author contrasts four methods for comparing survival time distributions when the distributions are exponential and uniform entry is assumed. The four methods are maximum likelihood estimators used in a normal test statistic, paired sign test, a comparison of proportion surviving, and an actuarial method.

Armitage, P. (1960), *Sequential Medical Trials*, Charles C. Thomas, Springfield, Ill.
An applied book on sequential analysis which includes examples and tables useful for the closed sequential designs.

Bain, L. J. and C. E. Antle (1967), Estimation of Parameters in the Weibull Distribution, *Technometrics*, Vol. 9, pp. 621–628.
The authors apply an estimation procedure of Gumbel to the parameters of the Weibull distribution. They obtain easily computed estimates that can be used in singly censored cases. They make a Monte Carlo comparison of four types of estimators for the Weibull distribution.

Barlow, R. E. and F. Proschan (1965), *Mathematical Theory of Reliability*, John Wiley & Sons, New York.
The authors direct this book to the mathematicians. The approach is nonparametric; however, assumptions are made concerning the hazard rate $h(t)$. They discuss IFR (increasing failure rate) and DFR (decreasing failure rate) distributions.

Barlow, R. E. and E. M. Scheuer (1966), Reliability Growth During a Development Testing Program, *Technometrics*, Vol. 8, pp. 53–60.
The authors develop a nonparametric reliability growth model assuming two types of system failure—inherent failures that cannot be eliminated without a major system change and assignable cause failures that can be so eliminated.

Bartholomew, D. J. (1957), A Problem in Life Testing, *Journal of the American Statistical Association*, Vol. 52, pp. 350–355.
The author obtains the maximum likelihood estimator for the parameter of the exponential survival distribution when the observations are censored. The assumption of uniform entry of observations is not made.

Bennett, C. and N. Franklin (1954), *Statistical Analysis in Chemistry and the Chemical Industry*, John Wiley & Sons, New York.
The authors discuss a general technique for determining variances of arbitrary functions.

Berger, A. and R. Z. Gold (1961), On Comparing Survival Times, *Proceedings of the Fourth Berkeley Symposium on Mathematical Statistics and Probability*, pp. 67–76.
This is a discussion of the problem of comparing two survival distributions at several different points in time. This solution to the problem is asymptotic.

Berkson, J. (1953), A Statistically Precise and Relatively Simple Method of Estimating the Bioassay with Quantal Response, Based on the Logistic Function, *Journal of the American Statistical Association*, Vol. 48, pp. 565–599.
The author applies the logistic function to quantal data (death or survival) in bioassay testing of animals, (i.e. the testing of a chemical compound in animals to determine lethal doses for various percentages of the particular animal population).

Berkson, J. and L. Elvebach (1960), Competing Exponential Risks with Particular Reference to the Study of Smoking and Lung Cancer, *Journal of the American Statistical Association*, Vol. 55, pp. 415–428.
The problem of competing risks in the exponential case is the subject of this paper. Maximum likelihood estimators of the requisite parameters and their large sample properties are discussed. Examples of patients with lung cancer versus other risks are given.

Berkson, J. and R. P. Gage (1950), Calculation of Survival Rates for Cancer, *Proceedings of the Staff Meetings of the Mayo Clinic*, Vol. 25, pp. 270–286. The article contains a discussion of T-year survival rates utilizing actuarial techniques.

Berkson, J. and R. P. Gage (1952), Survival Curve for Cancer Patients Following Treatment, *Journal of the American Statistical Association*, Vol. 47, pp. 501–515.
The authors estimate the parameters of a survival function when a proportion of the patients with a specific disease, say cancer, are cured. A constant hazard rate is assumed.

Berndt, G. D. (1966), *Estimating a Monotonically Changing Binomial Parameter*, Office of Operation Analysis, Strategic Air Command, Offutt Air Force Base, Nebraska.

The author considers a reliability growth model of the form

$$\frac{r_j}{1-r_j} = \frac{r}{1-r}\exp(\alpha/j),$$

where r_j is the success ratio for the jth stage. Maximum likelihood estimators are obtained for r and α.

Boag, J. W. (1949), Maximum Likelihood Estimates of the Proportion of Patients Cured by Cancer Therapy, *Journal of the Royal Statistical Society*, Vol. B11, pp. 15–53.
The author fits lognormal and exponential distributions to survival data. He presents a model for both cured and uncured patients; the uncured patients are assumed to have lognormal survival times. The solutions are iterative.

Boardman, T. J. and P. J. Kendell (1970), Estimation in Compound Exponential Failure Models, *Technometrics*, Vol. 12, pp. 891–900.
The authors discuss items on life-test that are subject to two independent single parameter causes of failure. The resulting failure times are used to estimate these parameters by the method of maximum likelihood. Both the censored and noncensored cases are considered.

Bradley, R. A. and J. Gart (1962), The Asymptotic Properties of ML Estimators When Sampling from Associated Populations, *Biometrika*, Vol. 49, pp. 205–213. The authors extend the basic theory of maximum likelihood estimation from a single population to associated populations.

Breslow, N. (1970), A Generalized Kruskal–Wallis Test for Comparing K Samples Subject to Unequal Patterns of Censorship, *Biometrika*, Vol. 57, pp. 579–594.
The author extends the results of Gehan's censored Wilcoxon test to the case of K samples with progressive right censoring.

Breslow, N. (1974), Covariance Analysis of Censored Survival Data, *Biometrics*, Vol. 30, pp. 89–99.
The author utilizes regression models for making covariance adjustments in comparison of survival curves. The paper compares the models of Glasser, Cox, and Feigl and Zelen.

Breslow, N. and J. Crowley (1974), A Large Sample Study of the Life Table and Product Limit Estimates Under Random Censorship, *Annals of Statistics*, Vol. 2, pp. 437–453.
Assuming progressive sampling, the necessary and sufficient conditions for the consistency of several life table estimates are derived.

Broadbent, S. (1958), Simple Mortality Rates, *Applied Statistics*, Vol. 7, pp. 86–95.
The author discusses four different hazard rates $\lambda(t)=\lambda$, $\lambda_0+2\lambda_1 t$, $\lambda^\gamma \gamma t^{\gamma-1}$, and $\lambda e^{\gamma t}$ with examples to illustrate the resulting distributions.

Buckland, W. R. (1964), *Statistical Assessment of the Life Characteristics*, No. 13 of Griffins Statistical Monographs, Charles Griffin and Company Limited, London.
This monograph presents a bibliographical survey of the literature of life testing. On one side of the page a concise development of life testing distributions is given and on the other side there is a bibliography for those requiring a more thorough treatment of each subject.

Chiang, C. L. (1960), A Stochastic Study of the Life Table and Its Applications: 1. Probability Distributions of the Biometric Function, *Biometrics*, Vol. 16, pp. 618–635.

The author defines cohort and current life tables. He considers life tables stochastically, treating all biometric functions as random variables.

Chiang, C. L. (1961), On the Probability of Death from Specific Causes in the Presence of Competing Risks, *Proceedings of the Fourth Berkeley Symposium on Mathematical Statistics and Probability*, pp. 169–180.
The author introduces the concept of proportionality in the theory of competing risks.

Chiang, C. L. (1966), On the Formula for the Variance of the Observed Expectation of Life—E. B. Wilson's Approach, *Human Biology*, Vol. 38, pp. 318–319.
The author derives the correct formula for the variance of the observed expectation of life. This formula was ingeniously derived by Wilson, who made a slight error in reasoning.

Chiang, C. L. (1967), Variance and Covariance of Life Table Functions Estimated from a Sample of Deaths, *Vital and Health Statistics*, Series 2, No. 20.
The author derives formulas for the variance and covariance of functions of abridged and complete life tables when a sample of deaths is available. He accounts for the component of sampling variation incurred by having the sample rather than the total of deaths available.

Chiang, C. L. (1968), *Introduction to Stochastic Processes in Biostatistics*, John Wiley & Sons, New York.
The author devotes the first half of this book to the study of birth and death processes. The second half reviews his work on life tables. The author also discusses his work on competing risks.

Chiang, C. L. (1970), Competing Risks and Conditional Probabilities, *Biometrics*, Vol. 26, pp. 767–776.
The author shows that the general model of competing risks under the proportionality assumption satisfies the criteria of internal consistency and reasonableness in describing survival data.

Choi, S. C. and V. A. Clark (1970), Sequential Decision for a Binomial Parameter with Delayed Observations, *Biometrics*, Vol. 26, pp. 411–420.
The authors consider a procedure based on Anderson [1964] whereby the decision based on a sequential probability ratio test may be altered by the information contained in m delayed observations. The data are assumed to follow a binomial distribution with either m known, m having a binomial distribution, or m having a Poisson distribution.

Cohen, A. C. (1950), Estimating the Mean and Variance of Normal Populations from Singly Truncated and Doubly Truncated Samples, *Annals of Mathematical Statistics*, Vol. 21, pp. 557–569.
The author provides formulas for estimating the mean and variance of a normal distribution when the total number of observations is known or when only the number of observations measured is known.

Cohen, A. C. (1951), Estimating Parameters of the Logarithmic Normal Distribution by Maximum Likelihood, *Journal of the American Statistical Association*, Vol. 46, pp. 206–212.
Maximum likelihood equations for estimating the population parameters for a three-parameter lognormal distribution are obtained. An iterative solution is outlined. The variance–covariance matrix for the maximum likelihood estimators is obtained. An example is included.

Cohen, A. C. (1965), Maximum Likelihood Estimation in the Weibull Distribution Based on Complete and on Censored Samples, *Technometrics*, Vol. 7, pp. 579–588.
The author obtains maximum likelihood estimates of the parameters (two-parameter case) either when the data are singly censored or when the data are progressively censored.

Cohen, A. C. (1966), Life Testing and Early Failure (Query), *Technometrics*, Vol. 8, pp. 539–545.
The author illustrates methods for analyzing progressively censored data for the exponential and Weibull distributions. He uses the maximum likelihood method for estimating the parameters.

Cohen, J. (1969), *Statistical Power Analysis for the Behavioral Sciences*, Academic Press, New York.
The author provides discussion and tables for determining sample sizes required for given levels of α and β for conventional tests of hypotheses.

Colton, T. (1963), A Model for Selecting One of Two Medical Treatments, *Journal of the American Statistical Association*, Vol. 58, pp. 388–400.
The author proposes a cost function model for designing an optimal clinical trial when N patients with a certain disease are to be allocated between two treatment groups. Both fixed sample and sequential sample plans are given.

Cook, A. A., Jr. and A. J. Gross, (1968), Estimation Techniques for Dependent Logit Models, *RAND Memorandum* RM-5734, RAND Corporation, Santa Monica, Calif.
The authors use a bivariate logistic model to develop the prediction of failure trends within aircraft as well as the prediction of the reenlistment rates among first-term airmen.

Cornfield, J. (1957), The Estimation of the Probability of Developing a Disease in the Presence of Competing Risks, *Journal of the American Public Health Association*, Vol. 47, pp. 601–607.
The author treats the problem of calculating and interpreting life table probabilities in the presence of competing risks.

Cornfield, J., M. Halperin and S. W. Greenhouse (1969), An Adaptive Procedure for Sequential Clinical Trials, *Journal of the American Statistical Association*, Vol. 64, pp. 759–770.
The authors consider an adaptive procedure for sequential clinical trials in which an increasing proportion of patients is assigned to the better of two treatments as evidence for it accumulates.

Cox, D. F. and O. Kempthorne (1963), Randomization Tests for Comparing Survival Curves, *Biometrics*, Vol. 19, pp. 307–317.
The authors describe the application of a permutation test of survival data for two treatment groups.

Cox, D. R. (1953), Some Simple Tests for Poisson Variates, *Biometrika*, Vol. 40, pp. 354–360.
The author develops a test for the equality of two exponential survival distributions when the data are censored.

Cox, D. R. (1959), The Analysis of Experimentally Distributed Life Times with Two Types of Failure, *Journal of the Royal Statistical Society*, Vol. B21, pp. 411–421.
The author presents a number of alternative probability models for the interpretation of failure data when there are two or more types of failure. Some of the techniques are illustrated on real data.

Cox, D. R. (1972), Regression Models and Life Tables, *Journal of the Royal Statistical Society*, Vol. B34, pp. 187–220.
The analysis of censored failure times is considered where the hazard function is a function of concomitant variables and unknown regression coefficients. The author assumes that the hazard function is given by $\lambda(t; Z) = \exp(Z\beta)\lambda_0(t)$, where $\lambda_0(t)$ is the hazard function for the standard set of conditions on Z the concomitant variable and β is the $p \times 1$ vector of unknown parameters.

Cramér, H. (1946), *Mathematical Methods of Statistics*, Princeton University Press, Princeton, N. J.
This standard mathematical statistics textbook provides much of the framework for modern theoretical statistics.

Cutler, S. J. and F. Ederer (1958), Maximum Utilization of the Life Table Method in Analyzing Survival, *Journal of Chronic Diseases*, Vol. 8, pp. 699–713.
The authors estimate a survival curve using an actuarial or life method. This paper is a standard reference in clinical life table studies.

Dahiya, R. C. and J. Gurland (1972), Pearson Chi-square Test of Goodness of Fit with Random Intervals, *Biometrika*, Vol. 59, pp. 147–153.
A modified form of the Pearson chi-square test statistic is considered where estimators based on the ungrouped sample are used in the test statistic as well as in determining the class interval end points.

Daniel, C. and F. S. Wood (1971), *Fitting Equations to Data*, John Wiley & Sons, New York.
The authors look at the aspects of fitting equations both linear and nonlinear to data that an experimenter is likely to collect.
Computer programs are available for fitting both linear and nonlinear equations to data.

David, H. A. (1970), On Chiang's Proportionality Assumption in the Theory of Competing Risks, *Biometrics*, Vol. 26, pp. 336–337.
The author obtains conditions on survival distributions so that the proportionality assumption is satisfied. The exponential and Weibull distributions satisfy these conditions.

Dixon, W. J., Ed. (1973), *Biomedical Computer Programs*, University of California Press, Berkeley.
This book includes two programs to compute clinical life tables and survival rates.

Dixon, W. J. and F. J. Massey (1969), *Introduction to Statistical Analysis*, McGraw-Hill Book Company, New York.
This book is an excellent introductory textbook. Extensive tables are included.

Draper, N. R. and H. Smith, Jr. (1968), *Applied Regression Analysis*, John Wiley & Sons, New York.
The authors discuss a variety of linear and nonlinear regression techniques and their associated estimation procedures. This book has become a standard reference.

Dubey, S. D. (1967), On Some Permissible Estimators of the Location Parameter of the Weibull and Certain Other Distributions, *Technometrics*, Vol. 9, pp. 293–307.
The author gives a graphical procedure for estimating location parameter of the Weibull distribution that is independent of the scale and shape parameters.

Efron, B. (1967), The Two Sample Problem with Censored Data, *Proceedings of the Fifth Berkeley Symposium on Mathematical Statistics and Probability*, Vol. 4, pp. 831–854.
The author proposes a modification of Gehan's generalization of the Wilcoxon test to increase its power.

Ehrentheil, O. F. and H. Muench (1959), Survival Rate of Patients Hospitalized with the Diagnosis of General Paresis, *Journal of Chronic Diseases*, Vol. 9, pp. 41–54.
   The authors investigate the survival rate in cases of general paresis, pointing out that the life expectancy has increased, especially with the advent of penicillin as a treatment.

Elvebach, L. (1958), Actuarial Estimation of Survivorship with Chronic Disease, *Journal of the American Statistical Association*, Vol. 53, pp. 420–440.
   The author assumes a constant force of mortality (hazard rate) that may change from interval to interval. Patients who are lost to follow-up are assumed to be constant. She also discusses different methods of follow-up.

Epstein, B. (1958), The Exponential Distribution and Its Rate in Life Testing, *Industrial Quality Control*, Vol. 15, No. 5, pp. 4–9.
   The author discusses conditions that would result in an exponential death density, using examples from life testing.

Epstein, B. and M. Sobel (1953), Life Testing, *Journal of the American Statistical Association*, Vol. 48, pp. 486–502.
   The authors assume a constant hazard rate. All observations have the same starting point; however, only the first r out of n are observed. The maximum likelihood estimator for the mean time between failures is obtained, and its distributional properties are discussed.

Feigl, P. and M. Zelen (1965), Estimation of Exponential Survival Probabilities with Concomitant Information, *Biometrics*, Vol. 21, pp. 826–838.
   The mean survival time for the ith patient in the study is assumed to be a linear function of a concomitant variate x_i. Analysis of the survival data of leukemia patients is presented.

Feinleib, M. (1960), A Method of Analyzing Lognormally Distributed Survival Data with Incomplete Follow-up, *Journal of the American Statistical Association*, Vol. 55, pp. 534–545.
   The author extends Boag's [1949] work to include survival data with incomplete follow-up. He uses estimates from life tables.

Fix, E. and J. Neyman (1951), A Simple Stochastic Model of Recovery, Relapse, Death and Loss of Patients, *Human Biology*, Vol. 24, pp. 205–241.
   Using cancer as their model disease, the authors describe a four-state stochastic model that characterizes a patient's condition at any point in time. The states are: $S_0 =$ under treatment, $S_1 =$ dead, $S_2 =$ recovery, and $S_3 =$ lost after recovery (includes death due to other causes).

Flehinger, B. J. and T. A. Louis (1971), Sequential Treatment Allocation in Clinical Trials, *Biometrika*, Vol. 58, pp. 419–426.
   The authors are concerned with allocating patients to two different treatments in a sequential clinical trial. They assume an exponential survival distribution for patients on each treatment and try to minimize the number of patients who receive the inferior treatment.

Flehinger, B. J., T. A. Louis, H. Robbins, and B. H. Singer (1972), Reducing the Number of Inferior Treatments in a Clinical Trial, *Proceedings of the National Academy of Sciences*, Vol. 69, pp. 2993–2994.
   The authors use a data-dependent allocation rule for deciding between two treatments when response is assumed to be normally distributed. The aim in using this rule is to minimize the number of patients who receive the inferior treatment.

Fletcher, R. and M. J. D. Powell (1963), A Rapidly Convergent Descent Method for Minimization, *Computer Journal*, Vol. 6, pp. 163–168.
A powerful iterative descent method for finding a local minimum of a function of several variables is described and its convergent properties are given.

Forsythe, A. B. and H. S. Frey (1970), Tests of Significance from Survival Data, *Computers and Biomedical Research*, Vol. 3, pp. 124–132.
The standard tests of significance applied to survival curves assume conditions that often are not satisfied (e.g., the data are exponentially distributed). The authors use a computer simulation that allows the distribution of any statistic to be determined from and for the existing data.

Frei, E., III, et al. (1961), Studies of Sequential and Combination Antimetabolite Therapy in Acute Leukemia: 6-Mercaptopurine and Methotrexate, *Blood*, Vol. 18, pp. 431–454.
The authors study the effects of 6-Mercaptopurine alone, Methotrexate alone, and the two drugs in combination in samples of children and adults with acute leukemia.

Freireich, E. J. et al. (1963), The Effect of 6-Mercaptopurine on the Duration of Steroid-induced Remission in Acute Leukemia, *Blood*, Vol. 21, pp. 699–716.
The authors present data that serve as an example in testing the equality of two survival distributions (6-MP patients vs. placebo patients) when censoring occurs.

Frome, E. L. and J. J. Beauchamp (1968), Maximum Likelihood Estimation of Survival Curve Parameters, *Biometrics*, Vol. 24, pp. 595–606.
The authors estimate the parameters of a survival curve that is used in the quantitative investigation of cytological damage resulting from ionizing radiation. This is done by maximum likelihood and weighted least squares. These two procedures are shown to be equivalent. The observations are assumed to be independent Poisson variables.

Furth, J., A. C. Upton, and A. W. Kimball (1959), Late Pathologic Effects of Atomic Detonation and Their Pathogenesis, *Radiation Research*, Supplement 1, pp. 243–264.
The authors present extensive data on survival of mice exposed to radiation from a nuclear detonation.

Gary, M. L., B. R. Rao, and C. K. Redmond (1970), Maximum Likelihood Estimation of the Parameters of the Gompertz Survival Function, *Applied Statistics*, Vol. 19, pp. 152–159.
The authors discuss maximum likelihood estimates of the parameters and include an example with five treatment groups.

Gehan, E. A. (1965a), A Generalized Wilcoxon Test for Comparing Arbitrarily Singly-censored Samples, *Biometrika*, Vol. 52, pp. 203–223.
The author presents a nonparametric two-sample test that is an extension of the Wilcoxon test to samples with arbitrary censoring on the right.

Gehan, E. A. (1965b), A Generalized Two-sample Wilcoxon Test for Doubly Censored Data, *Biometrika*, Vol. 52, pp. 650–652.
Gehan extends his earlier paper [1965a] to obtaining a nonparametric two-sample test for doubly censored data.

Gehan, E. A. (1967), Some Considerations in the Planning of a Clinical Trial. Talk at the Sixth International Biometric Conference, Sydney, Australia.
The author discusses various strategies in planning clinical trials. He compares the use of survival distributions to simply finding the proportion who survive T years,

and shows how using survival distributions can reduce the sample size need for fixed levels of type I and type II errors.

Gehan, E. A. (1969), Estimating Survival Functions from the Life Table, *Journal of Chronic Diseases*, Vol. 21, pp. 629–644.
The author develops methodology for estimating hazard rates, density functions, and survival functions from life tables. Formulas for estimating these functions and the approximate variances of these estimates are given.

Gehan, E. A. (1970), *Unpublished Notes on Survivability Theory*, University of Texas, M. D. Anderson Hospital and Tumor Institute, Houston.
The author presents a set of notes outlining new as well as old methods for analyzing survival data. Included in these notes are many of his own methods.

Gehan, E. A. and M. M. Siddiqui (1973), Simple Regression Methods for Survival Time Studies, *Journal of the American Statistical Association*, Vol. 68, pp. 848–856.
Least squares estimates are proposed for the parameters of the exponential, linear hazard, Gompertz, and Weibull distributions for grouped survival data. Monte Carlo comparisons are made with maximum likelihood estimates, and the least square estimates appear to be nearly as efficient.

Gehan, E. A. and D. G. Thomas (1969), The Performance of Some Two-sample Tests in Small Samples With and Without Censoring, *Biometrika*, Vol. 56, pp. 127–132.
The authors compare the power curves of three two-sample tests in small samples from the exponential and four forms of the Weibull distribution with and without sampling. The tests discussed are: *F* ratio, modified *F* ratio, and the generalized Wilcoxon test.

Gershenson, H. (1961), *Measurement of Mortality*, Society of Actuaries, Chicago.
This handbook, written for insurance actuaries, details methods of measuring mortality based on exposure to risk.

Gibbons, J. D. (1971), *Nonparametric Statistical Inference*, McGraw-Hill Book Company, New York.
This is an intermediate level textbook covering the better-known nonparametric techniques.

Glasser, M. (1965), Regression Analysis with Dependent Variable Censored, *Biometrics*, Vol. 21, pp. 300–307.
The author estimates multiple regression parameters where the underlying distribution is a censored normal distribution. The method is illustrated and applied to censored survival times.

Glasser, M. (1967), Exponential Survival with Covariance, *Journal of the American Statistical Association*, Vol. 62, pp. 561–568.
The author assumes a constant hazard rate that includes age as a covariable. He obtains iteratively the maximum likelihood estimates for two comparison groups.

Greenwood, J. A. and D. Durand (1960), Aids for Fitting the Gamma Distribution by Maximum Likelihood, *Technometrics*, Vol. 2, pp. 55–65.
The authors develop the maximum likelihood estimate equations for the two-parameter gamma distribution when the sample is not censored. They also present tables that are very useful in solving these equations for specific numerical cases.

Greenwood, M. (1926), The Natural Duration of Cancer, *Reports on Public Health and Medical Subjects* (Her Majesty's Stationery Office, London), Vol. 33, pp. 1–26.
The author presents a number of variance and covariance formulas for life table functions.

Grenander, U. (1956), On the Theory of Mortality Measurement, *Skandinavisk Aktuarie-tidskrift*, Vol. 39, Parts I and II, pp. 70–96 and pp. 125–153.
The author discusses various estimation procedures to estimate the requisite mortality parameters. Included among these procedures are maximum likelihood, minimum chi-square, and the method of moments.

Gross, A. J. (1971a), Monotonicity Properties of the Moments of Truncated Gamma and Weibull Density Functions, *Technometrics*, Vol. 13, pp. 851–858.
The author uses the monotonicity of the mean of the truncated exponential density to estimate its parameter by maximum likelihood.

Gross, A. J. (1971b), The Application of Exponential Smoothing to Reliability Assessment, *Technometrics*, Vol. 13, pp. 877–884.
The author develops two methods, one empirical and one Bayesian, to estimate the smoothing parameters, when exponential smoothing is applied to stagewise reliability growth.

Gross, A. J., R. L. Brunelle, and M. Huang (1975), Monte Carlo Simulation Comparisons of Weibull Distribution Parameters (submitted for publication).
The authors compare, by means of a Monte Carlo simulation, biases and their standard deviations of the maximum likelihood estimators and the two Bain-Antle estimators (Bain and Antle [1967]) for samples of size 25, 50, and 100.

Gross, A. J., V. A. Clark, and V. Liu (1971), Estimation of Survival Parameters When One of Two Organs Must Function for Survival, *Biometrics*, Vol. 27, pp. 369–377.
The authors consider a two-organ system (e.g., the lungs or kidneys of the human body) in which one organ must function for survival. They use maximum likelihood procedures to estimate the requisite parameters.

Gross, A. J., and M. Kamins (1968), Reliability Assessment in the Presence of Reliability Growth, *Annals of Assurance Sciences: 1968 Annual Symposium on Reliability*, pp. 406–416.
The authors consider a number of parametric reliability growth models, to assess and predict reliability growth when reliability grows stagewise.

Gumbel, E. J. (1958), *Statistics of Extremes*, Columbia University Press, New York.
This book contains a discussion of probability paper, a graph of numerous hazard functions, as well as a discussion of the extreme value distribution. It includes a method for estimating parameters of the Weibull distribution.

Halley, E. (1693), An Estimate of the Degrees of the Mortality of Mankind, Drawn from Curious Tables of the Births and Funerals of the City of Breslau, *Philosophical Transactions of the Royal Society of London*, Vol. 17, pp. 596–610.
The author makes one of the earliest rigorous presentations of the life table, similar to what is used today.

Halperin, M. (1952), Maximum Likelihood Estimation in Truncated Samples, *Annals of Mathematical Statistics*, Vol. 23, pp. 226–238.
The author discusses the properties and techniques for obtaining maximum likelihood estimates of parameters of distributions when the values of the first r ordered observations are known out of a total fixed sample size of n (with n known).

Halperin, M., E. Rogot, J. Gurian, and F. Ederer (1968), Sample Sizes for Medical Trials with Special Reference to Long-term Therapy, *Journal of Chronic Diseases*, Vol. 21, pp. 13–24.

The authors develop a model for estimating sample sizes for a control group and an experimental group when knowledge of the success probabilities for the two groups is assumed known, the type I and type II errors are specified, and a dropout rate for patients in the experimental group is assumed.

Halperin, M. and J. Ware (1974), Early Decision in a Censored Wilcoxon Two-sample Test for Accumulating Survival Data, *Journal of the American Statistical Association*, Vol. 69, pp. 414–422.
For both singly and progressively censored data, the authors explore the use of a Wilcoxon test statistic when the experiment is terminated and a specified percentage of the deaths has occurred in either of two groups.

Hankey, B. F. and M. H. Myers (1971), Evaluating Differences in Survival Between Two Groups of Patients, *Journal of Chronic Diseases*, Vol. 24, pp. 523–531.
The authors give a procedure for adjusting and comparing survival distributions for two groups of patients subject to two different initial conditions.

Harris, T. E., P. Meier, and J. W. Tukey (1950), The Timing of the Distribution of Events Between Observations, *Human Biology*, Vol. 22, pp. 249–270.
The authors look at a certain number of cases observed at irregular intervals to determine whether a specific event has occurred. Lost cases are also considered. A model is developed along with the maximum likelihood estimators of the parameter. The model is of the exponential form.

Harter, H. L. and A. H. Moore (1965), Maximum Likelihood Estimation of the Parameters of Gamma and Weibull Populations from Complete and Censored Samples, *Technometrics*, Vol. 7, pp. 639–643.
The authors provide iterative procedures for joint maximum likelihood estimation, from complete and censored samples, of the three parameters of gamma and Weibull populations. Numerical examples are given.

Harter, H. L. and A. H. Moore (1967), Asymptotic Variances and Covariances of Maximum Likelihood Estimators, from Censored Samples, of the Parameters of Weibull and Gamma Parameters, *Annals of Mathematical Statistics*, Vol. 38, pp. 557–570.
The authors describe conditions on censored Weibull and gamma samples that are needed to satisfy regularity conditions that in turn ensure the asymptotic multivariate normality of the maximum likelihood estimators of the parameters.

Hartley, H. O. (1958), Maximum Likelihood Estimation from Incomplete Data, *Biometrics*, Vol. 14, pp. 174–194.
The author discusses obtaining maximum likelihood estimators of parameters for truncated and censored samples.

Hartley, H. O. (1961), The Modified Gauss-Newton Method for Fitting of Nonlinear Regression Functions by Least Squares, *Technometrics*, Vol. 3, pp. 269–280.
The author considers a modification of the Gauss-Newton least squares nonlinear estimation procedure, which converges to a solution for most cases one is apt to encounter in practice.

Herman, R. J. and R. K. N. Patell (1971), Maximum Likelihood Estimation for Multi-risk Model, *Technometrics*, Vol. 13, pp. 385–397.
The authors discuss a failure model that accounts for competing causes of failure on a unit during its lifetime. Examples are given using the exponential and Weibull distributions.

Hoel, D. G. (1972), A Representation of Mortality Data by Competing Risks, *Biometrics*, Vol. 28, pp. 475–488.

The author depicts mortality data as a probabilistic combination of competing risks. Each risk is described by an age-at-death distribution and a net probability of occurrence. Both parametric and nonparametric models are considered, the nonparametric model being an extension of the Kaplan-Meier model.

Hoel, P. G. (1962), *Introduction to Mathematical Statistics*, John Wiley & Sons, New York.

A standard introductory text in statistical theory which requires only knowledge of elementary calculus.

Irwin, J. O. (1942), The Distribution of the Logarithm of Survival Times When the True Law is Exponential, *Journal of Hygiene*, Vol. 42, pp. 328–333.

The author discusses the difficulty in distinguishing a constant hazard function from the hazard function associated with a logarithmic survival distribution.

Irwin, J. O. (1949), The Standard Error of an Estimate of Expectation of Life, with Special Reference to Expectation of Tumorless Life in Experiments with Mice, *Journal of Hygiene*, Vol. 47, pp. 188–189.

The title for this paper is self-descriptive. The estimate given here is similar to Greenwood's estimate.

Jaffe, A. J. (1951), *Handbook of Statistical Methods for Demographers*, U.S. Bureau of the Census, Washington, D.C.

The author gives a detailed discussion of population life tables, in particular, analysis of early and late years of life. If unavailable, see Shryock and Siegel.

Jennrich, R. I. and P. F. Sampson (1968), An Application of Stepwise Regression to Nonlinear Estimation, *Technometrics*, Vol. 10, pp. 63–72.

The article describes the calculations made in stepwise regression and in the BMDX85 nonlinear regression program.

Kale, B. K. (1961), On the Solution of the Likelihood Equation by Iteration Process, *Biometrika*, Vol. 48, pp. 452–456.

The author discusses the Newton-Raphson and scoring methods for obtaining the maximum likelihood estimate iteratively for the single-parameter case.

Kale, B. K. (1962), On the Solution of Likelihood Equations by Iteration Processes. The Multiparametric Case, *Biometrika*, Vol. 49, pp. 479–486.

The author extends his previous results on Newton-Raphson and scoring to the multiparameter case.

Kaplan, E. L. and P. Meier (1958), Nonparametric Estimation from Incomplete Observations, *Journal of the American Statistical Association*, Vol. 53, pp. 457–481.

The authors obtain the nonparametric maximum likelihood estimate for $P(t)$ the proportion of individuals in a population whose life time exceeds t. They take into account the censored nature of survival data.

Kemeney, J. G. and J. L. Snell (1960), *Finite Markov Chains*, D. Van Nostrand Company, Princeton, N.J.

The authors present an excellent development of the theory of finite Markov chains. In particular they deal with the theory of absorbing and transient states within finite Markov chains.

Kendell, M. G. and A. Stuart (1963), *The Advanced Theory of Statistics*, Vol. 1, Hafner and Company, New York.

This standard reference for theoretical statistics includes various formulas for Mills' ratio or inverse of the hazard function for a normal distribution. The determination of standard errors of functions of random variables is discussed.

Kimball, A. W. (1960), Estimation of Mortality Intensities in Animal Experiments, *Biometrics*, Vol. 16, pp. 505–521.
The author compares nonparametric methods for estimating mortality intensities from small samples without censoring. Emphasis is on the construction of Gompertz plots and on techniques that would be useful in controlled animal studies.

Kimball, A. W. (1969), Models for Estimation of Competing Risks from Grouped Data, *Biometrics*, Vol. 25, pp. 329–337.
The author investigates populations exposed to several competing causes of death (i.e., competing risks). He is especially concerned with the probability of death from a given cause when one or more causes have been eliminated. The article compares two models proposed for estimating this probability based on several criteria, including biological relevance.

Kodlin, D. (1967), A New Response Time Distribution, *Biometrics*, Vol. 23, pp. 227–239.
The author considers a linear hazard rate $h(t)=\lambda+\mu t$. He obtains the likelihood function for both the censored and uncensored cases. The maximum likelihood estimates must be obtained iteratively.

Kowalik, J. and M. R. Osborne (1968), *Methods for Unconstrained Optimization Problems*, American Elsevier Publishing Company, New York.
The authors explain and contrast principal techniques used in computing the minimum of a function of several variables.

Kramer, M. and S. W. Greenhouse (1955), Determination of Sample Size and Selection of Cases, Psychopharmacology, *Problems in Evaluation*, Publication 583, National Academy of Sciences, pp. 356–371.
The authors consider some of the problems involved in determining the size of both experimental and control groups (i.e., the number of cases required) in experiments designed to detect the effects of tranquilizing agents. The approach is based on dichotomizing responses and using a simple contingency table approach. Note that in equation (ii) of the article the $1/n$ should have a plus sign which yields the expression for sample size used here in Chapter 8.

Krane, S. A. (1963), Analysis of Survival Data by Regression Techniques, *Technometrics*, Vol. 5, pp. 161–174.
The author assumes that the cumulative hazard function $\int_0^t \lambda(x)\,dx$ has a polynomial form from which he estimates the parameters. He applies this model to estimating the "survival" of automobiles and trucks of a public utility firm.

Kuzma, J. (1967), A Comparison of Two Life Table Methods, *Biometrics*, Vol. 23, pp. 51–64.
The author gives a numerical comparison of Chiang's and actuarial life table results.

Littell, A. S. (1952), Estimation of the *T*-year Survival Rate from Follow-up Studies over a Limited Period of Time, *Human Biology*, Vol. 24, pp. 87–116.
The author assumes a constant failure rate λ (which leads to an exponential survival distribution). Under this assumption he finds the maximum likelihood estimate of λ and its asymptotic variance when the observation time on each individual was censored and the observation times followed a rectangular distribution.

Lloyd, D. L. and M. Lipow (1962), *Reliability: Mathematics, Management, and Method*, Prentice-Hall, Englewood Cliffs, N.J.
The authors provide a fundamental book in the theory of reliability, covering such topics as distributions, estimation, reliability growth, and stochastic processes.

MacDonald, E. J. (1963), The Epidemiology of Melanoma, *Annals of the New York Academy of Sciences*, Vol. 100, pp. 4–15.
The author reports an epidemiological study of the etiology of melanoma, presenting survival data from the M. D. Anderson Hospital and Tumor Institute for males and females.

Mann, N. R., R. D. Schafer, and N. D. Singpurwalla (1974), *Methods for Statistical Analysis of Reliability and Life Data*, John Wiley & Sons, New York.
The authors cover a variety of methods of analyzing failure and life data from known distributions, including the exponential, gamma, lognormal, extreme value, and Weibull distributions. Their presentation of the Weibull distribution is particularly excellent.

Mantel, N. (1967), Ranking Procedures for Arbitrarily Restricted Observations, *Biometrics*, Vol. 23, pp. 65–78.
The author describes a modification of Gehan's generalization of the Wilcoxon test to include arbitrarily restricted observations on either the left- or right-hand side of the distribution.

Mantel, N. and M. Myers (1971), Problems of Convergence of Maximum Likelihood Procedures in Multiparameter Situations, *Journal of the American Statistical Association*, Vol. 66, pp. 484–491.
The authors discuss problems of convergence in the solution of maximum likelihood iterative equations. They point out that convergence may be highly dependent on initial estimates, particularly when one is working with censored data.

Marquardt, D. W. (1963), An Algorithm for Least Squares Estimation of Nonlinear Parameters, *Journal of SIAM*, Vol. 2, pp. 431–441.
The author discusses a modified gradient method, a compromise between the Newton-Raphson and the ordinary gradient method.

Maurice, R. J. (1957), A Minimax Procedure for Choosing Between Two Populations Using Sequential Sampling, *Journal of the Royal Statistical Society*; Series B, Vol. 19, pp. 255–261.
The author obtains a minimax solution for the loss function described by Colton [1963]. Colton and Maurice both note that the expected loss cannot be minimized for the most unfavorable case.

Mendenhall, W. and R. J. Hader (1958), Estimation of Parameters of Mixed Exponentially Distributed Failure Time Distributions from Censored Life Test Data, *Biometrika*, Vol. 45, pp. 504–520.
The authors separate a failure population into subpopulations, each representing a different type of failure. They obtain maximum likelihood estimators in the case where the subpopulations are exponential and the data are censored at a specified point in time.

Menon, M. V. (1963), Estimation of the Shape and Scale Parameters of the Weibull Distribution, *Technometrics*, Vol. 5, pp. 175–182.
This article, one of the earliest, presents a method for estimating the shape and scale parameters of the Weibull distribution which is not maximum likelihood.

Merrell, M. and L. E. Shulman (1955), Determination of Prognosis in Chronic Disease, *Journal of Chronic Diseases*, Vol. 1, pp. 12–32.
The authors develop a standard reference on clinical life tables.

Miller, I. and J. E. Freund (1965), *Probability and Statistics for Engineers*, Prentice-Hall, Inc., Englewood Cliffs, N.J.
A useful book for engineers and physical scientists who require statistics. Along

with the usual statistical topics, there are two chapters on quality assurance and reliability.

Miller, R. G. (1974), *Least Squares Regression with Censored Data*, Technical Report No. 3, Division of Biostatistics, Stanford University, Stanford, California.
The author describes and contrasts the approach of Cox [1972] and Kaplan and Meier [1958] in estimating the regression coefficients. He also presents a modification of the Kaplan and Meier procedure that can be readily used with more than one independent variable for censored data.

Moeschberger, M. (1974), Life-Tests Under Dependent Competing Causes of Failure, *Technometrics*, Vol. 16, pp. 39–47.
The author discusses some properties of the bivariate Weibull distribution and how it is applied to models where the competing risks are not assumed independent.

Moeschberger, M. and H. A. David (1971), Life-Tests Under Competing Causes of Failure and the Theory of Competing Risks, *Biometrics*, Vol. 27, pp. 909–933.
The authors obtain a general model for handling competing risks which allows for dependence of causes for both censored and uncensored data. The method of maximum likelihood is used to estimate parameters when the survival distributions of the competing risks are independent Weibull distributions.

Mood, A. M. (1950), *Introduction to the Theory of Statistics*, McGraw-Hill Book Company, New York.
The author discusses elliptical confidence regions for joint maximum likelihood estimates.

Mood, A. M. and F. A. Graybill (1963), *Introduction to the Theory of Statistics*, McGraw-Hill Book Company, New York.
The authors discuss large sample properties of maximum likelihood estimates, giving variance–covariance matrices in general situations.

Murthy, V. K., G. Swartz, and K. Yuen (1973), Realistic Models for Mortality Rates and their Estimation, I and II, Department of Biomathematics Technical Report, University of California at Los Angeles.
The article includes a discussion of so-called bathtub hazard rates.

Nam, J. M. (1973), Optimum Sample Sizes for the Comparison of the Control and Treatment, *Biometrics*, Vol. 29, pp. 101–108.
The optimum sample sizes for the control and treatment groups for testing the difference between the two mean survival times are investigated. It is assumed that the costs of testing the two groups differ.

Nelson, W. (1972), Theory and Application of Hazard Plotting for Censored Failure Data, *Technometrics*, Vol. 14, pp. 945–965.
The author presents theory and application of plotting cumulative hazard rates. This technique is very useful in deciding among distributions.

Parzen, E. (1967), *Analysis and Synthesis of Forecasting Models by Time Series Methods*, Technical Report 7, Stanford University, Stanford, California.
The author uses times series procedures to estimate the smoothing constant in an exponential smoothing model.

Pearson, E. S. and H. O. Hartley (1966), *Biometrika Tables for Statisticians*, Vol. 1, 3rd ed., Cambridge University Press, Cambridge, England.
This is a standard reference book on tables for statisticians.

Pearson, K. (1922), *Tables of the Incomplete Γ - function*, Cambridge University Press for the Biometrika Trustees, Cambridge, England. (Reprinted 1957.)

The author gives extensive tables for the incomplete gamma function.

Peto, R. and J. Peto (1972), Asymptotically Efficient Rank Invariant Procedures, *Journal of the Royal Statistical Society*, Series A. Vol. 135, pp. 185–207.
The authors present the logrank test that can be used to test the equality of two or more groups of observations of survival times. It has greater local power than any other rank-invariant test procedure for detecting Lehmann-type differences between groups of independent observations. It is an easily used test and can be performed if there is censoring.

Peto, R. and M. C. Pike (1973), Conservatism of the Approximation $\Sigma(0-E)^2/E$ in the Logrank Test for Survival Data or Tumor Incidence Data, *Biometrics*, Vol. 29, pp. 579–584.
The authors describe the procedure for performing this rank test for detecting multiplicative differences in failure rates with censored or noncensored data. They compare the power of alternative procedures to the logrank test.

Rao, C. R. (1952), *Advanced Statistical Methods in Biometric Research*, John Wiley & Sons, New York.
In this book the author discusses a large number of procedures which are applicable to biometry problems such as discriminant analysis. In particular he discusses the Method of Scoring which is a variation of the Newton-Raphson procedure.

Robbins, H. and D. Siegmund (1972), On a Class of Stopping Rules for Testing Parametric Hypotheses, *Sixth Berkeley Symposium on Mathematical Statistics and Probability* Vol. 5 *Biology and Health*, pp. 37–41.
The authors present a theoretical development for determining stopping rules for parametric hypotheses.

Saaty, T. L. (1959), *Mathematical Methods of Operation Research*, McGraw-Hill Book Company, New York.
The author includes extensive discussion of methods of optimization with and without constraints.

Saaty, T. L. and J. Bram (1964), *Nonlinear Mathematics*, McGraw-Hill Book Company, New York.
The authors discuss, generally, problems in nonlinear equation solutions. In particular maximization techniques such as the gradient method are discussed.

Sacher, G. A. (1956), On the Statistical Nature of Mortality, with Special Reference to Chronic Radiation Mortality, *Radiation*, Vol. 67, pp. 250–257.
The author develops an estimate of the hazard function in clinical life tables based on the assumption of constant hazard rate within but not necessarily between intervals.

Sampford, M.R. (1952), The Estimation of Response Time Distribution, I and II, *Biometrics*, Vol. 8, pp. 13–32, 307–369.
The author discusses the fundamentals and general methods for estimating response time distributions for both the single- and multi-stimulus situations.

Sampford, M. R. (1954), The Estimation of Response Time Distribution, III, *Biometrics*, Vol. 10, pp. 531–561.
The author discusses problems of survival data in the presence of truncation.

Scheuer, E. M. (1968), Testing Grouped Data for Exponentiality, Rand Memorandum RM-5692-PR, Santa Monica, Calif.
The author discusses testing the fit of grouped data to an exponential distribution;

a JOSS computer program is provided to implement the test.

Schork, M. A. and R. D. Remington (1967), The Determination of Sample Size in Treatment–Control Comparisons for Chronic Disease Studies in which Drop-out or Non-Adherence is a Problem, *Journal of Chronic Diseases*, Vol. 20, pp. 233–239.
The authors assume that a fixed proportion of the patients shift from the treatment group to the control per unit of time.

Seal, H. L. (1954), The Estimation of Mortality and Other Decremental Probabilities, *Skandinavisk Aktuarietidskrift*, Vol. 37, pp. 137–162.
The author reviews the various estimates for the probability of surviving T years and discusses their properties.

Sheppard, W. F. (1939), *The Probability Integral*, British Association of Mathematics Tables, Vol. 7, Cambridge University Press, Cambridge, England.
This book contains tables of the inverse of the hazard rate for the normal distribution.

Shryock, H. S. and J. S. Siegel (1973), *The Methods and Materials of Demography*, Vol. I and II, U. S. Bureau of Census, Government Printing Office, Washington, D. C.
The authors have written a very comprehensive explanation of population life tables and the data that comprise them.

Siegel, S. (1956), *Nonparametric Statistics for the Behavioral Sciences*, McGraw-Hill Book Company, New York.
The author provides examples of the use of many nonparametric tests. On the inside cover is printed an index based on Stevens' classification of data that allows the reader to quickly find an appropriate test. No proofs are provided.

Sobel, M. and G. H. Weiss (1970), Play-the-Winner Sampling for Selecting the Better of Two Binomial Populations, *Biometrika*, Vol. 57, pp. 357–365.
The authors consider the sequential allocation of treatments by an experimenter for determining which of two binomial populations has the larger mean. In particular, they consider a play-the-winner rule.

Spiegelman, M. (1968), *Introduction to Demography*, rev. ed., Harvard University Press, Cambridge, Mass.
In Chapter 5 the author presents life table construction techniques.

Taylor, A. E. (1955), *Advanced Calculus*, Ginn and Company, Boston.
The author presents a thorough treatment of important topics in advanced calculus, including determining maxima and minima of functions of two-dimensional variables.

Thoman, D. R. and L. J. Bain (1969), Two Sample Tests in the Weibull Distribution, *Technometrics*, Vol. II, pp. 805–815.
The authors propose a test for the equality of the shape parameters in two Weibull distributions when the scale parameters are unknown. Their test is based on maximum likelihood estimators. Tests for the equality of the scale parameters are also presented, along with a procedure for selecting the Weibull process with the larger mean life.

Thoman, D. R., L. J. Bain, and C. E. Antle (1969), Inferences on the Parameters of the Weibull Distribution, *Technometrics*, Vol. II, pp. 445–460.
The problems of estimation and testing hypotheses regarding the parameters in the Weibull distribution are considered in this paper. Among other results, the authors obtain exact confidence intervals for parameters based on maximum likelihood

322

estimators, as well as a table of unbiasing factors (depending on sample size) for the maximum likelihood estimator of the shape parameter.

Turnbull, B. W. (1974), Nonparametric Estimation of a Survivorship Function with Doubly Censored Data, *Journal of the American Statistical Association*, Vol. 69, pp. 169–173.
The author presents an iterative procedure for obtaining estimates of a response time distribution with double censored data based on the product-limit method of Kaplan and Meier.

Turnbull, B. W., B. W. Brown, and M. Hu (1974), Survivorship Analysis of Heart Transplant Data, *Journal of the American Statistical Association*, Vol. 69, pp. 74–80.
The authors discuss a number of techniques for analyzing heart transplant data when information is available on length of time from acceptance to transplant for some of the patients but only acceptance to death or date last seen for others.

Wald, A. (1947), *Sequential Analysis*, John Wiley & Sons, New York.
This classic book describes the sequential procedure of testing hypotheses. The author was the principal contributor to this theory.

Wasan, M. T. (1970), *Parametric Estimation*, McGraw-Hill Book Company, New York.
This is an intermediate-level text on properties of various statistical estimation procedures.

Watson, G. S., and M. R. Leadbetter (1964), Hazard Analysis, I, *Biometrika*, Vol. 51, pp. 175–184.
The authors define the hazard rate, obtain estimators of it, and conclude the article with numerical examples. They also discuss some applications to life tables.

Weibull, W. (1939), A Statistical Theory of the Strength of Materials, *Ingeniors Vetenskaps Akademien Handlingar*, No. 151; The Phenomenon of Rupture in Solids, *ibid.*, No. 153.
The author presents the survival distribution that has since borne his name—the Weibull distribution.

Weibull, W. (1951), A Statistical Distribution of Wide Applicability, *Journal of Applied Mechanics*, Vol. 18, pp. 293–297.
The author gives some empirical justifications for using the Weibull distribution in applications.

Weiner, J. M., C. E. Hopkins, and J. Marmorston (1965), Use of Correlation Structure Analysis in Estimating Survival Benefit, *Annals of the New York Academy of Sciences*, Vol. 126, pp. 743–757.
The authors present a method for analyzing survival data in the presence of concomitant variables. They separate out those covariates and combinations of covariates which are associated with survival.

Weinman, D. G., G. Dugger, W. E. Franck, and J. E. Hewett (1973), On a Test for the Equality of Two Exponential Distributions, *Technometrics*, Vol. 15, pp. 177–182.
The authors discuss the power of likelihood ratio tests for exponential distribution when a common, unknown scale parameter is present.

Wetherill, G. B. (1966), *Sequential Methods in Statistics*, Methuen and Company, London.
The book surveys the sequential methods literature.

White, J. S. (1969), The Moments of Log-Weibull Order Statistics, *Technometrics*, Vol. II, pp. 373–386.
The author obtains the means and variances of the order statistics of the log-Weibull distribution $F(x) = 1 - \exp(-\exp x)$. Tables of these means and variances

are given for $n = 1, \ldots, 50$; $n = 55, 60, 65, \ldots, 500$. Examples of the use of these tables in obtaining weighted least squares estimates from censored samples from a Weibull distribution are also given.

Whitman, S. (1969), *The Gompertz Function as a Conditional Mortality Function*, Unpublished doctoral dissertation, Yale University, New Haven, Conn.
The author presents an extensive discussion of the historical background of the Gompertz distribution. He also describes the properties of this distribution and gives algorithms for obtaining maximum likelihood estimates of the parameters.

Wilde, D. J. (1964), *Optimum Seeking Methods*, Prentice-Hall, Englewood Cliffs, N. J.
The author provides an excellent discussion of strategies for seeking the optimum of functions whose characteristics are unknown. The book contains many examples and diagrams.

Wilde, D. J. and C. S. Beightler (1967), *Foundations of Optimization*, Prentice-Hall, Englewood Cliffs, N. J.
The authors present a unified theory of optimization. The book includes many examples and diagrams.

Wilk, M. B., R. Gnanadesikan, and M. J. Huyett (1962a), Estimation of Parameters of the Gamma Distribution Using Order Statistics, *Biometrika*, Vol. 49, pp. 525–545.
The authors provide tables for estimating the parameters when either a complete sample is available or data on the m smallest values of the sample, where $m < n$, the complete sample size, are known.

Wilk, M. B., R. Gnanadesikan, and M. J. Huyett (1962b), Probability Plots for the Gamma Distribution, *Technometrics*, Vol. 4, pp. 1–20.
The authors give a procedure for probability plots for random samples from the gamma distribution for any specified value of the shape parameter between 0.1 and 20.0.

Zahl, S. (1955), A Markov Process Model for Follow-up Studies, *Human Biology*, Vol. 27, pp. 90–119.
The author introduces a time-stationary, continuous Markov process with a finite number of states as a model for the follow-up study of a disease.

Zelen, M. (1969), Play the Winner Rule and the Controlled Clinical Trial, *Journal of the American Statistical Association*, Vol. 64, pp. 131–146.
The author considers a play-the-winner rule for allocating patients to two different treatment groups when the response to treatment is binomial (i.e., success or failure).

Zellner, A. and T. H. Lee (1965), Joint Estimation of Relationships Involving Discrete Random Variables, *Econometrica*, Vol. 33, pp. 382–394.
The authors take advantage of linear regression techniques to estimate parameters of the logistic function.

Zippin, C. and P. Armitage (1966), Use of Concomitant Variables and Incomplete Survival Information with Estimation of an Exponential Survival Parameter, *Biometrics*, Vol. 22, pp. 665–672.
The authors extend the results of Feigl and Zelen where a single covariate is introduced to the progressively censored case for the exponential distribution.

Index